ANIMATION MATHS

eCAMPUS
LEARN

IVO DE PAUW    BIEKE MASSELIS

# Animation
# Maths

Lannoo
Campus

eCampusLearn

Go to **www.ecampuslearn.com.**

eCAMPUS
LEARN

Fill in the following code: 1f2d2f3c9baa3d6c

Good luck!

Chapter 1, David Ritter; 2, John Evans; 3, Wouter Verweirder; 4, Daryl Beggs, Juan Pablo Arancibia Medina; 5, Stephanie Berghaeuser; 6, Martin Walls; 7, 12, Wouter Tansens; 8, Danie Pratt; 9, Ivo De Pauw; 10, Caetano Lacerda; 11, Ken Munyard; 13, Bieke Masselis; 14, Boke Haide; 15, Waldemar Zielinski; 16, Detje Holger; 17, Cornelia Roessing; p.25, p.103, Wouter Tansens; p.48, Wouter Verweirder; p.52, Leo Storme; p.185, Bieke Masselis; p.268, Yu-Sung Chang; p.343, Angelo Fallein.

D/2021/45/69 – ISBN 978 94 014 7495 5 – NUR 918

Cover design: Keppie & Keppie
Interior design: Ivo De Pauw, Bieke Masselis

LannooCampus Publishers
Vaartkom 41 box 01.02        P.O. Box 23202
3000 Leuven                  1100 DS Amsterdam
Belgium                      Netherlands
www.lannoocampus.com

This book is dedicated to Malaika.


*"Sometimes I'm black, sometimes I'm white*
*it all depends on who is on the other side*
*there are things they cannot see*
*and there are things I can not hide"*

*Bruno Deneckere (Someday, June 2006)*

# Contents

Chapter 7 · Vectors                                                                           121

# Acknowledgements

We hereby insist on thanking a lot of people who made this book possible: Prof. Dr. Leo Storme, Wim Serras, Wouter Tansens, Wouter Verweirder, **Koen Samyn** [14] (credited for the chapters Transformation Analysis, Scene Graphs, View Transformation), Hilde De Maesschalck, Ellen Deketele, Conny Meuris, Hans Ameel, Dr. Rolf Mertig, Dick Verkerk, ir. Gose Fischer, Prof. Dr. Fred Simons, Sofie Eeckeman, Dr. Luc Gheysens, Dr. Bavo Langerock, Wauter Leenknecht, Marijn Verspecht, Sarah Rommens, Prof. Dr. Marcus Greferath, Dr. Cornelia Roessing, Tim De Langhe, Niels Janssens, Peter Flynn, Jurgen Leemans, Stef Lantsoght, Hilde Vanmechelen, Jef De Langhe, Ann Deraedt, Rita Vanmeirhaeghe, Prof. Dr. Jan Van Geel, Dr. Ann Dumoulin, Bart Uyttenhove, Rik Leenknegt, Peter Verswyvelen, Roel Vandommele, ir. Lode De Geyter, Bart Leenknegt, Olivier Rysman, ir. Johan Gielis, Frederik Jacques, Kristel Balcaen, ir. Wouter Gevaert, Bart Gardin, Dieter Roobrouck, Dr. Yu-Sung Chang (*Wolfram Demonstrations* [24]), Prof. Dr. Sy Blinder (*Wolfram Demonstrations*), Steven De Keninck, Prof. Dr. Mark McClure (*Wolfram Demonstrations*), Dr. Felipe Dimer de Oliveira (*Wolfram Demonstrations*), Steven Verborgh, Ingrid Viaene, Kayla Chauveau, Angelika Kirkorova, Thomas Vanhoutte, Fries Carton, Jef Daels, Andries Geens, Angelo Fallein, Jolan Plaum, Charles Derre, Anna Rich and whomever we might have forgotten!

Hereby our special thanks to **Dick Verkerk** for supporting our *Wolfram Application Server* [25]) server in the Netherlands at can.nl

# Chapter 1 · Arithmetic refresher

As this chapter offers all the necessary mathematical skills for the full mastery of all further topics explained in this book, we strongly recommend it. To serve its purpose, the successive paragraphs below refresh some required aspects of mathematical language as used on the applied level.

## 1.1 Algebra

REAL NUMBERS

We typeset the set of:

$\triangleright$   natural numbers (unsigned integers) as $\mathbb{N}$ including zero,

$\triangleright$   integer numbers as $\mathbb{Z}$ including zero,

$\triangleright$   rational numbers as $\mathbb{Q}$ including zero,

$\triangleright$   real numbers (floats) as $\mathbb{R}$ including zero.

All the above make a chain of subsets: $\mathbb{N} \subset \mathbb{Z} \subset \mathbb{Q} \subset \mathbb{R}$.

To avoid possible confusion, we outline a brief glossary of mathematical terms. We recall that using the correct mathematical terms reflects correct mathematical thinking. Putting down ideas in the correct words is of major importance for profound insight.

Sets

$\triangleright$   We recall writing all **subsets** in between braces, e.g. the **empty set** appears as $\{\}$.

$\triangleright$   We define a **singleton** as any subset containing only one element, e.g. $\{5\} \subset \mathbb{N}$, as a subset of natural numbers.

$\triangleright$   We define a **pair** as any subset containing just two elements, e.g. $\{115, -4\} \subset \mathbb{Z}$, as a subset of integers. In programming the boolean values *true* and *false* make up a pair $\{true, false\}$ called the boolean set which we typeset as $\mathbb{B}$.

$\triangleright$   We define $\mathbb{Z}^- = \{\ldots, -3, -2, -1\}$ whenever we need negative integers only. We express symbolically that $-1234$ is an **element** of $\mathbb{Z}^-$ by typesetting $-1234 \in \mathbb{Z}^-$.

$\triangleright$   We typeset the **set minus** operator to delete elements from a set by using a backslash, e.g. $\mathbb{N} \setminus \{0\}$ reading all natural numbers except zero, $\mathbb{Q} \setminus \mathbb{Z}$ meaning all pure rational numbers after all integer values left out and $\mathbb{R} \setminus \{0, 1\}$ expressing all real numbers apart from zero and one.

Calculation basics

| operation | expression | $a$ | $b$ | $c$ |
|---|---|---|---|---|
| to add | $a+b=c$ | term | term | sum |
| to subtract | $a-b=c$ | term | term | difference |
| to multiply | $a \cdot b = c$ | factor | factor | product |
| to divide | $\frac{a}{b}=c,\, b \neq 0$ | numerator | divisor or denominator | quotient or fraction |
| to exponentiate | $a^b = c$ | base | exponent | power |
| to take the root | $\sqrt[b]{a}=c$ | radicand | index | radical |
| return factorial | $n! = c$ | | | $n$ factorial |

We define the **factorial** of a natural argument as the returned product of this argument multiplied with all natural numbers from this number $n$ down to 1. Put in symbols:

$$n! = n \cdot (n-1) \cdot (n-2) \cdot \ldots \cdot 3 \cdot 2 \cdot 1 \quad \text{restricted to } n \in \mathbb{N}$$

Furthermore we define $1! = 1$ and as well $0! = 1$.

*Examples*:

$$2! = 2 \cdot 1 = 2, \quad 3! = 3 \cdot 2 \cdot 1 = 6, \quad 4! = 4 \cdot 3 \cdot 2 \cdot 1 = 24.$$

We write the **opposite** of a real number $r$ as $-r$, defined by the sum $r + (-r) = 0$. We typeset the **reciprocal** of a nonzero real number $r$ as $\frac{1}{r}$ or $r^{-1}$, defined by the product $r \cdot r^{-1} = 1$.

We define **subtraction** as equivalent to adding the opposite: $a - b = a + (-b)$. We define **division** as equivalent to multiplying with the reciprocal: $a : b = a \cdot b^{-1}$.

When we mix operations we need to apply priority rules for them. There is a fixed priority list 'PEMDAS' in performing mixed operations in $\mathbb{R}$ that can easily be memorised by 'Please Excuse My Dear Aunt Sally'.

▷ First process all that is delimited in between Parentheses,

▷ then Exponentiate,

▷ then Multiply and Divide from left to right,

▷ finally Add and Subtract from left to right.

Now we discuss the **distributive law** ruling within $\mathbb{R}$, which we define as threading a 'superior' operation over an 'inferior' operation. In conclusion, distributing requires two *different* operations.

Hence we distribute *exponentiating* over *multiplication* as in $(a \cdot b)^3 = a^3 \cdot b^3$. Likewise rules *multiplying* over *addition* as in $3 \cdot (a+b) = 3 \cdot a + 3 \cdot b$.

However we should never stumble on this 'Staircase of Distributivity' by going too fast:

$$(a+b)^3 \neq a^3 + b^3,$$

$$\sqrt{a+b} \neq \sqrt{a} + \sqrt{b},$$

$$\sqrt{x^2 + y^2} \neq x + y.$$

Fractions

A **fraction** is what we call any rational number written as $\frac{t}{n}$ given $t, n \in \mathbb{Z}$ and $n \neq 0$, wherein $t$ is called the **numerator** and $n$ the **denominator**. We define the reciprocal of a nonzero fraction $\frac{t}{n}$ as $\frac{1}{\frac{t}{n}} = \frac{n}{t}$ or as the power $\left(\frac{t}{n}\right)^{-1}$. We define the opposite fraction as $-\frac{t}{n} = \frac{-t}{n} = \frac{t}{-n}$. We summarise fractional arithmetic:

| | |
|---|---|
| sum | $\frac{t}{n} + \frac{a}{b} = \frac{t \cdot b + n \cdot a}{n \cdot b}$ |
| difference | $\frac{t}{n} - \frac{a}{b} = \frac{t \cdot b - n \cdot a}{n \cdot b}$ |
| product | $\frac{t}{n} \cdot \frac{a}{b} = \frac{t \cdot a}{n \cdot b}$ |
| division | $\frac{\frac{t}{n}}{\frac{a}{b}} = \frac{t}{n} \cdot \frac{b}{a}$ |
| exponentiation | $\left(\frac{t}{n}\right)^m = \frac{t^m}{n^m}$ |
| singular fractions | $\frac{1}{0} = \pm\infty$ **infinity** (see page 76) |
| | $\frac{0}{0} = ?$ **indeterminate** |

Powers

We define a **power** as any real number written as $g^m$, wherein $g$ is called its **base** and $m$ its **exponent**. The opposite of $g^m$ is simply $-g^m$. The reciprocal of $g^m$ is $\frac{1}{g^m} = g^{-m}$, given $g \neq 0$.

According to the exponent type we distinguish between:

$$g^3 = g \cdot g \cdot g \qquad\qquad 3 \in \mathbb{N},$$
$$g^{-3} = \frac{1}{g^3} = \frac{1}{g \cdot g \cdot g} \qquad\qquad -3 \in \mathbb{Z},$$
$$g^{\frac{1}{3}} = \sqrt[3]{g} = w \Leftrightarrow w^3 = g \qquad\qquad \frac{1}{3} \in \mathbb{Q},$$
$$g^0 = 1 \qquad\qquad g \neq 0.$$

Whilst calculating powers we may have to:

multiply $\qquad g^3 \cdot g^2 = g^{3+2} = g^5,$

divide $\qquad \frac{g^3}{g^2} = g^3 \cdot g^{-2} = g^{3-2} = g^1,$

exponentiate $\quad \left(g^3\right)^2 = g^{3 \cdot 2} = g^6$ them.

We insist on avoiding typesetting radicals like $\sqrt[7]{g^3}$ and strongly recommend their contemporary notation using radicand $g$ and exponent $\frac{3}{7}$, consequently exponentiating $g$ to $g^{\frac{3}{7}}$. We recall the fact that all square roots are non-negative numbers, $\sqrt{a} = a^{\frac{1}{2}} \in \mathbb{R}^+$ for $a \in \mathbb{R}^+$.

As well as knowing the above exponent types, understanding the above rules to calculate them is necessary for using powers successfully. We advise memorising the integer squares running from $1^2 = 1$, $2^2 = 4$, ..., up to $15^2 = 225$, $16^2 = 256$ and the integer cubes running from $1^3 = 1$, $2^3 = 8$, ..., up to $7^3 = 343$, $8^3 = 512$ in order to easily recognise them.

Recall that the only way out of any power is exponentiating with its reciprocal exponent. For this purpose we need to exponentiate both left hand side and right hand side of any given relation (see also paragraph 1.2).

*Example*: Find $x$ when $\sqrt[7]{x^3} = 5$ by exponentiating this power.

$$x^{\frac{3}{7}} = 5 \Longleftrightarrow \left(x^{\frac{3}{7}}\right)^{\frac{7}{3}} = (5)^{\frac{7}{3}} \Longleftrightarrow x \approx 42.7494.$$

We emphasise the above strategy as the only successful one to free base $x$ from its exponent, yielding its correct expression numerically approximated if we wish to.

*Example*: Find $x$ when $x^2 = 5$ by exponentiating this power.

$$x^2 = 5 \Longleftrightarrow \left(x^2\right)^{\frac{1}{2}} = (5)^{\frac{1}{2}} \text{ or } -(5)^{\frac{1}{2}} \Longleftrightarrow x \approx 2.23607 \text{ or} -2.23607.$$

We recall the above double solution whenever we free base $x$ from an *even* exponent, yielding their correct expression as accurately as we wish to.

Mathematical expressions

Composed mathematical expressions can often seem intimidating or cause confusion. To gain transparency in them, we firstly recall indexed variables which we define as subscripted to count them: $x_1, x_2, x_3, x_4, \ldots, x_{99999}, x_{100000}, \ldots$, and $\alpha_0, \alpha_1, \alpha_2, \alpha_3, \alpha_4, \ldots$ . It is common practice in industrial research to use thousands of variables, so just picking unindexed characters would be insufficient. Taking our own alphabet as an example, it would only provide us with 26 characters.

We define finite expressions as composed of (mathematical) operations on objects (numbers, variables or structures). We can for instance analyse the expression $(3a + x)^4$ by drawing its **tree form**. This example reveals a Power having exponent 4 and a subexpression in its base. The base itself yields a sum of the variable $x$ Plus another subexpression. This final subexpression shows the product 3 Times $a$.

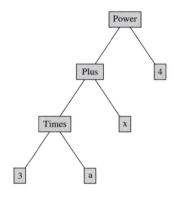

Let us also evaluate this expression $(3a + x)^4$. Say $a = 1$, then we see our expression partly collapse to $(3 + x)^4$. If, on top of this, we assign $x = 2$, our expression then finally turns to the numerical value $(3 + 2)^4 = 5^4 = 625$.

When we expand this power to its **pure sum expression** $81a^4 + 108a^3x + 54a^2x^2 + 12ax^3 + x^4$, we did nothing but *reshape* its **pure product expression** $(3a + x)^4$.

We warn that trying to solve this expression – which is not a relation – is completely in vain. Recall that inequalities, equations and systems of equations or inequalities are the only objects in the universe we can (try to) solve mathematically.

Relational operators

We also refresh the use of correct terms for inequalities and equations.

We define an **inequality** as any *variable* expression comparing a left hand side to a right hand side by applying the 'is-(strictly)-less-than' or by applying the 'is-(strictly)-greater-than' operator. For example, we can read $(3a + x)^4 \leqslant (b + 4)(x + 3)$ containing variables $a$, $x$, $b$. Consequently we may solve such inequality for any of the unknown quantities $a, x$ or $b$.

We define an **equation** as any *variable* expression comparing a left hand side to a right hand side by applying the 'is-equal-to' operator. For example $(3a + x)^4 = (b + 4)(x + 3)$ is an equation containing variables $a$, $x$, $b$. Consequently we also may solve equations for any of the unknown quantities $a, x$ or $b$.

We define an **equality** as a constant relational expression being *true*, e.g. $7 = 7$. We define a **contradiction** as a constant relational expression being *false*, e.g. $-10 > 5$.

## REAL POLYNOMIALS

We elaborate upon the mathematical environment of polynomials over the real numbers in their variable or indeterminate $x$, a set we denote with $\mathbb{R}_{[x]}$.

▷ Monomials

We define a **monomial** in $x$ as any product $ax^n$, given $a \in \mathbb{R}$ and $n \in \mathbb{N}$. We can extend this concept to several indeterminates $x, y, z, \ldots$ like the monomials $3(xy)^6$ and $3(x^2 y^3 z^6)$ are.

We define the **degree** of a monomial $ax^n$ as its natural exponent $n \in \mathbb{N}$ to the **indeterminate part** $x$. We say constant numbers are monomials of degree 0 and linear terms are monomials of degree 1. We say squares have degree 2 and cubes have degree 3, followed by monomials of higher degree.

For instance, the real monomial $-\sqrt{12}x^6$ is of degree 6. Extending this concept, the monomial $3(xy)^6$ is of degree 6 in $xy$ and the monomial $3(x^2 y^3 z^6)^9$ is of degree 9 in $x^2 y^3 z^6$.

We define **monomials of the same kind** as those having an identical indeterminate part. For instance, both $\frac{5}{7}x^6$ and $-\sqrt{12}x^6$ are of the same kind. Extending the concept, likewise $\frac{5}{7}x^3 y^5 z^2$ and $-\sqrt{12}x^3 y^5 z^2$ are of the same kind.

All basic operations on monomials emerge simply from applying the calculation rules of fractions and powers.

▷ Polynomials

We define a **polynomial** $V(x)$ as any sum of monomials. We define the **degree** of $V(x)$ as the maximal exponent $m \in \mathbb{N}$ to the indeterminate variable $x$. For instance, the real polynomial

$$V(x) = 17x^2 + \frac{1}{4}x^3 + 6x - 7x^2 - \sqrt{12}x^6 - 13x - 1,$$

is of degree 6.

Whenever monomials of the same kind appear in it, we can simplify the polynomial. For instance, our polynomial simplifies to $V(x) = 10x^2 + \frac{1}{4}x^3 - 7x - \sqrt{12}x^6 - 1$.

Moreover, we can sort any given polynomial either in an ascending or descending order according to its powers in $x$. Sorting our polynomial $V(x)$ in an ascending order yields $V(x) = -1 - 7x + 10x^2 + \frac{1}{4}x^3 - \sqrt{12}x^6$. Sorting $V(x)$ in a descending order yields $V(x) = -\sqrt{12}x^6 + \frac{1}{4}x^3 + 10x^2 - 7x - 1$.

Eventually we are able to evaluate any polynomial, getting a numerical value from it. For instance evaluating $V(x)$ in $x = -1$, yields $V(-1) = -\sqrt{12}(-1)^6 + \frac{1}{4}(-1)^3 + 10(-1)^2 - 7(-1) - 1 = -\sqrt{12} - \frac{1}{4} + 16 = \frac{63}{4} - 2\sqrt{3} \in \mathbb{R}$.

▷ **Basic operations**

Adding two monomials of the same kind: we add their coefficients and keep their indeterminate part

$$5a^2 - 3a^2 = (5-3)a^2 = 2a^2.$$

Multiplying two monomials of any kind: we multiply both their coefficients and their indeterminate parts

$$-5ab \cdot \frac{7}{4}a^2b^3 = -5 \cdot \frac{7}{4} \cdot a^{1+2}b^{1+3} = \frac{-35}{4}a^3b^4.$$

Dividing two monomials: we divide both their coefficients and their indeterminate parts

$$\frac{-8a^6b^4}{-4a^4} = \frac{-8}{-4}a^{6-4}b^{4-0} = 2a^2b^4.$$

Exponentiating a monomial: we exponentiate each and every factor in the monomial

$$\left(-2a^2b^4\right)^3 = (-2)^3(a^2)^3(b^4)^3 = -8a^6b^{12}.$$

Adding or subtracting polynomials: we add or subtract all monomials of the same kind

$$(x^2 - 4x + 8) - (2x^2 - 3x - 1) = x^2 - 4x + 8 - 2x^2 + 3x + 1 = -x^2 - x + 9.$$

Multiplying two polynomials: we multiply each monomial of the first polynomial with each monomial of the second polynomial and simplify all those products to the resulting product polynomial

$$\begin{aligned}
(2x^2 + 3y) \cdot (4x^2 - y) &= 2x^2(4x^2 - y) + 3y(4x^2 - y) \\
&= 2x^2 \cdot 4x^2 + 2x^2 \cdot (-y) + 3y \cdot 4x^2 \\
&\quad + 3y \cdot (-y) \\
&= 8x^4 - \boxed{2x^2y} + \boxed{12x^2y} - 3y^2 \\
&= 8x^4 + 10x^2y - 3y^2.
\end{aligned}$$

## 1.2 Equations in one variable

In anticipation of this section, we will refresh the required vocabulary. A **solution** is any value assigned to the variable that turns the given equation into an *equality* (being *true*). The **scope** of an equation is any number set in which the equation resides, realising it will most likely be $\mathbb{R}$. We define the **solution set** as the set containing all legal solutions to an equation. This solution set is always a subset of the scope of the equation.

### LINEAR EQUATIONS

A **linear equation** is an algebraic equation of degree one, referring to the maximum natural exponent of the unknown quantity. By simplifying we can always standardise any linear equation to

$$ax + b = 0, \tag{1.1}$$

given $a \in \mathbb{R} \setminus \{0\}$ and $b \in \mathbb{R}$. We cite $3x + 7 = 22, 5x - 9d = c$ and $5(x-4) + x = -2(x+2)$ as examples of linear equations, and $3x^2 + 7 = 22$ and $5ab - 9b = c$ as counter examples. The adjective 'linear' originates from the Latin word 'linea' meaning (straight) line as referring to the graph of a linear function (see chapter 4).

We solve a linear equation for its unknown part by rewriting the entire equation until its shape exposes the solution explicitly.

We recall easily the required rules for rewriting a linear equation by the metaphor denoting a linear equation as a 'pair of scales'. This way we should never forget to keep the equation's balance: whatever operation we apply, it has to act on both sides of the equals-sign. If we add to (or subtract from) the left hand 'scale' than we are obliged to add the same term to (or subtract it from) the right hand 'scale'. If we multiply (or divide) the left hand side, than we

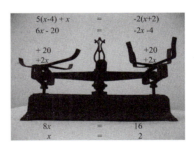

are likewise obliged to multiply (or divide) the right hand side with the same factor. If not, our equation would lose its balance just like a pair of scales would. We realise that our metaphor covers all usual 'rules' to handle linear equations.

The reason we perform certain rewrite steps depends on which variable we are aiming for. This is called *strategy*. Solving the equation for a different variable implies a different sequence of rewrite steps.

*Example*:   We solve the equation $5(x-4)+x=-2(x+2)$ for $x$. Firstly, we apply the distributive law: $5x-20+x=-2x-4$. Secondly, we put all terms dependent of $x$ to the left hand side and the constant numbers to the right hand side $5x+x+2x=-4+20$. Thirdly, we simplify both sides $8x=16$. Finally, we find $x=2$ leading to the solution singleton $\{2\}$.

## QUADRATIC EQUATIONS

Handling quadratic expressions and solving quadratic equations are useful basics in order to study topics in multimedia, digital art and technology.

▷  Expanding products

We refresh **expanding** a product as (repeatedly) applying the distributive law until the initial expression ends up as a pure *sum* of terms. Note that our given polynomial $V(x)$ itself does not change: we just shift its appearance to a pure sum. We illustrate this concept through $V(x)=(2x-3)(4-x)$.

$$\begin{aligned}(2x-3)(4-x) &= (2x-3)\cdot 4+(2x-3)\cdot(-x)\\ &= (8x-12)+(-2x^2+3x)\\ &= -2x^2+11x?12.\end{aligned}$$

Other examples are

$$5a(2a^2-3b)=5a\cdot 2a^2-5a\cdot 3b=10a^3-15ab$$

and

$$\begin{aligned}4\left(x-\frac{1}{2}\right)\left(x+\frac{13}{2}\right) &= (4x-2)\left(x+\frac{13}{2}\right)\\ &= (4x-2)\cdot x+(4x-2)\cdot\frac{13}{2}\\ &= 4x^2-2x+26x-13=4x^2+24x-13.\end{aligned}$$

▷  Factoring polynomials

We define **factoring** a polynomial as decomposing it into a pure *product* of (as many as possible) factors. Note that our given polynomial $V(x)$ itself does not change: we just shift its appearance to a pure product. Our **trinomial** $V(x)=-2x^2+11x-12$ just shifts its appearance to the pure product $V(x)=(2x-3)(4-x)$ when factored. It merely shows that the product $(2x-3)(4-x)$ is a factorisation of the trinomial $-2x^2+11x-12$.

Imagine we had to factor the trinomial $-2x^2 + 11x - 12$ without any hint. This way we realise that factoring generally is a hard job to do. Especially because we do not have any clue about which factors build up the pure product for a polynomial. Many questions arise: *how many* factors to expect, *where* to start from, *what* is the opening step towards factorisation?

We observe the need for at least a minimum asset of factoring methods. As an extra motivation, we emphasise the importance of factoring as it reveals all essential building blocks of any polynomial. Knowing the roots of a polynomial gives us a deeper insight. We therefore introduce some factoring basics in the next paragraphs.

## Common Factor

We show how to separate common factors if they appear.

For instance $6 + 12x = 6 \cdot 1 + 6 \cdot (2x) = 6 \cdot (1 + 2x)$ results in a pure product of a number and a linear factor by separating the common factor 6. Another polynomial like $5x + x^2 = 5x + xx = (5 + x) \cdot x$ separates into two linear factors by the use of the common factor $x$. An example expression like $39x + 3xy = 3 \cdot 13x + 3xy = 3 \cdot x \cdot (13 + y)$ yields a pure product of a number factor, a linear factor in $x$ and a linear factor in $y$ by separating the common factors 3 and $x$. Occasionally we may have to factor by grouping. For instance

$$\begin{aligned} 1 + x + x^2 + x^3 &= (1 + x) + (x^2 + x^3) = (1 + x) + (x^2 \cdot 1 + x^2 \cdot x) \\ &= (1 + x) + x^2(1 + x) = 1 \cdot (1 + x) + x^2(1 + x) \\ &= (1 + x^2) \cdot (1 + x) \end{aligned}$$

results stepwise into a pure product of a quadratic and a linear factor in $x$.

## Perfect powers

Expanding the natural powers of the **binomial** $A + B$ reveals their corresponding pure sum shapes.

$$\begin{aligned} (A + B)^2 &= (A + B)\ (A + B) = A^2 + 2AB + B^2 \\ (A + B)^3 &= (A + B)^2(A + B) = A^3 + 3A^2B + 3AB^2 + B^3 \\ (A + B)^4 &= (A + B)^3(A + B) = A^4 + 4A^3B + 6A^2B^2 + 4AB^3 + B^4 \end{aligned}$$

We define a **perfect power** as any natural exponentiation of a binomial. The important power is $(A + B)^2$ which we define as the **perfect square** of its binomial $A + B$.

Those perfect powers of $A + B$, when ordered to ascending natural exponents, display **Pascal's Triangle** for all $n \in \mathbb{N}$.

$$1$$
$$1A + 1B$$
$$1A^2 + 2AB + 1B^2$$
$$1A^3 + 3A^2B + 3AB^2 + 1B^3$$
$$1A^4 + 4A^3B + 6A^2B^2 + 4AB^3 + 1B^4$$
$$1A^5 + 5A^4B + 10A^3B^2 + 10A^2B^3 + 5AB^4 + 1B^5$$
$$\vdots$$

Notice how a **coefficient** is produced as a sum of its upper two, leading to a symmetric triangle of numbers with the constant '1' on both edges. This 'triangle' is named after its explorer **Blaise Pascal** (1623 –1662).

Despite the diminishing need for perfect power formulas in this century of ruling computing power, we do advise you to know at least the perfect square by heart.

To put the perfect square into words: *'The square of a binomial equals the sum of both squares plus two times the product'*.

$$(A+B)^2 = A^2 + 2AB + B^2. \tag{1.2}$$

We provide a visual aid to help you memorise it. The area of the total square equals $(A+B) \cdot (A+B)$. Alternatively we puzzle this area piece by piece, via adding both white square areas $A^2$ and $B^2$ plus the *two* grey rectangular areas $AB$, jointly equalling the perfect square as the trinomial $A^2 + B^2 + 2AB$.

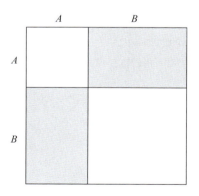

Consequently we now can explore a new factoring method. For instance, we intend to factor the trinomial $1 - 2x + x^2$, whilst we have no guarantee for its pure product shape to even exist.

Strategically we perform two subsequent checks.

1) Verify whether both squares carry the same sign.

2) Then find $2AB$ corresponding correctly to the given $A$ and $B$.

Only when both checks hold, are we able to shift the given trinomial to its perfect product $(A+B)^2$. We give an example of this strategy to the trionomial $1 - 2x + x^2$.

We rewrite the trinomial to $+(-1)^2 - 2x + (x)^2$ by assigning $A = -1$ and $B = x$. By substituting $A$ and $B$ into $2AB$, we find $+(-1)^2 + 2(-1)(x) + (x)^2$ equalling our trinomial. Therefore we confirm $A = -1$ and $B = x$, which allows the shift $1 - 2x + x^2 = (-1 + x)^2$. We realise that alternatively $(+1 - x)^2$ is a correct factorisation as well.

Perfect quotient

$$(A + B)(A - B) = A^2 - B^2 \qquad (1.3)$$

▷ Quadratic formula for quadratic equations

A **quadratic equation** is an algebraic equation of degree two in the unknown quantity $x$ that can be reduced to the default shape

$$ax^2 + bx + c = 0 \qquad (1.4)$$

given $a \in \mathbb{R} \setminus \{0\}$ and $b, c \in \mathbb{R}$.

To solve this equation for $x$ we firstly divide both sides of it by $a$. Dividing by $a$ is valid since $a \neq 0$. In case of $a = 0$ we would no longer have a quadratic but a linear equation.

$$ax^2 + bx + c = 0 \iff x^2 + \frac{b}{a}x + \frac{c}{a} = 0$$

Secondly, we aim for a perfect square by adding and subtracting the special term $\left(\frac{b}{2a}\right)^2$ which is again valid since this is equivalent to adding 0. This way we have created a perfect square $(A + B)^2$, assigning $A = x$ and $B = \frac{b}{2a}$.

$$x^2 + 2\left(\frac{b}{2a}\right)x + \left(\frac{b}{2a}\right)^2 - \left(\frac{b}{2a}\right)^2 + \frac{c}{a} = 0$$

$$\iff \left(x^2 + 2\left(\frac{b}{2a}\right)x + \left(\frac{b}{2a}\right)^2\right) - \left(\frac{b}{2a}\right)^2 + \frac{c}{a} = 0$$

$$\iff \left(x + \frac{b}{2a}\right)^2 - \left(\frac{b}{2a}\right)^2 + \frac{c}{a} = 0$$

$$\iff \left(x + \frac{b}{2a}\right)^2 = \left(\frac{b}{2a}\right)^2 - \frac{c}{a}$$

The left hand side of this equation is now a square. Before proceeding we make sure that all denominators of the right hand side are equal.

$$\left(x + \frac{b}{2a}\right)^2 = \frac{b^2}{4a^2} - \frac{c \cdot 4a}{a \cdot 4a} \iff \left(x + \frac{b}{2a}\right)^2 = \frac{b^2 - 4ac}{4a^2}$$

Arithmetically this holds: $L^2 = R \Leftrightarrow L = \sqrt{R}$ or $L = -\sqrt{R}$. Hence we reach two similar solutions to our equation:

$$x + \frac{b}{2a} = \sqrt{\frac{b^2 - 4ac}{4a^2}} \qquad \text{or} \qquad x + \frac{b}{2a} = -\sqrt{\frac{b^2 - 4ac}{4a^2}}$$

$$\Downarrow$$

$$x + \frac{b}{2a} = \frac{\sqrt{b^2 - 4ac}}{2a} \qquad \text{or} \qquad x + \frac{b}{2a} = -\frac{\sqrt{b^2 - 4ac}}{2a}$$

$$\Downarrow$$

$$x = -\frac{b}{2a} + \frac{\sqrt{b^2 - 4ac}}{2a} \qquad \text{or} \qquad x = -\frac{b}{2a} - \frac{\sqrt{b^2 - 4ac}}{2a}$$

The **discriminant** of a quadratic equation $ax^2 + bx + c = 0$, given $a \neq 0$, is the real number to be calculated as

$$D = b^2 - 4ac. \tag{1.5}$$

Furthermore, we solve $ax^2 + bx + c = 0$ for $x$ like this:

if $D < 0$ then there are no solutions in $\mathbb{R}$,

if $D = 0$ we find one real root $x_1 = \frac{-b}{2a}$,

if $D > 0$ we have two similar **roots**

$$x_1 = \frac{-b + \sqrt{D}}{2a} \text{ and } x_2 = \frac{-b - \sqrt{D}}{2a}. \tag{1.6}$$

As a spin-off these roots enable factoring the default left hand side as

$$ax^2 + bx + c = a(x - x_1)(x - x_2) \text{ when } D > 0 \tag{1.7}$$

and as

$$ax^2 + bx + c = a(x - x_1)^2 \text{ when } D = 0.$$

*Examples*:   Solving the quadratic equation $-2x^2 + 11x - 12 = 0$ for $x$, we firstly calculate its discriminant $D = 11^2 - 4 \cdot (-2) \cdot (-12) = 25$ to subsequently determine its roots as $x_1 = \frac{-11 + \sqrt{25}}{2 \cdot (-2)} = \frac{3}{2}$ and $x_2 = \frac{-11 - \sqrt{25}}{2 \cdot (-2)} = 4$. As a bonus, this allows us to factor $-2x^2 + 11x - 12$ as $(-2)\left(x - \frac{3}{2}\right)(x - 4)$. The solution set is the pair $\{\frac{3}{2}, 4\} \subset \mathbb{R}$.

We solve $25x^2 - 60x + 36 = 0$ for $x$ as a next example. In this case the discriminant equals zero, yielding a unique root of multiplicity 2 to be found as $x = \frac{-(-60) \pm \sqrt{0}}{2 \cdot 25} = \frac{6}{5}$ and thus leading to the solution singleton $\{\frac{6}{5}\} \subset \mathbb{R}$.

Finally solving also $25x^2 + 49x + 36 = 0$ for $x$, we calculate its discriminant as $D = -1199$. It is not possible for us to find any real root due to the fact $\sqrt{-1199} \notin \mathbb{R}$, which in this case leads to an empty solution set $\{\} \subset \mathbb{R}$.

Equations of higher degree

Solving the polynomial quadratic equation $ax^2 + bx + c = 0$ for $x$ by means of the Quadratic Formula

$$\frac{-b \pm \sqrt{D}}{2a}$$

dates back to Babylonian and Greek times. The next big leap forward for solving equations of higher degree had to wait until the $16^{th}$ century, till the time of the Renaissance.

▷ Cubic equations

> **Geronimo Cardano** (1501–1576) published a similar Cubic Formula for solving polynomial equations of degree three. Despite Cardano publishing it, the Cubic Formula was actually discovered by another Italian. Historians claim that this formula was discovered by the mathematician **Niccolo Fontana** (1499 –1557) (nicknamed Tartaglia or '*stutterer*').

▷ Quartic equations

> Shortly after the former formula, **Lodovico Ferrari** (1522 – 1565) a pupil of Cardano, and also an Italian mathematician, found the Quartic formula to solve polynomial equations of degree four.

▷ Quintic equations

> For an apotheosis one needed to wait until the $19^{th}$ century in France: the very young and brilliant mathematician **Evariste Galois** (1811–1832) proved the impossibility of finding a similar Quintic Formula for polynomial equations of larger than degree four. Meanwhile, as a workaround (from, among others, Isaac Newton, around 1676) we can solve any polynomial equation numerically, yielding approximations for its solutions. Apart from this modern numerical approach, special subtypes of polynomial equations of larger degree can also still be solved exactly by means of formulas of radical expressions.

## 1.3 Logarithms

▷ We define the **common or Briggsian logarithm** as an exponent to base 10,

$$\log_{10}(a) = x \Leftrightarrow 10^x = a$$

which satisfies the existence condition $a \in \mathbb{R}^+ \setminus \{0\}$ with base $10 \in \mathbb{R}^+ \setminus \{0, 1\}$.

*Examples*:

$$\begin{aligned} \log_{10}(100) &= 2 \\ \log_{10}(1000) &= 3 \\ \log_{10}(100000) &= 5 \end{aligned}$$

The decimal logarithm is an idea of the Englishman **Henry Briggs** (1561–1630), contemporary of the Scotsman **John Napier** (1550 – 1617). Around 1615, both mathematicians agreed that the logarithm base number 10 would offer the better future perspective. Among others, the famous scientist **Simon Stevin** (1548 – 1620) from Bruges, in Belgium, contributed a lot in establishing decimal numbers worldwide, in line with base number 10. In 1624, Briggs published the very first decimal logarithm table in his book 'Arithmetica Logaritmica'.

Back in 1618, Napier published (unknowingly) the **natural base**, which was later re-discovered as the transcendental limit value

$$2.718281828459\ldots \tag{1.8}$$

by the Swiss **Jakob Bernoulli** (1654 –1705) and – similarly to the transcendental number $\pi \approx 3.14$ for circles – by the next Swiss genius **Leonhard Euler** (1707–1783) named Euler's number, simply called by the symbol $e \approx 2.72$.

▷  We define the **natural logarithm** or **Napier's logarithm** as an exponent to base $e \approx 2.72$,

$$\log_e(a) = x \Leftrightarrow e^x = a$$

which satisfies the existence condition $a \in \mathbb{R}^+ \setminus \{0\}$ with base $e \in \mathbb{R}^+ \setminus \{0,1\}$.

*Examples*:

$$\begin{aligned} \log_e(e^1) &= 1 \\ \log_e(e^2) &= 2 \\ \log_e(e^5) &= 5 \end{aligned}$$

John Napier conceived of the natural logarithm around 1594. After decades of calculation, he finally published the first natural logarithm table in his book 'Mirifici Logarithmorum Canonis Descriptio' in 1614. New mathematical ideas acquire a common status proportional to their ease of use, but Napier's first design based on $\frac{1}{e} \approx 0.368$ was seemingly less practical. In general, Napier contributed substantially to the popularity and adoption of decimal numbers and the decimal logarithm.

AR I T H M E T I C   R E F R E S H E R

▷ We define the **binary logarithm** as an exponent to base 2,

$$\log_2(a) = x \Leftrightarrow 2^x = a$$

which satisfies the existence condition $a \in \mathbb{R}^+ \setminus \{0\}$ with base $2 \in \mathbb{R}^+ \setminus \{0, 1\}$.

*Examples*:

$$
\begin{aligned}
\log_2(4) &= 2 \\
\log_2(8) &= 3 \\
\log_2(32) &= 5 \\
\log_2(1024) &= 10
\end{aligned}
$$

This binary logarithm is especially applicable in data communication and other binary environments.

## 1.4 Exercises

**Exercise 1** Simplify the following expressions.

1) $a^5 \cdot a^5$

2) $a^5 + a^5$

3) $a^5 - a^5$

4) $(a^5)^3$

5) $a^{10} \cdot a^{-3}$

6) $\frac{a^8}{a^2}$

7) $\frac{a^6}{a^{-2}}$

8) $(2a^{-3})^{-2}$

**Exercise 2** Tick every correct answer, given $a, b, c, d, m, n \in \mathbb{Z} \setminus \{0\}$.

| | | | | | |
|---|---|---|---|---|---|
| 1) | $-(a+b) =$ | $-a-b$ | $a-b$ | $-a+b$ | $a+b$ |
| 2) | $-\frac{a+b}{c} =$ | $\frac{a+b}{-c}$ | $\frac{-a+b}{c}$ | $\frac{-a-b}{c}$ | $\frac{-a-b}{-c}$ |
| 3) | $\frac{c}{d} + \frac{3}{4} =$ | $\frac{c+3}{d+4}$ | $\frac{c+3}{4d}$ | $\frac{4c+3d}{4d}$ | $\frac{4c+3d}{d+4}$ |
| 4) | $\frac{c}{d} \cdot \frac{3}{4} =$ | $\frac{c+3}{d+4}$ | $\frac{4d}{3c}$ | $\frac{3c}{4d}$ | $\frac{12c}{4d}$ |
| 5) | $\left(\frac{-1}{7}\right)^{-1} =$ | $\frac{1}{7}$ | $-7$ | $\frac{-1}{7}$ | $7$ |
| 6) | $a^2 \cdot a^n \cdot a^m =$ | $a^{2nm}$ | $a^{2+n+m}$ | $a^{2+nm}$ | $a^{2n+m}$ |
| 7) | $(-2a^2b^3)^3 =$ | $-6a^6b^9$ | $-6a^2b^6$ | $-8a^6b^9$ | $-8a^8b^{27}$ |
| 8) | $\frac{c^6d^3}{c^2d} =$ | $c^3d^3$ | $c^4d^3$ | $c^4d^2$ | $c^3d^2$ |
| 9) | $\left(\frac{a^5}{b}\right)^{-2} =$ | $\frac{-a^7}{b^2}$ | $\frac{b^2}{a^{10}}$ | $\frac{b^2}{a^{-10}}$ | $\frac{-a^{10}}{b^2}$ |
| 10) | $a^{12} : a^3 =$ | $a^9$ | $a^4$ | $a^{-4}$ | $a^{-9}$ |

**Exercise 3** Solve both linear equations for $x$.

1) $2 \cdot \left(\frac{x}{3} + 1\right) = \frac{x}{4} - \frac{1}{2}$

2) $\left(3 + \frac{2}{7}\right) \cdot \left(\frac{x}{3} - 1\right) = \frac{x}{7} + \frac{5}{3}$

**Exercise 4** The sum of three successive odd integers is 141. Find these three numbers.

**Exercise 5** Simplify the following expressions by using appropriate rules and use exclusively positive exponents to answer. All variables are nonzero rational numbers.

1)  $(a^3)^{-4}$

2)  $a^{12} : a^{13}$

3)  $(a^{-4})^5$

4)  $\left(\frac{a^2}{b^3}\right)^3$

5)  $b^3 : b^{-5}$

6)  $(-2 \cdot a^3 \cdot b^2)^4$

7)  $((b^5)^5)^2$

8)  $((a^{-5})^3 \cdot a^{11}) : (a^9)^{-3}$

9)  $\frac{(-4 \cdot a^4 \cdot b)^3}{(-2 \cdot a^3 \cdot b)^2}$

10)  $(2 \cdot a^2 \cdot b^3)^4 \cdot (-3 \cdot a \cdot b^4)^2$

**Exercise 6** Simplify as much as possible.

1)  $-4x^2y^2 + (2xy)^2$

2)  $-4x^2y^2(+2xy)^2$

3)  $-4x^2y^2 : (2xy)^2$

4)  $(x^2y)^3(-x^6y^3)$

5)  $(x^2y)^3(-x^6y)^3$

6)  $(x^2y)^3 - (x^6y^3)$

7)  $x(x-1) + (2x+1)(x-3)$

8)  $6x(x+5) - 8x(x^2 - 4x + 3)$

9)  $(3x-2)(x-4) - 3x^2 + 7x(2x-1)$

**Exercise 7** Solve these quadratic equations for their unknown quantity (variable) within the scope of $\mathbb{R}$.

1)  $6\delta^2 + 5\delta = 0$

2)  $4x^2 - 25 = 0$

3)  $2t^2 - t - 6 = 0$

4)  $4x^2 + 5x + 1 = 0$

5)  $(t+1)^2 + (t+2)^2 + (t+3)^2 = 2$

6)  $t(t+1) = 6$

**Exercise 8** Was the polynomial $V(x,y) = 2(x-2y)(a-3b) + 5(a+3b)(2y-x)$ factored? If so, why is it a factorisation? If not so, then factor $V(x,y)$.

**Exercise 9** Factor the quadratic trinomial $K(x) = 1 + 6x + 9x^2$ using the Quadratic Formula.

**Exercise 10** Provide each expression with the most simplified answer relying on the definition of the logarithm used.

1) $\log_e(1)$

2) $\log_2(16)$

3) $\log_4(16)$

4) $\log_8(16)$

5) $\log_e\left(\frac{1}{e\sqrt{e}}\right)$

6) $\log_{\frac{1}{2}}(16)$

# Chapter 2 · Linear systems

Though it is not a skill that is inevitably acquired, solving linear systems is a particularly useful skill for the expert in multimedia topics. As chosen in this book, the elimination of variables for solving linear systems is as straightforward to implement as to scale. Scalability is crucial in situations where variables might run into thousands.

## 2.1 Definitions

We define a **linear system** as a collection of $n$ linear equations involving the same set of $m$ variables, hence it is often called an $n$ by $m$ system.

*Example*:

$$\begin{cases} 3x + 2y = 6 \\ 5x + 2y = 4 \end{cases}$$

Finding the solution to a linear system does not mean separately solving each equation. It implies the *simultaneous* solving of all equations.

We define **equivalent systems** as systems having the same solution.

*Example*: The 3 by 3 system

$$\begin{cases} -4y + z + 7 = -2x \\ x + y - 3z = 6 \\ -y + 3x + z = 0 \end{cases}$$

is equivalent to

$$\begin{cases} 2x + 2y - 6z = 12 \\ - 6y + 7z = -19 \\ 3x - y + z = 0 \end{cases}$$

and equivalent to

$$\begin{cases} x + y - 3z = 6 \\ 2y - 5z = 9 \\ 8z = -8. \end{cases}$$

We link equivalent systems by inserting the equivalent-sign '$\Longleftrightarrow$'.

$$\begin{cases} -4y + z + 7 = -2x \\ x + y - 3z = 6 \\ -y + 3x + z = 0 \end{cases} \Longleftrightarrow \begin{cases} x = 1 \\ y = 2 \\ z = -1 \end{cases}$$

The solution set to the latter equivalent system is a singleton containing one triple: $\{(1,2,-1)\} \subset \mathbb{R} \times \mathbb{R} \times \mathbb{R} = \mathbb{R}^3$.

We define the **solution to an $n$ by $m$ linear system**, if it exists, as the list of $m$ values that satisfies all $n$ equations when assigned to the variables.

*Example*: We solve the linear system

$$\begin{cases} 3x + 2y = 6 \\ 5x + 2y = 4 \end{cases}$$

as

$$\begin{cases} x = -1 \\ y = \frac{9}{2} \end{cases}$$

often typeset in a **solution set** $\{(-1, \frac{9}{2})\} \subset \mathbb{R} \times \mathbb{R} = \mathbb{R}^2$.

We define a **square system** as any system having as many equations as unknown quantities and often referred to as an $n$ by $n$ system.

*Example*: This 3 by 3 system

$$\begin{cases} 2x & + & 2y & - & 6z & = 12 \\ & & -6y & + & 7z & = -19 \\ 3x & - & y & + & z & = 0 \end{cases}$$

is a square system.

We define a **triangular system** as a system with all of its equations ordered to their variables, showing a triangle of coefficients 0 under the diagonal.

*Example*:

$$\begin{cases} x & + & y & - & 3z & = 6 \\ & & 2y & - & 5z & = 9 \\ & & & & 8z & = -8 \end{cases}$$

Notice how this diagonal runs along the ordered variables $x, y, z$ under which exclusively coefficients 0 appear.

We define a linear system as **underdetermined** when its solution contains one (or more) free **parameter(s)**. An underdetermined linear system has a **degree of freedom** equal to the number of free parameters in its solution.

*Example*:

$$\begin{cases} -3x + 2y & = -4 \\ 9x - 6y & = 12 \end{cases}$$

We notice that both equations are dependent. Dividing both the left hand side and right hand side of the second equation by $-3$ simplifies the system to

$$\begin{cases} -3x + 2y = -4 \\ -3x + 2y = -4. \end{cases}$$

It seems that we actually apply just one condition to two variables $x$ and $y$. Consequently we need to set free one of both variables, either $x$ or $y$. Let us for instance choose to set free $x \in \mathbb{R}$. The degree of freedom of this system is 1 and we simply proceed by solving this system for $y$.

$$\left\{ \begin{array}{l} y = \frac{-4+3x}{2} \\ x \in \mathbb{R} \text{ (free parameter)} \end{array} \right. \iff \left\{ \left( x, \frac{-4+3x}{2} \right) \text{ with } x \in \mathbb{R} \right\} \subset \mathbb{R} \times \mathbb{R}$$

However this answer admits an infinite number of solutions, they are all according to the solution template as dictated by the free parameter.

$$\left\{ \ldots, \left( -1, \frac{-7}{2} \right), (0, -2), \left( 1, \frac{-1}{2} \right), (2, 1), \left( 3, \frac{5}{2} \right), \ldots \right\} \subset \mathbb{R} \times \mathbb{R}$$

To distinguish the free parameter from the ordinary variable $x$ we typeset it by a Greek character such as $\lambda$. We summarise the above solution set as

$$\left\{ \left( \lambda, \frac{-4+3\lambda}{2} \right) \text{ with } \lambda \in \mathbb{R} \right\} \subset \mathbb{R} \times \mathbb{R}.$$

We define a linear system as **inconsistent** when two (or more) of its equations cause a conflict. Consequently, inconsistent systems have an empty solution set.

*Example*:

$$\left\{ \begin{array}{ll} 3x - 4y & = 2 \\ -6x + 8y & = 6 \end{array} \right. \iff \left\{ \begin{array}{l} 3x - 4y = 2 \\ 0x + 0y = 10 \end{array} \right.$$

Note that one of these equations holds a contradiction, leading to an inconsistent system. Hence no value set can ever make this system *true*. As solutions cannot exist, this system has an empty solution set $\{\} \subset \mathbb{R} \times \mathbb{R}$.

## 2.2 Methods for solving linear systems

### SOLVING BY SUBSTITUTION

Substitution is a method involving systematically replacing the variables of the system. We solve one of the equations for one variable of our choice. For instance, we choose to solve the second equation for $y$.

$$\left\{ \begin{array}{ll} -2x + y & = -2 \\ x + y & = 10 \end{array} \right. \iff \left\{ \begin{array}{ll} -2x + y & = -2 \\ y & = 10 - x \end{array} \right.$$

We then replace or 'substitute' $y$ by its expression $y = 10 - x$ in the first equation $-2x + y = -2$. We are now able to solve the resulting equation for $x$ only.

$$\begin{cases} -2x + (10 - x) &= -2 \\ y &= 10 - x \end{cases} \iff \begin{cases} -3x + 10 &= -2 \\ y &= 10 - x \end{cases}$$

$$\iff \begin{cases} -3x &= -12 \\ y &= 10 - x \end{cases}$$

$$\iff \begin{cases} x &= 4 \\ y &= 10 - x \end{cases}$$

Finally we replace the obtained value $x = 4$ back in the second equation.

$$\begin{cases} x &= 4 \\ y &= 10 - x \end{cases} \iff \begin{cases} x &= 4 \\ y &= 10 - 4 = 6 \end{cases}$$

Solving linear systems by substitution has three disadvantages.

1) Substitution becomes digressive in solving systems of larger $n \times m-$dimensions.

2) As a consequence, solving by substitution becomes increasingly error prone.

3) Lacking a clear strategy, solving by substitution may end up in a vicious circle.

## SOLVING BY ELIMINATION

We strongly recommend solving any *linear* system by elimination. This method involves a stepwise renewing of the initial system, dictated by the algorithm of the famous German mathematical genius **Carl Friedrich Gauss** (1777–1855), until the complete solution appears. Be aware that this method is valid for *linear* systems only. Solving by elimination is as straightforward to use with pen and paper as to implement in software.

Gaussian algorithm

**Step 1**: We **trim** the initial system for each equation to its variables. Remember to maintain the same order of appearance for the variables in each equation. Leave a blank space in case of an absent variable.

**Step 2**: Renew this trimmed version to an equivalent **triangular** system. To achieve this, only three **elementary row operations** can be used:

▷ swapping two equations,

▷ multiplying an equation with a nonzero number,

▷ adding two equations.

Strategically we aim at setting coefficients to zero below each targeted variable.

**Step 3**: Finally we **replace** from the bottom equation to the top equation each determined value for the subsequent variable. At the final stage, we find the complete solution to the linear system.

*Example*:    We solve a 3 by 3 linear system for $x, y$ and $z$ using the above Gaussian algorithm.

$$\begin{cases} -4y + z + 7 & = -2x \\ x + y - 3z & = 6 \\ -y + 3x + z & = 0 \end{cases}$$

**Step 1**: We trim the initial system.

$$\begin{cases} 2x & - & 4y & + & z & = -7 \\ x & + & y & - & 3z & = 6 \\ 3x & - & y & + & z & = 0 \end{cases}$$

**Step 2**: We renew the trimmed system to an equivalent triangular system.

Whenever it suits us, we may swap equations in order to get an 'easier' one on top. In this example, we switch the first and second equation, what we denote by $R_1 \leftrightarrow R_2$.

$$\begin{cases} x & + & y & - & 3z & = 6 \\ 2x & - & 4y & + & z & = -7 \\ 3x & - & y & + & z & = 0 \end{cases}$$

We will use the top equation $x + y - 3z = 6$ as a tool for eliminating the first variable $x$ in the underlying equations. Meanwhile we will keep copying our unaffected 'tool' $x + y - 3z = 6$ to the very end. We eliminate $x$ in the second equation by adding $-2$ times the top equation to it.

$$\begin{cases} x & + & y & - & 3z & = 6 & | -2 \\ 2x & - & 4y & + & z & = -7 & | 1 \\ 3x & - & y & + & z & = 0 \end{cases}$$

We denote the above action by $R_2 \rightarrow -2R_1 + R_2$. We recall that we keep the top equation and only renew the second equation by our strategic combination. This leads to

$$\begin{cases} x & + & y & - & 3z & = 6 \\ 0x & - & 6y & + & 7z & = -19 \\ 3x & - & y & + & z & = 0. \end{cases}$$

The equation $0x - 6y + 7z = -19$ is the result of adding the multiples $(\mathbf{-2})(x + y - 3z = 6)$ $+ \mathbf{1}(2x - 4y + z = -7)$.

Likewise, we eliminate $x$ in the third equation via the renewal $R_3 \to -3R_1 + R_3$.

$$\begin{cases} x & + & y & - & 3z & = 6 & \quad | -3 \\ 0x & - & 6y & + & 7z & = -19 & \quad | \\ 3x & - & y & + & z & = 0 & \quad | 1 \end{cases}$$

$$\Longleftrightarrow \begin{cases} x & + & y & - & 3z & = 6 \\ 0x & - & 6y & + & 7z & = -19 \\ 0x & - & 4y & + & 10z & = -18 \end{cases}$$

The equation $0x - 4y + 10z = -18$ is the result of adding the multiples $(\mathbf{-3})(x + y - 3z = 6)$ $+ \mathbf{1}(3x - y + z = 0)$.

We now acquire a decreased 2 by 2 system in $y$ and $z$. We now make this 2 by 2 system subject to the Gaussian algorithm.

$$\begin{cases} x & + & y & - & 3z & = 6 \\ & & - & 6y & + & 7z & = -19 \\ & & - & 4y & + & 10z & = -18 \end{cases}$$

After simplifying and omitting obsolete zeros we reach the linear system above. Whenever it suits us, we may multiply any equation with a nonzero value in order to simplify it. We multiply the second equation with the factor $-1$, logging this action as $R_2 \to -R_2$. We likewise multiply the third equation with the factor $-\frac{1}{2}$, logging this action as $R_3 \to -\frac{1}{2}R_3$.

$$\begin{cases} x & + & y & - & 3z & = 6 \\ & & 6y & - & 7z & = 19 \\ & & 2y & - & 5z & = 9 \end{cases}$$

We proceed by swapping $R_2 \leftrightarrow R_3$.

$$\begin{cases} x & + & y & - & 3z & = 6 \\ & & 2y & - & 5z & = 9 \\ & & 6y & - & 7z & = 19 \end{cases}$$

We keep the first equation. We adopt the second equation $2y - 5z = 9$ as the next tool to eliminate $y$ in the underlying equation. Consequently we now keep $2y - 5z = 9$ till the algorithm ends. We eliminate the variable $y$ by adding $-3$ times $2y - 5z = 9$ to the third equation.

$$\begin{cases} x & + & y & - & 3z & = 6 \\ & & 2y & - & 5z & = 9 & \quad | -3 \\ & & 6y & - & 7z & = 19 & \quad | 1 \end{cases}$$

We add $(-3)(2y - 5z = 9) + 1(6y - 7z = 19)$, logging it as $R_3 \rightarrow -3R_2 + R_3$, which yields $8z = -8$.

$$\begin{cases} x & + & y & - & 3z & = 6 \\ & & 2y & - & 5z & = 9 \\ & & & & 8z & = -8 \end{cases}$$

We simplify the third equation to $z = -1$.

$$\begin{cases} x & + & y & - & 3z & = 6 \\ & & 2y & - & 5z & = 9 \\ & & & & z & = -1 \end{cases}$$

**Step 3**: This triangular system allows for substituting values for the variables, from the bottom equation up to the top equation. Substituting $z = -1$ in the above equations, we find the value for $y$.

$$\begin{cases} x & + & y & - & 3(-1) & = 6 \\ & & 2y & - & 5(-1) & = 9 \\ & & & & z & = -1 \end{cases} \iff \begin{cases} x & + & y & + & 3 & = 6 \\ & & y & & & = 2 \\ & & & & z & = -1 \end{cases}$$

Substituting also $y = 2$ in the above equations, we find the value for $x$.

$$\begin{cases} x & + & 2 & + & 3 & = 6 \\ & & y & & & = 2 \\ & & & & z & = -1 \end{cases} \iff \begin{cases} x & = 1 \\ y & = 2 \\ z & = -1 \end{cases}$$

We solved this system yielding the solution set $\{(1, 2, -1)\} \subset \mathbb{R} \times \mathbb{R} \times \mathbb{R} = \mathbb{R}^3$.

Verifying the solution is straightforward and can be done in no time:

$$\begin{cases} 2(1) & - & 4(2) & + & (-1) & = -7 \\ 1 & + & 2 & - & 3(-1) & = 6 \\ 3(1) & - & 2 & + & (-1) & = 0. \end{cases}$$

Corresponding to the disadvantages of solving by substitution, solving by elimination has three advantages.

1) Solving by elimination scales properly, even for very large $n$ by $m$ dimensions.

2) This scalability firmly reduces the error rate when performing on large systems.

3) Based upon a transparent algorithm with a clear ending phase (announced by the triangular shape), it is unlikely to end up in a vicious circle.

## 2.3 Exercises

**Exercise 11** Solve this linear system in $\mathbb{R}^4$:

$$\begin{cases} x + y + 2z + 3v & = 9 \\ 2x + z + 4v & = 11 \\ -y + 2z + v & = 4 \\ 2x + 7y + 8z & = 10. \end{cases}$$

**Exercise 12** Solve this linear system in $\mathbb{R}^3$:

$$\begin{cases} 2x_2 + x_3 + x_1 & = 4 \\ x_1 + 2x_3 + 2x_2 & = 3 \\ x_3 + x_1 + 6x_2 & = 1. \end{cases}$$

**Exercise 13** A house contains four identical wooden windows. In total it took 32 metres of wood to construct them. The length of each window equals three times the width. Find the dimensions of each window.

**Exercise 14** The sum of two real numbers is 210 and their difference is 36. Find these numbers.

**Exercise 15** A father tells his son: 'Three years ago I used to be four times your age.' Then 'Indeed,' his son answers, 'but next year you will be three times my age.' Find today's ages of both the father and son.

**Exercise 16** After finishing exams, Adison and Valence compare their marks. Adison says: 'I was graded 50 points more than Valence'. Valence replies: 'If they had graded me only 310 points more, I would have twice the points of Adison'. Find the graded points, both for Adison and for Valence.

**Exercise 17** The main component of a computer is its motherboard: it determines the processor type and frequency, and therefore the computer architecture. A hardware constructor uses two different assembly robots for producing 29 000 motherboards. The fastest robot assembles 10 000 motherboards in 1 hour, the other robot 6000. Due to repairs the fastest robot remains out of service for the first 30 minutes. Determine how long it then will take to assemble the total batch of 29 000 motherboards and how many of them were made by each robot.

**Exercise 18** For manufacturing 3 types of laptop components (I, II and III) we need 3 machines ($A, B$ and $C$).

We detail the involved production times (in minutes):

 ▷ component I: 3 minutes using machine $A$, 11 minutes using machine $B$ and 27 minutes using machine $C$,

 ▷ component II: 15 minutes using machine $A$, 5 minutes using machine $B$ and 6 minutes using machine $C$,

 ▷ component III: 12 minutes using machine $A$, 14 minutes using machine $B$ and 5 minutes using machine $C$.

How many laptop components of each type should we produce in an 8 hour working day in order to use each machine full time?

**Exercise 19** Teddy and Phiona each cut a rectangular sheet of paper into two equal parts. Phiona has cut her sheet into two rectangles, each having a circumference of 40 cm. The same applies to Teddy, though with a rectangular circumference of 50 cm. We emphasize that both of them cut an identical rectangular sheet of paper. Find the initial rectangular circumference of the sheet both girls cut.

**Exercise 20** Two masses, one of 155 kg and one of 264 kg, are put on a pair of scales. When doubling the smallest mass, we need to shift the centre on the arm by 7 m in order to restore balance. Find the arm length. Hint: $\text{mass}_1 \cdot \text{distance}_1 = \text{mass}_2 \cdot \text{distance}_2$.

**Exercise 21** It takes an aeroplane 3 hours to fly the 2 000 km from Brussels to Athens, benefiting from a tailwind. During the return flight, the aeroplane now faces the same wind as a headwind, which causes a return duration of 4 hours. Find both the speed of the plane and the speed of this wind.

**Exercise 22** On a hot summer day, Edmond, Edward and Evelyn are enjoying a drink outdoors. Due to the heat, they first order four beers and two sodas and Evelyn pays 8 EUR for it. Edmond pays 7.5 EUR for the next order of three beers and three sodas. Although there is no price list outside, Edward manages to figure out the price of one beer. Can you?

**Exercise 23** Rita wraps a parcel tying it with 16 cm of ribbon. Since it is a brick-shaped parcel, she has three possibilities in doing so and respectively she is left with 2 cm, 5 cm or just 1 cm for the knot. Find the parcel's dimensions.

# Chapter 3 · Trigonometry

Computer animation often deals with variables representing position, distance, angle, ...
The major mathematical concepts that cover these measures are sine and cosine. All the
remaining concepts in this chapter recur frequently in the subsequent chapters.

Trigonometry originated from measuring our surrounding space and definitely leads to
a multitude of applications such as coordinate systems, rotation, periodical phenomena
(light and sound), .... We initially define what an angle is in order to turn to triangles.
These triangles allow us to define the trigonometric concepts which we then extend to the
unit circle. We finalise this chapter by outlining the inverse trigonometric functions.

## 3.1 Angles

We consider the cone that connects the pupil of our eye to a golf ball at arm's length.
Cutting this 3D-cone vertically, we obtain a plane angle formed by its two rays or sides,
sharing the common endpoint or vertex $O$ (at our pupil). We now try to measure the size
of this spacial quantity 'angle' as the inner sector bordered by both intersecting sides.

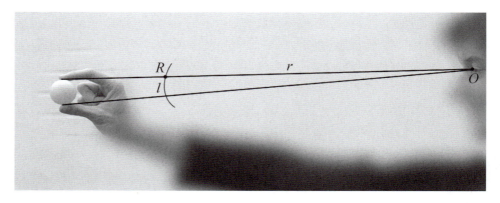

*Figure 3.1*: Definition of an angle

We catch this sector using a pair of compasses and define an **angle** as the circular arc
length $l$ to its radius $|OR| = r$:

$$\alpha = \frac{l}{r}. \tag{3.1}$$

As the above fraction is insensitive to scaling, it indeed guarantees a reliable measurement
for any angle. For instance, when we measure the size of an angle as a circular arc of 6
cm to a radius of 5 cm, then it will remain as a circular arc of 24 cm to its proportionally
adjusted radius of 20 cm. We then calculate $\alpha = \frac{6\,\text{cm}}{5\,\text{cm}} = \frac{24\,\text{cm}}{20\,\text{cm}} = 1.2$. The unit for this angle
$\alpha$, based upon the dimensionless fraction $\frac{\text{circular arc length}}{\text{radius}}$, is called **radians** and commonly

typeset as rad. Hence the above discussion led to an angle of 1.2 rad.

We may be more familiar with **degrees** as a common unit for angles. This ancient Sumerian measure refers to the approximately 360 days for the Earth to complete its 'circular' orbit around the Sun, stating $\alpha_{\text{max}} = 360°$.

Converting angles from radians to degrees and the other way round can be done easily by applying the **Rule Of Three**. We obtain the maximal plane angle by revolving one ray until $\alpha_{\text{max}} = \frac{l_{\text{max}}}{r} = \frac{\text{circle circumference}}{r} = \frac{2\pi r}{r} = 2\pi$ rad. Hence we can rely on the **full angle** equality $2\pi$ rad $= 360°$ given $\pi \approx 3.14$.

*Example*:  Conversion between radians and degrees.

| from degrees to radians | from radians to degrees |
|---|---|
| $360° = 2\pi$ rad | $2\pi$ rad $= 360°$ |
| $1° = \frac{2\pi \text{ rad}}{360}$ | $1$ rad $= \frac{360°}{2\pi}$ |
| $30° = \frac{2\pi \text{ rad}}{360} \cdot 30 = \frac{\pi}{6}$ rad | $\frac{\pi}{6}$ rad $= \frac{360°}{2\pi} \cdot \frac{\pi}{6} = 30°$ |

The fractional part of an angle represented in degrees can be expressed in two different base forms. On the one hand, we can use the popular DD-form (Duo Decimal) for it, writing the fraction in the decimal number base 10. On the other hand, we might use the ancient DMS-form (Degrees Minutes Seconds) for it, writing the fraction in the sexagesimal number base 60 using accents for separating minutes from seconds. Our contemporary time keeping still uses minutes and seconds as parts of one hour. We might for instance express an angle of $180.5°$ as $180°30'00''$ in the latter system. Converting fractional parts of angles from DD to DMS and the other way round goes the same way as converting numbers from decimal base to number base $B = 60$.

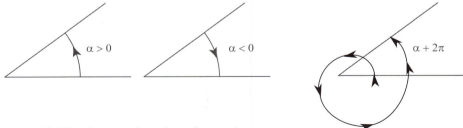

*Figure 3.2*: The sign or orientation of an angle

We can represent the same angle in different ways. We even may attribute a sign to an angle. We define **positive angles** by revolving a ray counter clockwise. Alternatively, we define **negative angles** by revolving a ray clockwise. Furthermore, all angles $\alpha$ and $\alpha'$ for which $\alpha' = \alpha + k\,2\pi$ with parameter $k \in \mathbb{Z}$ (or alternatively $\alpha' = \alpha + k\,360°$) are equivalent due to the full angle being their **elementary period**.

We differentiate the types of angles in some more detail. A **zero** angle equals exactly $0°$. An **acute** angle is larger than $0°$ and smaller than $90°$. A **right** angle equals exactly $90°$ or the quarter of a circle. Both sides of a right angle are said to be orthogonal or perpendicular. An **obtuse** angle is larger than $90°$ and smaller than $180°$. A **straight** angle equals exactly $180°$ or the half of a circle. A **reflex** angle is larger than $180°$ and smaller than $360°$ (full circle).

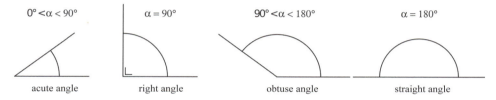

*Figure 3.3*: Types of angles

## 3.2 Triangles

Every three points or vertices $A, B$ and $C$ that do not lie on the same straight line make up a **scalene triangle** $ABC$ when it has no two sides of equal length (see figure 3.10). The measures of the interior angles of a triangle $ABC$ always add up to the straight angle, $\hat{A} + \hat{B} + \hat{C} = 180° = \pi$ rad.

A **right triangle** is a triangle having one of its interior angles equal to $90° = \frac{\pi}{2}$ rad. Consequently, the sum of both acute angles of any right triangle equals $90° = \frac{\pi}{2}$ rad. Its largest side is the edge opposite the right angle and is called the **hypotenuse**.

An **isosceles triangle** is a triangle having two equal sides through the apex and their corresponding base angles having the same measure. An **equilateral triangle** is a triangle having three equal sides, their equal interior angles measuring $60°$.

We recall some geometric concepts as they are defined and used in triangles. An **angle bisector** of a triangle is the straight line through a vertex which cuts the corresponding angle in half. A **median** or **side bisector** of a triangle is the straight line through a vertex and the midpoint of the opposite side. An **altitude** of a triangle is the straight line through a vertex and perpendicular to the opposite side. This opposite side is called the **base** of the altitude and its intersection point is called the **foot**. A **perpendicular bisector** of a triangle is the straight line through the midpoint of a side and being perpendicular to it. In a scalene triangle (see figure 3.4) these four geometric lines differ clearly. In isosceles triangles the four geometric lines through the apex coincide. In equilateral triangles the four geometric lines always coincide.

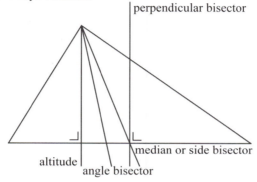

*Figure 3.4*: Geometric lines in a triangle

Two plane polygons are **similar** in case

  ▷  the ratio of their corresponding edges is constant, and

  ▷  their corresponding angles measure the same size.

We recall more specifically a similarity criterion for triangles: *two triangles are similar whenever two of their corresponding angles measure the same size.*

Before proceeding with the discussion on triangles we explain the ruler-and-compass method to construct a perpendicular bisector on a given line segment, as it will be useful in subsequent chapters.

  ▷  We construct the perpendicular bisector $m$ of a segment $[AB]$ by drawing two circles $C(A,r)$ and $C(B,r)$ with the same radius $r$. These circles intersect, when applying a sufficiently large radius $r > 0$, in two points spanning the perpendicular bisector of the segment $[AB]$. Consequently we acquire the midpoint $M$ of the segment $[AB]$ as a bonus.

  ▷  We construct the altitude $l$ from apex $D$ to a base $k$ by drawing a circle $C(D,s)$ which intersects the base $k$ in two distinguished points $P$ and $Q$. Finally we apply the above construction for the perpendicular bisector $m$ of the segment $[PQ]$. Consequently we acquire the foot as the midpoint $D'$ of the segment $[PQ]$.

We refresh the **Pythagorean theorem**, named after the Greek **Pythagoras of Samos** (582–507 Before Christ), which states in any right triangle: the square of the length of the

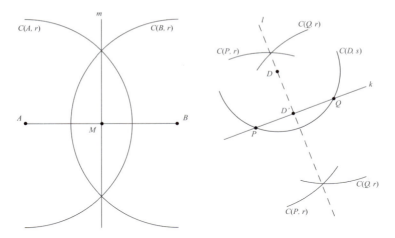

*Figure 3.5*: Constructions of the perpendicular bisector *m* and the altitude *l*

hypotenuse equals the sum of the squares of the lengths of the two other sides.

$$(\text{side}_1)^2 + (\text{side}_2)^2 = (\text{hypotenuse})^2 \tag{3.2}$$

*Figure 3.6*: Statue of Pythagoras on the isle of Samos

*Proof*:  One way to prove this famous theorem goes via the area of this tiled square. In this square we count four identical right triangles featuring hypotenuse $c$ and other sides $a$ and $b$.

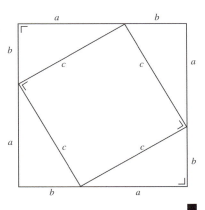

$$\frac{\text{large}}{\text{square area}} = \frac{\text{inner}}{\text{square area}} + 4 \text{ times right triangle area}$$

$$(a+b)^2 = c^2 + 4\frac{ab}{2}$$

$$a^2 + 2ab + b^2 = c^2 + 2ab$$

$$a^2 + b^2 = c^2 \qquad \blacksquare$$

*Example*:  A staircase consists of 17 steps of height 19 cm and depth 15 cm. Find the length of the entire banisters.

The Pythagorean theorem yields $\sqrt{19^2 + 15^2}$ for the length of the hypotenuse of a single step. In conclusion, for the length of the banisters we multiply $17 \cdot \sqrt{19^2 + 15^2} = 411.53$ cm.

Game programming often involves calculating the distance between two points on the screen, e.g. between the anchor points of two colliding objects or two interacting personages. The programmed game may for instance respond as soon as its player moves sufficiently near the enemy. We realise now that various screen situations require a fast calculation of the distance between two points. Applying the Pythagorean theorem is the straightforward way to do so.

Assuming we know two points $P(x_1, y_1)$ and $Q(x_2, y_2)$ allows us to draw a right triangle with hypotenuse $[PQ]$. Figure 3.7 reveals $(x_2 - x_1)$ and $(y_2 - y_1)$ for the lengths of the other sides. In conclusion, the Pythagorean theorem states for the distance between $P$ and $Q$, typeset as $d(P, Q)$, the formula

$$d(P, Q) = \sqrt{(x_2 - x_1)^2 + (y_2 - y_1)^2}. \qquad (3.3)$$

*Example*:  Find the distance between the points $A(1, 2)$ and $B(5, 6)$.
Their distance yields $d(A, B) = \sqrt{(5 - 1)^2 + (6 - 2)^2} = \sqrt{32}$.

We can extend the plane distance formula to three dimensions. We just need to take the $z$-coordinate into account to realise it. Consequently, the distance between two spacial points $P(x_1, y_1, z_1)$ and $Q(x_2, y_2, z_2)$ then equals

$$d(P, Q) = |PQ| = \sqrt{(x_2 - x_1)^2 + (y_2 - y_1)^2 + (z_2 - z_1)^2}. \qquad (3.4)$$

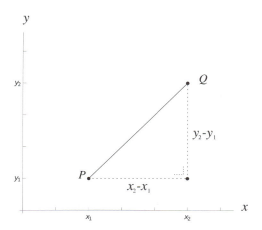

*Figure 3.7*: Distance between two points

## 3.3 Right triangle

We recall measuring the size of angles in radians as defined by the ratio $\alpha = \frac{l}{r}$. Alternatively we define more such ratios as based upon a right triangle. We replace the drawing of a circular arc on radius $r$ (using a pair of compasses) by the drawing of an opposite side perpendicular to $r$ (using a square). We realise the elegance of using a square over the use of a pair of compasses, as the length of a side is far easier to determine than the length of a circular arc.

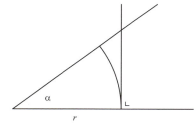

*Figure 3.8*: Definition of trigonometric ratios

This way we acquire a new measure for the interior angle $\alpha$ as the ratio $\frac{\text{opposite side}}{r}$. Such a measure is a 'trigonometric ratio'. We define the above ratio of sides as the **tangent** of the acute angle $\alpha$. As the above fraction is insensitive to scaling, it again guarantees a reliable measurement for any angle.

The total number of trigonometric ratios is six and they are called **sine**, **cosine**, tangent, **cotangent**, cosecant and secant. They are of fundamental importance to study and practice the more advanced trigonometry.

We hereby list all correct definitions for the ratios sine (sin), cosine (cos) and tangent (tan), as for their reciprocals cosecant (csc), secant (sec) and cotangent (cot).

$$\sin \alpha = \frac{\text{opposite side}}{\text{hypotenuse}} \qquad\qquad \csc \alpha = \frac{1}{\sin \alpha}$$

$$\cos \alpha = \frac{\text{adjacent side}}{\text{hypotenuse}} \qquad\qquad \sec \alpha = \frac{1}{\cos \alpha}$$

$$\tan \alpha = \frac{\text{opposite side}}{\text{adjacent side}} \qquad\qquad \cot \alpha = \frac{1}{\tan \alpha}$$

As a useful mnemonic for the above ratios we make acronyms using their starting characters: 'SOH, CAH, TOA'. We realise that these definitions, as based upon right triangles, are limited to acute angles $\alpha$ only. We recall that each of both acute angles of a right triangle measures a size smaller than $90°$.

Based upon the above trigonometric ratios, we immediately discover our first trigonometric formula as

$$\tan \alpha = \frac{\sin \alpha}{\cos \alpha}. \tag{3.5}$$

## 3.4  Unit circle

The **unit circle** is a circle within the orthonormal (see paragraph 6.1) frame. The centre of the unit circle is the origin $O$ and its radius equals one. We consider the unit circle as being our trigonometric 'dashboard' because for any angle $\alpha$ revolved from the horizontal $x$-axis (reference side), we find a corresponding point $E_\alpha$ on it (see figure 3.9).

We subdivide this unit circle in four equal parts that we index counter clockwise, starting from the zero angle (on the positive $x$-axis). Consequently, all acute angles between $0°$ and $90°$ lie in the first quadrant and all obtuse angles are in the second quadrant bordered by $90°$ and $180°$. The reflex angles fall in the third quadrant as bordered by $180°$ and $270°$ or eventually in the fourth quadrant between $270°$ and $360°$.

We simplify the former trigonometric ratios sine, cosine and tangent for a hypotenuse corresponding to the radius of the unit circle, in other words for the hypotenuse set to one. Note that we redefine the trigonometric tangent via a larger right triangle bordering the vertical tangent line to the unit circle at the point $E_0(1,0)$.

$$\cos \alpha = \frac{\text{adjacent side}}{1}$$

$$\sin \alpha = \frac{\text{opposite side}}{1}$$

$$\tan \alpha = \frac{\text{opposite side at the vertical tangent line } x = 1}{1}$$

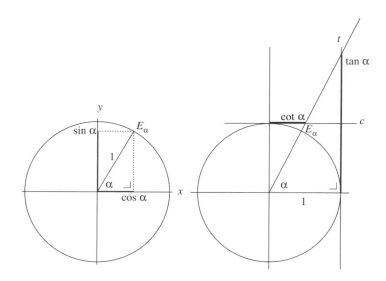

*Figure 3.9*: Trigonometric ratios in the unit circle

In the unit circle we interpret $\cos \alpha$ as the horizontal shadow on the $x$-axis caused by the revolving side of angle $\alpha$. Likewise, we meet $\sin \alpha$ as the vertical shadow on the $y$-axis caused by the revolving side of angle $\alpha$. Finally, we encounter $\tan \alpha$ as the opposite side to the angle $\alpha$ on the vertical tangent line $x = 1$. In other words, for the revolving side of an angle $\alpha$ intersecting the unit circle in the corresponding point $E_\alpha$ we define the $x$-coordinate of $E_\alpha$ as the **cosine** of $\alpha$ and the $y$-coordinate of $E_\alpha$ as the **sine** of $\alpha$. Any angle $\alpha$ therefore determines unambiguously its cosine and sine. But for the other way round, every couple of cosine and sine values determines its angle $\alpha$ only to an integer multiple of $2\pi$ rad or $360°$. This is due to the periodicity of each angle with its **elementary period** of $2\pi$ rad or $360°$.

The redefined sine, cosine and tangent in a unit circle overcome their previous limitation to acute angles in such a way that they now are defined for any angle size.

Applying the Pythagorean theorem (3.2) in the unit circle, yields

$$(\sin \alpha)^2 + (\cos \alpha)^2 = 1. \tag{3.6}$$

This important formula linking sine and cosine is known as the trigonometric **Pythagorean Identity**.

The Pythagorean theorem applies to right triangles only. Scalene triangles are ruled by the **Law of Cosines** and the **Law of Sines** to link their interior angles to the length of their sides. For a scalene triangle $ABC$ featuring interior angles $\alpha, \beta$ and $\gamma$ at corresponding vertices $A, B$ and $C$ and opposite sides $a, b$ and $c$, we state both laws (omitting their proofs):

$$\frac{a}{\sin \alpha} = \frac{b}{\sin \beta} = \frac{c}{\sin \gamma} \qquad \text{Law of Sines,} \qquad (3.7)$$

$$a^2 = b^2 + c^2 - 2bc \cos \alpha \qquad \text{Law of Cosines.} \qquad (3.8)$$

The Law of Cosines for a right angle $\alpha$ implies $\cos \alpha = 0$ and thus simplifies to the Pythagorean theorem. In addition, we refresh the general formula to calculate the area of a scalene triangle as

$$\text{area}_{triangle} = \frac{1}{2} a\, b \sin \gamma \qquad \text{Area of triangles.} \qquad (3.9)$$

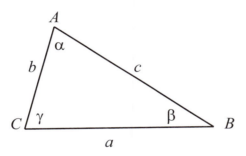

*Figure 3.10*: Scalene triangle formulas

## 3.5 Special angles

Given their definitions, the trigonometric ratios are limited to these intervals:

$$\sin \alpha \in [-1, 1] \qquad\qquad \cos \alpha \in [-1, 1]$$
$$\tan \alpha \in\, ]-\infty, +\infty[ \qquad\qquad \cot \alpha \in\, ]-\infty, +\infty[$$

### Trigonometric ratios for an angle of $45° = \frac{\pi}{4}$ rad

As the interior angles of the triangle $OAE$ add up to $180°$, we find the angle $\hat{E}$ equals $45°$. As this triangle features equal angles $\hat{O} = \hat{E} = 45°$ we conclude the triangle $OAE$ to be isosceles with apex $A$ and for its corresponding length of sides $|OA| = |AE|$. This geometric reasoning leads finally to $\cos 45° = \sin 45°$. Substituting the above geometric conclusion in the Pythagorean Identity $(\sin 45°)^2 + (\cos 45°)^2 = 1$ yields $(\sin 45°)^2 + (\sin 45°)^2 = 1$. Solving this equality for $\sin 45°$ leads to $\sin 45° = \pm\sqrt{\frac{1}{2}}$. Since the angle of $45°$ resides in the first quadrant, it limits $\sin 45° \geqslant 0$, and so we

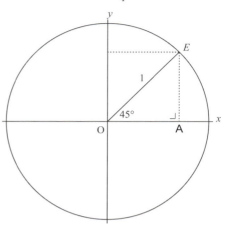

conclude $\sin 45° = \frac{1}{\sqrt{2}} = \frac{\sqrt{2}}{2}$. As an immediate consequence we conclude $\cos 45° = \frac{\sqrt{2}}{2}$.

### Trigonometric ratios for an angle of $30° = \frac{\pi}{6}$ rad

Firstly we **reflect** the point $E_{30°}$ over the $x$-axis to its image point $E'$. Reflections are isometric mappings: they leave distances and angle sizes unchanged. Therefore we conclude the triangle $EOE'$ to have an apex angle $\hat{O} = 60°$ and two equal sides $|OE| = |OE'|$. As the interior angles of the triangle $EOE'$ add up to $180°$, we find equal angles $\hat{E} = \hat{E}' = 60°$. Since $\hat{E} = \hat{O} = \hat{E}' = 60°$ we conclude the triangle $EOE'$ to be equilateral with length of sides $|OE| = |OE'| = |EE'| = 1$. Based on the coinciding of all geometric lines in equilateral triangles, we consider the altitude $OA$

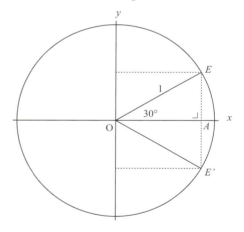

on the base $[EE']$ also to be the median that bisects the segment $[EE']$. Therefore we conclude $|EA| = \sin 30° = \frac{1}{2}$. Substituting this geometric conclusion in the Pythagorean Identity straightforwardly yields the corresponding $\cos 30° = \frac{\sqrt{3}}{2}$.

## TRIGONOMETRIC RATIOS FOR AN ANGLE OF $60° = \frac{\pi}{3}$ RAD

For the special angle of $60°$, we consider the triangle $OEC$ as the start of a geometric reasoning. Given its vertices $E$ and $C$ lying on the unit circle, we know that $|OE| = |OC| = 1$ is leading us to an isosceles triangle $OEC$ with apex $O$ and consequently equal base angles $\hat{E} = \hat{C}$. As the interior angles of the triangle $OEC$ add up to $180°$, we find equal angles $\hat{E} = \hat{C} = 60°$. Based on the coinciding of all geometric lines in equilateral triangles, we consider the altitude $EA$ on the base $[OC]$ also to be the median that bisects the segment $[OC]$. Therefore we conclude that $|OA| = \cos 60° = \frac{1}{2}$. Substituting this

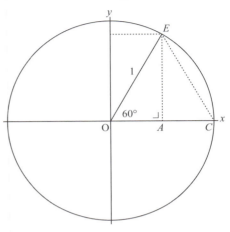

geometric conclusion in the Pythagorean Identity yields the corresponding $\sin 60° = \frac{\sqrt{3}}{2}$.

## OVERVIEW

Given the above sine and cosine values, we can easily calculate the corresponding tangents via their quotient (see formula (3.5)). We hereby draw for some special angles $\alpha$ their corresponding values for sine and cosine within the unit circle and their tangents on the vertical tangent line $x = 1$.

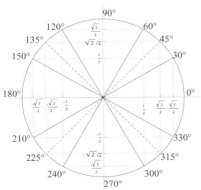

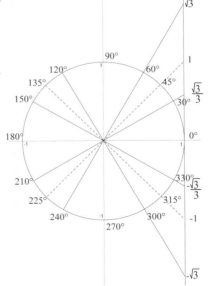

*Figure 3.11*: Graphical overview via the unit circle

## 3.6 Pairs of angles

We briefly explain the properties of two pairs of angles that are useful in this book.

**Coterminal angles** or oppositely signed angles; their measurements add up to $0°$. In other words, if $\alpha$ and $\beta$ are coterminal then $\alpha + \beta = 0°$ or $\beta = -\alpha$. The corresponding figure shows how the cosines of coterminal angles remain invariant, while their sines receive opposite signs. This leads to the trigonometric formulas $\cos(-\alpha) = \cos\alpha$ and $\sin(-\alpha) = -\sin\alpha$.

**Complementary angles**; their measurements add up to $90°$. In other words, if $\alpha$ and $\beta$ are complementary then $\alpha + \beta = 90°$ or $\beta = 90° - \alpha$. The corresponding figure shows how the sine of $\alpha$ equals the cosine of $90° - \alpha$ and the cosine of $\alpha$ equals the sine of $90° - \alpha$. This leads to the trigonometric formulas $\cos(90° - \alpha) = \sin\alpha$ and $\sin(90° - \alpha) = \cos\alpha$.

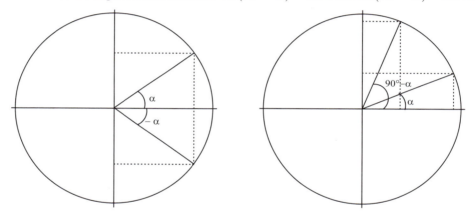

*Figure 3.12*: Coterminal and complementary angles

## 3.7 Sum identities

In this paragraph we state and prove all trigonometric ratios of a sum of two angles. We firstly emphasise the non-linearity of all trigonometric ratios: e.g. for the sine we encounter $\sin(\alpha + \beta) \neq \sin\alpha + \sin\beta$. Indeed, e.g. for angles $\alpha = 60°$ and $\beta = 30°$ the value $\sin 90° = 1$ does not equal the sum $\sin 60° + \sin 30° = \frac{\sqrt{3}+1}{2}$. Given the above inequality, we realise the need for the correct formulas which are stated below.

$$\sin(\alpha + \beta) = \sin \alpha \cos \beta + \cos \alpha \sin \beta \qquad \sin(\alpha - \beta) = \sin \alpha \cos \beta - \cos \alpha \sin \beta$$

$$\cos(\alpha + \beta) = \cos \alpha \cos \beta - \sin \alpha \sin \beta \qquad \cos(\alpha - \beta) = \cos \alpha \cos \beta + \sin \alpha \sin \beta$$

$$\tan(\alpha + \beta) = \frac{\tan \alpha + \tan \beta}{1 - \tan \alpha \tan \beta} \qquad \tan(\alpha - \beta) = \frac{\tan \alpha - \tan \beta}{1 + \tan \alpha \tan \beta}$$

*Proof*: First of all we prove the formula for the cosine of a subtraction of angles, using the included right triangles and the Law of Cosines. Thereafter we will prove all other Sum Identities based upon this formula for the cosine of a subtraction.

▷ $\cos(\alpha - \beta) = \cos \alpha \cos \beta + \sin \alpha \sin \beta$

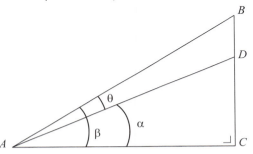

*Figure 3.13*: Proving the formula for the cosine of a subtraction

The right triangle $ACD$ yields $\cos \alpha = \frac{|AC|}{|AD|}$ and $\sin \alpha = \frac{|DC|}{|AD|}$ and the right triangle $ABC$ yields $\cos \beta = \frac{|AC|}{|AB|}$ and $\sin \beta = \frac{|BC|}{|AB|}$. Substituting the sines and cosines in the right hand side of the formula leads to

$$\cos \alpha \cos \beta + \sin \alpha \sin \beta = \frac{|AC|}{|AD|} \cdot \frac{|AC|}{|AB|} + \frac{|DC|}{|AD|} \cdot \frac{|BC|}{|AB|}$$

$$= \frac{|AC|^2 + |DC| \cdot |BC|}{|AD| \cdot |AB|}.$$

Secondly, we interpret the left hand side $\cos(\alpha - \beta) = \cos(\beta - \alpha)$ as $\cos \theta$ in the scalene triangle $BAD$. We determine $\cos \theta$ using the Law of Cosines (3.8) for the angle $\theta$ and its opposite side $|BD|$ as $|BD|^2 = |AB|^2 + |AD|^2 - 2|AB| \cdot |AD| \cdot \cos \theta$.

To finalise the proof, eliminating $\cos \theta$ from both steps should lead to an equality. Substituting the factor $\cos \theta$ in the second step results in:

$$|BD|^2 = |AB|^2 + |AD|^2 - 2|AB| \cdot |AD| \cdot \left( \frac{|AC|^2 + |DC| \cdot |BC|}{|AD| \cdot |AB|} \right)$$

$$= |AB|^2 + |AD|^2 - 2(|AC|^2 + |DC| \cdot |BC|)$$

We may replace the length $|BD|$ by the difference $|BC| - |DC|$ and apply the Perfect Square:

$$|BD|^2 = |AB|^2 + |AD|^2 - 2(|AC|^2 + |DC| \cdot |BC|)$$
$$(|BC| - |DC|)^2 = |AB|^2 + |AD|^2 - 2(|AC|^2 + |DC| \cdot |BC|)$$
$$|BC|^2 - 2|BC| \cdot |DC| + |DC|^2 = |AB|^2 + |AD|^2 - 2|AC|^2 - 2|DC| \cdot |BC|$$
$$|BC|^2 + |DC|^2 = |AB|^2 + |AD|^2 - 2|AC|^2.$$

We finally obtain the equality $(|BC|^2 + |AC|^2) + (|DC|^2 + |AC|^2) = |AB|^2 + |AD|^2$ after grouping terms and applying the Pythagorean Theorem twice.

▷ Based upon $\cos(\alpha - \beta) = \cos \alpha \cos \beta + \sin \alpha \sin \beta$, we have less difficulties in proving the five remaining Sum Identities

$$\cos(\alpha + \beta) = \cos(\alpha - (-\beta))$$
$$= \cos \alpha \cos(-\beta) + \sin \alpha \sin(-\beta)$$
$$= \cos \alpha \cos \beta - \sin \alpha \sin \beta \quad \text{(coterminal angles)}$$

$$\sin(\alpha - \beta) = \cos(90° - (\alpha - \beta)) \quad \text{(complementary angles)}$$
$$= \cos((90° - \alpha) + \beta)$$
$$= \cos(90° - \alpha)\cos \beta - \sin(90° - \alpha)\sin \beta$$
$$= \sin \alpha \cos \beta - \cos \alpha \sin \beta \quad \text{(complementary angles)}$$

$$\sin(\alpha + \beta) = \sin(\alpha - (-\beta))$$
$$= \sin \alpha \cos(-\beta) - \cos \alpha \sin(-\beta)$$
$$= \sin \alpha \cos \beta + \cos \alpha \sin \beta \quad \text{(coterminal angles)}$$

$$\tan(\alpha \pm \beta) = \frac{\sin(\alpha \pm \beta)}{\cos(\alpha \pm \beta)}$$

$$= \frac{\sin \alpha \cos \beta \pm \cos \alpha \sin \beta}{\cos \alpha \cos \beta \mp \sin \alpha \sin \beta}$$

$$= \frac{\frac{\sin \alpha \cos \beta}{\cos \alpha \cos \beta} \pm \frac{\cos \alpha \sin \beta}{\cos \alpha \cos \beta}}{\frac{\cos \alpha \cos \beta}{\cos \alpha \cos \beta} \mp \frac{\sin \alpha \sin \beta}{\cos \alpha \cos \beta}}$$

$$= \frac{\tan \alpha \pm \tan \beta}{1 \mp \tan \alpha \tan \beta}$$

When applying the quotient formula of tangent we should never divide by zero! This would occur in case $\tan \alpha = \pm \frac{1}{\tan \beta}$ or equivalently whenever $\alpha \pm \beta = 90° + k180°$ given $k \in \mathbb{Z}$. These angular values would cause $\tan(\alpha \pm \beta) = \pm \infty$.

## 3.8 Inverse trigonometric functions

When an angle $\alpha$ returns $\sin \alpha = \frac{1}{2}$ then we may conclude, apart from the obvious $\alpha = \frac{\pi}{6}$, for infinitely more angular sizes $\alpha$ to return $\sin \alpha = \frac{1}{2}$. This is due to the periodicity of the function sine and the sine values (except for $-1$ and $1$) to appear twice within one period. We have $\sin \alpha = \frac{1}{2}$ if $\alpha = \frac{\pi}{6}$ but also if $\alpha = \frac{5\pi}{6}$ and we may add multiples of $2\pi$ to each of them (due to the periodicity of the plane angle $\alpha$). We avoid that an infinite number of angles $\alpha$ are the solution to $\sin \alpha = \frac{1}{2}$ when we restrict the angle $\alpha$ to the radian interval $[\frac{-\pi}{2}, \frac{\pi}{2}]$. A similar issue also applies for the cosine and for the tangent, solved in a similar way. We agree on the restricted interval $[0, \pi]$ for cosine and on the restricted interval $]\frac{-\pi}{2}, \frac{\pi}{2}[$ for tangent.

For any $x$ being the result of a sine, cosine or tangent, we find one unique angle $\alpha$ in their restricted interval giving $\sin \alpha = x$, $\cos \alpha = x$ or $\tan \alpha = x$ respectively. The corresponding functions to trace this unique angle $\alpha$ are called the **arcsine**, the **arccosine** and the **arctangent**. We can typeset the arcsine either as arcsin, asin or $\sin^{-1}$.

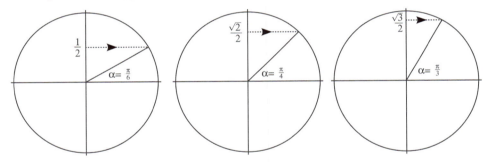

*Figure 3.14*: The arcsine

Arcsine returns for a given sine value its corresponding circular arc (in radians restricted to quadrants I and IV). For instance $\arcsin \frac{\sqrt{2}}{2} = \frac{\pi}{4}$ because $\sin \frac{\pi}{4} = \frac{\sqrt{2}}{2}$.

$$\alpha = \arcsin x \Leftrightarrow x = \sin \alpha \text{ with } \frac{-\pi}{2} \leqslant \alpha \leqslant \frac{\pi}{2}$$

Similarly, arccosine returns for any given cosine value its corresponding circular arc (in radians restricted to quadrants I and II).

$$\alpha = \arccos x \Leftrightarrow x = \cos\alpha \text{ with } 0 \leqslant \alpha \leqslant \pi$$

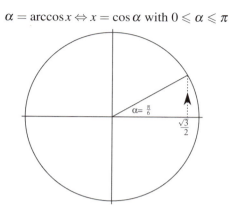

*Figure 3.15*: The arccosine

Finally, arctangent returns for a given tangent value its corresponding circular arc (in radians restricted to quadrants I and IV).

$$\alpha = \arctan x \Leftrightarrow x = \tan\alpha \text{ with } \frac{-\pi}{2} < \alpha < \frac{\pi}{2}$$

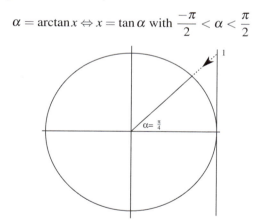

*Figure 3.16*: The arctangent

*Example*: What size is the (smallest) angle made by the straight line through the points $O(0,0)$ and $P\left(\frac{1}{2}, \frac{\sqrt{3}}{2}\right)$ and the $x$-axis?

Drawing the lines shows an opposite side of length $\frac{\sqrt{3}}{2}$ and an adjacent side of length $\frac{1}{2}$. This allows us to calculate the size of angle $\alpha$ made by the line $OP$ and the $x$-axis via the tangent: $\tan\alpha = \frac{\frac{\sqrt{3}}{2}}{\frac{1}{2}} = \sqrt{3}$. We need to restrict the size of angle $\alpha$ to the radian interval $\left]\frac{-\pi}{2}, \frac{\pi}{2}\right[$ for the arctangent to return the corresponding size of the targeted angle as $\alpha = \arctan\sqrt{3} = \frac{\pi}{3}$.

## 3.9 Exercises

**Exercise 24**  Calculate the missing side lengths and/or angle sizes in the triangle $ABC$ as typeset in Figure 3.10 for
1) $\alpha = 48°$, $b = 29$ cm and $\gamma = 90°$,
2) $a = 10$ cm, $\beta = 65°$ and $c = 12$ cm.

**Exercise 25**  When Vian drives a car on the highway his eyes are 1 metre above ground level. He approaches a major traffic sign which is hanging above his head. The traffic sign hangs 6 metres above the road surface and its height measures 3 metres. Find Vian's visual angle when he approaches 20 metres (measured along the road surface) from this traffic sign.

**Exercise 26**  The Belgian national railroad company renews the railroad from Kortrijk to Bruges. Accidently, ignoring the rail's expansion due to the summer heat, they miscalculated its length. It takes 50 km rail from Kortrijk to Bruges, but in summer this rail undergoes an overall expansion of 1 metre. The railroad authorities decide to incline the track instead of shortening the rail's length by 1 metre. Find the track's height at halfway along the line from Kortrijk to Bruges.

**Exercise 27**  A pilot receives flight data from a plane in order to intercept that plane. The situation below was shown on the pilot's display before the data signal was scrambled. Find the distance from the pilot to the point of interception.

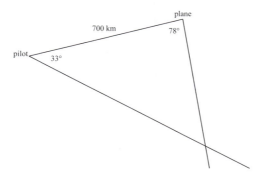

**Exercise 28**  A scouting plane takes off from a runway at the east coast, heading in the direction of 80°, being measured clockwise from the north direction of 0° onwards. Due to awful weather conditions, the flying scout returns to another east coast landing strip at a distance of 250 km north of its home base. Then the plane takes the direction of 283° for its return flight. What is the total distance this plane will travel?

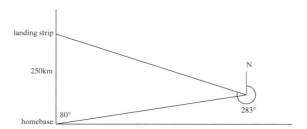

**Exercise 29**  Two friends work together to determine an aeroplane's height. They keep a distance of 180 metres in the open field and spot the plane simultaneously. Abid measures an angle of 60° and Isaac an angle of 30°. Together they are able to calculate the aeroplane's altitude. Are you?

**Exercise 30**  Find $\alpha$ in case of

1)  $\sin \alpha = \frac{\sqrt{3}}{2}$

2)  $\tan \alpha = 1$

3)  $\cos 2\alpha = \frac{-\sqrt{2}}{2}$

4)  $\tan \frac{\alpha}{2} = -\sqrt{3}$

**Exercise 31**  At each vertex of a square field $ABCD$ whose side measures 14 metres, stands a flagpole. The flagpole at vertex $A$ measures 7 metres, the flagpole at vertex $B$ has a length of 10 metres and in vertex $C$ the pole is 15 metres. Somewhere in the square's interior a point $P$ is located at equal distance to each flagpole top.

▷  Find the location of this point $P$. Therefore calculate distances $x$ and $y$, being the horizontal distance from $P$ to the side $[AD]$ and the vertical distance from $P$ to the side $[AB]$ respectively.

▷  Given the distance from $P$ to the flagpole top at vertex $D$ is the same distance from $P$ to the previous tops in vertices $A$, $B$ and $C$, find the height of the flagpole in $D$.

How does an application determine how characters move or whether their trajectories intersect? An answer to the latter and much more can be achieved by the use of functions. In this chapter, we outline some basic concepts on functions and we discuss the main prototypes of the polynomial, the logarithmic and the exponential functions. We also initiate collision prediction via intersection of trajectories. We study the trigonometric functions which are fit to describe periodical phenomena, but also the inverse trigonometric functions including the atan2-variant which return angular data. We complete all of these functions by their local maclaurin approximations, for their deeper understanding as well as for purely practical purposes later on.

## 4.1 Basic concepts on real functions

▷ We define a **function** as a mapping $f$ that for each argument $x$ returns at most one image $f(x)$.

▷ We define the **domain** of a function as the set of arguments $x$ which have exactly one image $f(x)$.

▷ We define the **range** of a function as the set of all images $f(x)$ returned by the function $f$.

▷ We define a **root** of a function $f$ as each argument $x_0$ that maps to $f(x_0) = 0$.

All four of the above definitions can be summarised in an explicit formula or recipe $f$ : *domain* $\rightarrow$ *range* $: x \mapsto y = f(x)$. The real function $f : \mathbb{R} \rightarrow \mathbb{R} : x \mapsto y = f(x)$ maps arguments $x$ of the domain in $\mathbb{R}$ onto images $f(x)$ of the range in $\mathbb{R}$. Therefore we can easily visualise the function in an $\mathbb{R} - \mathbb{R}$ graph.

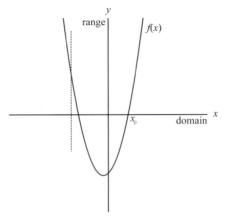

*Figure 4.1*: The graph of a function $f(x)$

We revisit the former concepts approaching them from the graph of the function $f$.

 ▷ We verify the defining property graphically via running a vertical line horizontally through the graph of $f$: at most one intersection point has to occur for $f$ being a function.

 ▷ We recognise the domain of a function $f$ graphically as the area on the $x$-axis where the function $f$ is defined.

 ▷ We recognise the range of a function $f$ graphically as the set of all returned images $f(x)$ on the $y$-axis.

 ▷ We graphically interpret a root of a function $f$ as an intersection point $x_0$ of the graph of the function and the $x$-axis.

## 4.2 Polynomial functions

We define a **polynomial function** $V$ via the recipe $y = V(x)$ given $V(x)$ a real polynomial. These functions have domain $\mathbb{R}$, a possibly limited range in $\mathbb{R}$ and up to maximal $n$ roots as solutions to the equation $V(x) = 0$. We call $n$ the degree of the polynomial $V(x)$ and we call $n$ the degree of the polynomial function $V$ as well.

### LINEAR FUNCTIONS

We define a **linear function** via the recipe

$$y = mx + c. \tag{4.1}$$

These functions have domain $\mathbb{R}$, a range in $\mathbb{R}$ which is limited only in case of a horizontal line and a (most likely) unique root as solution to the linear equation $mx + c = 0$.

We define the **slope $m$** as the coefficient to $x$ in the recipe. The slope means the inclination of the linear graph to the horizontal $x$-axis. A horizontal run of $k$ steps on the $x$-axis corresponds to a rise of $m \cdot k$ steps on the $y$-axis when $m > 0$ and to a descent of $m \cdot k$ steps on the $y$-axis when $m < 0$.

If both axes are perpendicular and equally calibrated, then the slope $m$ equals the tangent of the inclination angle $\alpha$ of the linear graph to the $x$-axis:

$$\text{slope } m = \frac{\Delta y}{\Delta x} = \tan \alpha. \tag{4.2}$$

The capital Greek character 'delta' $\Delta$ refers to 'd' as the starting character of 'difference'. Hence $\Delta x$ denotes the difference of two $x$-values while $\Delta y$ typesets the difference of two $y$-values.

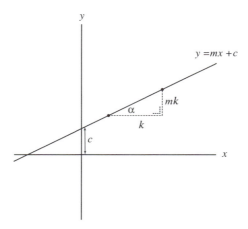

*Figure 4.2*: The linear function $y = mx + c$

In case $m < 0$ we have a descending line, in case $m = 0$, a horizontal line and in case $m > 0$, then the linear graph is ascending. Any two linear functions $y = mx + c$ and $y = nx + d$ draw **parallel lines** if their slopes simply equal: $m = n$. Any two linear functions $y = mx + c$ and $y = nx + d$ draw **perpendicular lines** if their slopes multiply to negative one:

$$m \cdot n = -1 \quad \text{(see exercise 32)}.$$

We define the **intercept** $c$ as the constant term in the recipe. The intercept means the position where the linear graph intersects the $y$-axis.

Any two points $Q(x_Q, y_Q)$ and $R(x_R, y_R)$ determine the recipe of the linear function they are spanning.

Based upon the formula (4.2) defining the slope, for any point $P(x, y)$ on a non-vertical straight line holds

$$\text{slope } m = \frac{\Delta y}{\Delta x} = \frac{y_R - y_Q}{x_R - x_Q} = \frac{y - y_Q}{x - x_Q}.$$

As a consequence of the above formulas, we derive an expression for the straight line stretching from point $Q(x_Q, y_Q)$ to point $R(x_R, y_R)$.

$$
\begin{aligned}
y - y_Q &= \frac{y_R - y_Q}{x_R - x_Q} \cdot (x - x_Q) \\
y - y_Q &= m \cdot (x - x_Q).
\end{aligned}
\tag{4.3}
$$

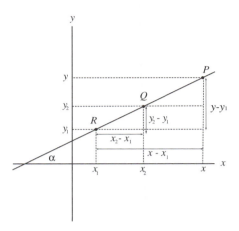

*Figure 4.3*: The straight line connecting points $Q$ and $R$

*Example*:    We determine the linear function $f(x)$ connecting two points $R(0,2)$ and $Q(11,23)$.

Formula (4.3) yields $y - 2 = \frac{23-2}{11-0}(x - 0)$ which we simplify to the recipe $y = \frac{21}{11}x + 2$.

As a bonus, we also aim for the recipe of the line being perpendicular to the line $y = \frac{21}{11}x + 2$ and passing through the point $(2, -7)$. We therefore make use of the property of the slopes of perpendicular lines multiplying to $-1$. Hence the slope of the perpendicular line we wish to construct equals $-\frac{11}{21}$. Eventually this perpendicular line is $y + 7 = -\frac{11}{21}(x - 2)$ which we simplify to the recipe $y = -\frac{11}{21}x - \frac{125}{21}$.

## QUADRATIC FUNCTIONS

We define the real **quadratic function** as a mapping $f$ which returns for any argument $x$ *maximal* one image $y = ax^2 + bx + c$, given $a, b, c$ real coefficients with $a \neq 0$.

Quadratic functions take $\mathbb{R}$ for their domain, limit their range to one side and can have up to two roots $x_1$ and $x_2$ which are traceable by the quadratic formulas (see page 30). We call the graph of the quadratic function a parabola.

*Example*:  We hereby present the elementary prototype

$f : \mathbb{R} \longrightarrow \mathbb{R} : x \mapsto x^2$, alternatively $y = x^2$.

We verify this parabola for being a function by running a vertical line horizontally through the $(x, y)$-frame and counting for the number of return values $f(x)$ corresponding to each argument $x$. We call $f(x) = x^2$ an **even function** because $f(-x) = f(x)$ for all $x$.

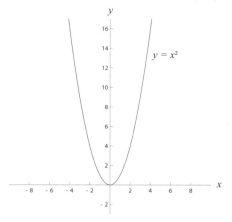

*Figure 4.4*: The graph of the elementary even function $y = x^2$ is symmetric

We see the **parabola opening up** in case $a > 0$ or the **parabola opening down** for $a < 0$. We call the lowest or highest point on it the **vertex of a parabola**. We calculate the $x$-coordinate of the vertex of the parabola as $x_{vertex} = \frac{-b}{2a}$. The vertical line through the vertex is the axis of symmetry of the parabola. The coefficient $a$ measures the width of the parabola: the larger the absolute value of $a$ the steeper the parabola. The coefficient $b$ is the slope of the tangent line to the parabola in its intersection point on the $y$-axis. The coefficient $c$ determines the parabola's intersection point on the $y$-axis, also known as its **intercept** $c$.

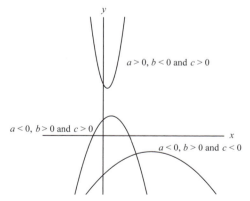

*Figure 4.5*: The effect of the coefficients on the quadratic function $y = ax^2 + bx + c$

We also look at the prototype of the **horizontal parabola** as it is defined by $y^2 = x$.

By reducing this equation to zero we can define this horizontal parabola equivalently by $y^2 - x = 0$. Such a horizontal parabola is clearly *no function* as running a vertical line horizontally through the $(x,y)$-frame reveals *two* return values $y$ for each argument $x$. Instead we generally call a horizontal parabola a **curve**.

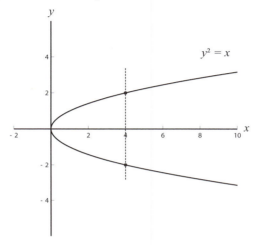

*Figure 4.6*: The graph of the curve $y^2 = x$

## 4.3 Intersection of functions

Sometimes animation programming requires the intersection of two lines. For this occasion those lines often model the edge of a building, the horizon or the trajectory of a moving object.

Let us take an easy start aiming for the intersection of just two straight lines. Two different straight lines in the same plane will either intersect in one point or they appear to be parallel. In the former case, how can we find their intersection point? We can do this by tracing the $x$-value that leads to identical return values. Because we try to locate where both straight lines *equal each other*, we have to compare their recipes in one equation.

*Example*: We intersect the two straight lines $f(x) = 2x - 2$ and $g(x) = -x + 10$. Therefore, we compare their recipes in one equation $2x - 2 = -x + 10$ which we solve for $x$. Doing so yields $x = 4$ as the unique solution. Substituting this $x$-value in one of both recipes yields the corresponding $y$-value, $y = f(4) = g(4) = 6$. In conclusion, we find the intersection point $(x,y) = (4,6)$.

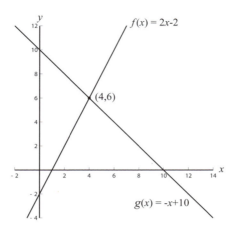

*Figure 4.7*: Intersecting two straight lines

Alternatively we can find the intersection point of these straight lines by solving the 2 by 2 system (see page 40)

$$\begin{cases} y = 2x - 2 \\ y = -x + 10. \end{cases}$$

Two straight lines may not always have one unique intersection point. They can be parallel without having any intersection point or they can coincide having an infinite number of them.

We apply the same method to determine whether a straight line and a parabola intersect or even to find out whether two parabolas intersect.

*Example*:  Consider two quadratic functions $f(x) = 2x^2 - 10x + 5$ and $g(x) = -x^2 + 2x - 4$. We impose $f(x) = g(x)$ for the $x$-coordinate of an intersection point, consequently $2x^2 - 10x + 5 = -x^2 + 2x - 4$ after substituting their recipes and equivalently $3x^2 - 12x + 9 = 0$. Solving this quadratic equation using the quadratic formulas (see page 30) yields $x_1 = 1$ and $x_2 = 3$. Calculating $f(1) = g(1) = -3$ and $f(3) = g(3) = -7$ we obtain the coordinates of both intersection points of these parabolas as $(1, -3)$ and $(3, -7)$.

We generalise the above method of 'comparing' recipes. In order to possibly trace the intersection points of two different functions $f(x)$ and $g(x)$, we solve the equation $f(x) = g(x)$ for the horizontal coordinate $x$. Equivalently we may solve the to zero reduced equation $f(x) - g(x) = 0$ for $x$. Substituting any found $x$-value in the recipe of either $f(x)$ or $g(x)$, yields the vertical coordinate $y$ of the corresponding intersection point.

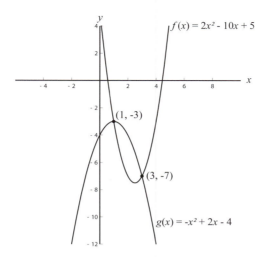

*Figure 4.8*: Determining intersection points

## 4.4 *Logarithmic functions*

We define the **logarithmic functions** by a logarithm in their recipe $y = \log_g(x)$ with base $g \in \mathbb{R}^+ \backslash \{0, 1\}$. Given the existence condition of the logarithmic functions their domain is $\mathbb{R}^+ \backslash \{0\}$ featuring the unique root $x_0 = 1$. Furthermore, their range is $\mathbb{R}$.

We define as an **asymptote** to a real function $f$ each line which approaches the graph of $f$ arbitrarily closely. The word 'asymptote' originates from the Greek language and literally means 'not coinciding'.

*Examples*:  We see how the function $f(x) = \log_2(x)$ is ascending. The graph of the function $\log_2(x)$ aims for the $y-$axis as its vertical asymptote when $x$ tends to 0.

We also examine the graph of a logarithmic function with base $g < 1$, amongst which is $y = \log_{\frac{1}{2}}(x)$. In this case, we see the function $f(x) = \log_{\frac{1}{2}}(x)$ descending. Once more the graph of the function $\log_{\frac{1}{2}}(x)$ aims for the $y-$axis as its vertical asymptote where we see $x$ striving for 0.

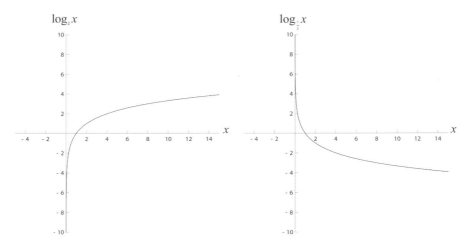

*Figure 4.9*: Displayed graphs of the ascending $\log_2(x)$ and the descending $\log_{\frac{1}{2}}(x)$

## 4.5  Exponential functions

We define **exponential functions** by the recipes $y = g^x$ with their base $g \in \mathbb{R}^+\backslash\{0,1\}$. Exponential functions are named after their variable expo-nent. Exponential functions are defined over the domain $\mathbb{R}$ without any root, and reach the range $\mathbb{R}^+\backslash\{0\}$. We il-lustrate these features through some various examples.

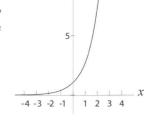

We define **positive infinity** typeset by the symbol $+\infty$ as a quantity boundlessly larger than any number and can therefore never be a number, in symbols $+\infty \notin \mathbb{R}$.

We define **negative infinity** typeset by the symbol $-\infty$ simply as the opposite of positive infinity, for which we conclude in symbols:

▷   $-\infty = -(+\infty)$

▷   $-\infty \notin \mathbb{R}$.

Negative infinity is considered the unbounded lowest pos-sible quantity and hence not a number. In conclusion, we keep in mind that $+\infty$ as well as $-\infty$ are not numbers.

*Examples*:   In the exponentially growing function $y = 2^x$ it is the base number 2 caus-ing the function's explosive 'accelerated yield'. This exponential function $2^x$ features the $x$−axis as its horizontal asymptote when $x$ tends to $-\infty$. In general, we should never confuse such an exponential function with any power function. We define power functions as a subtype of polynomial functions and hence they have a variable base. Let the elementary quadratic function $f(x) = x^2$ be the evergreen default species of power functions (see page 72). The concept of power functions (variable in their base) entirely opposes the concept of *exponential* functions (variable in their *exponent*).

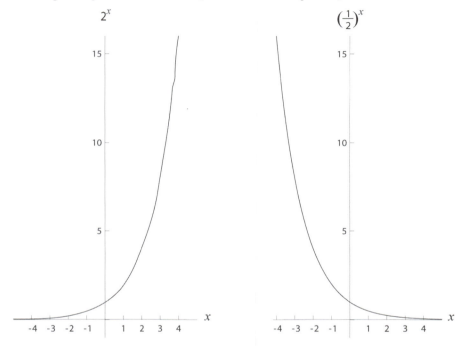

*Figure 4.10*: Displayed graphs of the exponential functions $2^x$ and as well $\left(\frac{1}{2}\right)^x$

By studying $y = \left(\frac{1}{2}\right)^x$ we also look into the graph of a descending exponential function based on a base $g < 1$. Because its base number $\frac{1}{2}$ is less than the constant base 1, we notice this function's so-called **asymptotic** decline towards the $x-$axis where $x$ strives for $+\infty$. These exponential functions decrease very rapidly and then level off to end asymptotic towards the horizontal $x-$axis.

As a concluding example, we refer to the previously portrayed exponential function based on the natural base $e \approx 2.72 \in \mathbb{R}^+$ (see formula (1.8)). Being the most essential species, this **natural exponential function** is typeset by a worldwide mathematical keyword **exp** $(x) = e^x$. The natural exponential function $\exp(x)$ obviously shows an accelerated ascending graph with increasing $x$, being powered by its base $e \approx 2.72$ which causes such an explosive 'accelerated yield'. The natural exponential function $\exp(x)$ asymptotically declines to the $x-$axis when its variable $x$ decreases down to $-\infty$.

## 4.6 Trigonometric functions

In a previous chapter, we met the trigonometric ratios. Amongst them, the trigonometric sine function models various applications which exhibit periodicity. We will need a more analytical insight on how the sine evolves over a time interval. Whenever we input an angle argument $x$ (in radians), which is variable in time, we call it a sine function $\sin x$.

### ELEMENTARY SINE FUNCTION

We first of all study the **elementary sine** function $f : \mathbb{R} \to [-1, 1] : x \mapsto \sin x$. Running through the unit circle we return the sine value for each angle.

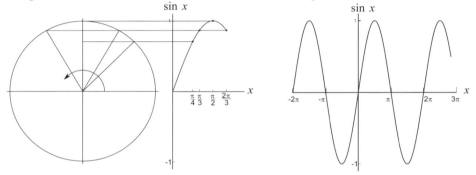

*Figure 4.11*: The graph of the elementary sine function is a **sinusoid**

The domain of this function is the horizontal $x$-axis or in other words the set $\mathbb{R}$. Due to the between $-1$ and $1$ limited sine values, the range of this function is the vertical window $[-1, 1]$. The roots of the elementary sine function are given by $k\pi$, for any $k \in \mathbb{Z}$. Since we can keep running through the unit circle, this function inherits the periodicity of its angle argument, demonstrating an **elementary period** of $2\pi$ rad. We call the elementary $\sin(x)$ an **odd function** because $\sin(-x) = -\sin(x)$ for all $x$ with a point reflected graph.

### GENERAL SINE FUNCTION

We generalise the above sine function by introducing four coefficients to it. This extended function may look like

$$f(x) = r\sin(\omega x + \theta_0) + c. \tag{4.4}$$

▷ The **amplitude** or **elongation** $r > 0$ is the maximum position from the equilibrium. We use the amplitude to stretch the sine function along the $y$-axis. We define the **peak-to-peak amplitude** as the distance between the maximum returned **crest** and the minimum returned **trough**, which equals $2r$. In figure 4.12(a) we show the effect of an amplitude $r$ that was brought to 4.

▷ The **angular speed** or **pulsation** $\omega > 0$ is the inner coefficient we use to stretch the sine function along the $x$-axis. We prove the constant $\omega = \frac{2\pi}{T}$ with $f = \frac{1}{T}$ the **frequency** in Hertz, named after the German physicist **Heinrich Rudolf Hertz** (1857–1894) who was the pioneer of producing and detecting radio waves. Therefore, the pulsation is also known as the **angular frequency** $\omega = 2\pi f$. Figure 4.12(b) demonstrates how a pulsation $\omega$ doubled to 2 affects the elementary period halved to the new $T = \pi$.

*Proof*: We recall the periodicity of the elementary sine function. For its elementary period equals the full angle $2\pi$ rad, we express this as $\sin x = \sin(x + 2\pi)$. This equality holds, even after attributing $x$ a positive coefficient $\omega$ which we express as $\sin(\omega x) = \sin((\omega x) + 2\pi)$.

Factoring this introduced $\omega$ leads to $\sin(\omega(x)) = \sin(\omega(x + \frac{2\pi}{\omega}))$ which reveals its **general period** as $T = \frac{2\pi}{\omega}$. Rewriting this formula for $\omega$ proves $\omega = \frac{2\pi}{T}$. ∎

▷ We practically define the **phase** as the argument $x_0$ of the 'first incoming crossing' at the baseline of the sine function. Therefore our crossing $\omega x + \theta_0 = 0$ corresponds to phase $x_0 = \frac{-\theta_0}{\omega}$ as indicated by index 0.

A positive phase $x_0 > 0$ of the general sine graph is the length by which its elementary version was shifted to the right. For a negative phase $x_0 < 0$, it is the length by which its elementary version was shifted to the left. Therefore, we appreciate the phase $x_0$ as the horizontal shift of an elementary sine graph. In figure 4.12(c) the elementary graph is shifted by $\frac{\pi}{2}$ radians to the right. This corresponds to the effect of a phase $x_0 = +\frac{\pi}{2}$.

Unlike in other resources, our practical definition in this book remains valid throughout various occurrences of general sine functions.

*Example:* For another popular version like $g(x) = r\sin(\omega(x - q))$ we accordingly solve $\omega(x - q) = 0$ for $x$ which consequently corresponds to phase $x_0 = q$.

According to their mutual phase difference, a sine function can be 'ahead of the other'. Sine functions with phase difference equal to zero are said to be 'inphase'. Sine functions with a phase difference equal to the straight angle are in 'antiphase'.

a)                                                          b)

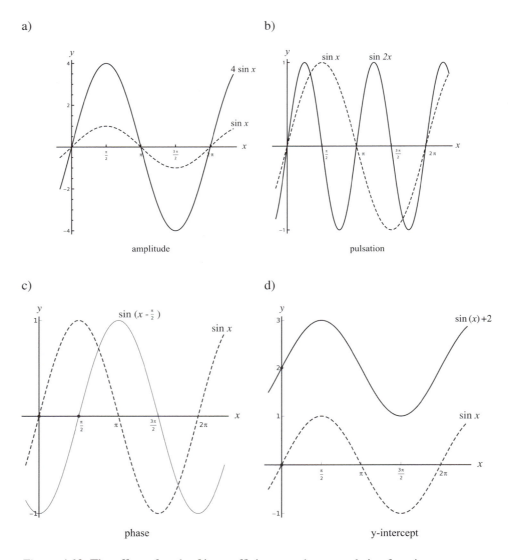

*Figure 4.12*: The effect of each of its coefficients on the general sine function

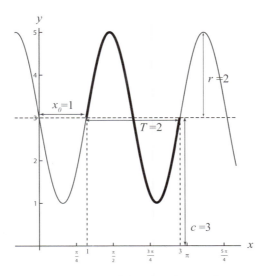

*Figure 4.13*: Graph of the sine function $f(x) = 2\sin(\pi x - \pi) + 3$

▷ The **intercept** is the constant term $c$ added to the general sine function. In case $c \neq 0$ the former roots of its elementary version are called **baseline crossings**, along the equilibrium line $y = c$ different from the $x$-axis. Graphically, the intercept $c$ is simply the value by which the elementary graph is shifted upwards ($c > 0$) or downwards ($c < 0$). Therefore, we recognise the intercept $c$ as the vertical position from the elementary equilibrium. In figure 4.12(d) the elementary graph is shifted upwards by 2, demonstrating the effect of $c = +2$.

*Example*: To retrieve the recipe of a general sine function as portrayed in figure 4.13, we need to determine all of its coefficients. Note that the order matters in doing so. The trough reads $y_{min} = 1$ and the crest $y_{max} = 5$. We calculate the amplitude by solving $2r = 5 - 1$ for $r = 2$. Consequently this sine function will oscillate around its horizontal baseline $y = \frac{1+5}{2} = 3$, which yields an intercept $c = +3$. We measure its general period $T$ for instance from $x = 1$ to $x = 3$. Equivalently $T = 3 - 1 = 2$ leads to the pulsation $\omega = \pi$. We spot the 'first incoming crossing' at its baseline seemingly shifted by 1 to the right. The corresponding phase $x_0 = +1$ yields the coefficient $\theta_0 = -\omega x_0 = -\pi(+1) = -\pi$. Assembling it all, we determine this general sine function as $f(x) = 2\sin(\pi x + (-\pi)) + 3$.

TRANSVERSAL OSCILLATIONS

The general sine function $f(x) = r\sin(\omega x + \theta_0) + c$ models oscillations, such as the motion of a spring. In nature, various of them exist, such as light (in vacuum), sound (although in reality a longitudinal phenomenon) or mechanical waves.

The amplitude of light is its intensity measured in *candela* and its frequency corresponds to colour. Although actually not a transversal oscillation, the general sine function may also model sound. The amplitude of modelled sound is its intensity measured in *decibel* and its frequency corresponds to pitch or tone.

## 4.7 Inverse trigonometric functions

We define the inverse trigonometric functions (alternatively called **cyclometric functions**) based on their former definitions (see page 63).

ARCSINE

Logically, the real range $[-1, 1]$ of the elementary sine function becomes the domain of its inverse function, called arcsine. We recall the periodicity of the elementary sine function (given its elementary period of $2\pi$). Due to this periodicity, the arcsine would not be a function, as to every argument between $-1$ and $1$, there corresponds an infinite number of returned angles. For this reason, we restrict the range of the arcsine to the real interval $\left[\frac{-\pi}{2}, \frac{\pi}{2}\right]$.

$$\begin{array}{rccc}
\sin: & \mathbb{R} & \longrightarrow & [-1, 1] \\
& x \text{ (in rad)} & \longmapsto & y = \sin x
\end{array}$$

$$\begin{array}{rccc}
\arcsin: & [-1, 1] & \longrightarrow & \left[\dfrac{-\pi}{2}, \dfrac{\pi}{2}\right] \\
& x & \longmapsto & y \text{ (in rad)} = \arcsin x
\end{array}$$

Note that sine acts as a function on angles (in radians) to return values in the real interval $[-1, 1]$, while arcsine acts as a function on numbers (between $-1$ and $1$) to return their corresponding angle in radians (between $\frac{-\pi}{2}$ rad and $\frac{\pi}{2}$ rad).

*Example*: We calculate the arcsine return for some arguments:

$$\text{arcsin}: \quad [-1,1] \quad \longrightarrow \quad \left[\frac{-\pi}{2}, \frac{\pi}{2}\right]$$

$$x \qquad \mapsto \qquad y \text{ (in rad)} = \text{arcsin}\, x$$

$$-1 \qquad \mapsto \qquad \text{arcsin}(-1) = \frac{-\pi}{2}$$

$$-\frac{1}{2} \qquad \mapsto \qquad \text{arcsin}\left(-\frac{1}{2}\right) = \frac{-\pi}{6}$$

$$0 \qquad \mapsto \qquad \text{arcsin}\,0 = 0$$

$$\frac{1}{2} \qquad \mapsto \qquad \text{arcsin}\left(\frac{1}{2}\right) = \frac{\pi}{6}$$

$$\frac{\sqrt{2}}{2} \qquad \mapsto \qquad \text{arcsin}\left(\frac{\sqrt{2}}{2}\right) = \frac{\pi}{4}$$

$$\frac{\sqrt{3}}{2} \qquad \mapsto \qquad \text{arcsin}\left(\frac{\sqrt{3}}{2}\right) = \frac{\pi}{3}$$

$$-2 \quad \mapsto \quad \text{arcsin}(-2) \text{ is not defined because } -2 \notin [-1,1]$$

Doing so for all arguments $x$ running through the real interval $[-1,1]$ leads to the graph of the arcsine function.

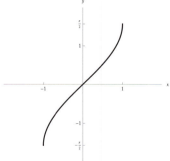

*Figure 4.14*: The arcsine function

ARCCOSINE

In the same way, arccosine is the inverse function of cosine which returns for any given argument (between $-1$ and $1$) its corresponding angle (restricted to the closed interval $[0, \pi]$) in radians.

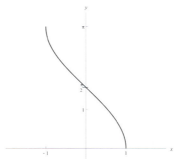

*Figure 4.15*: The arccosine function

ARCTANGENT

Arctangent is the inverse function of tangent which returns for any given argument its corresponding angle restricted to the open interval $\left] \frac{-\pi}{2}, \frac{\pi}{2} \right[$ in radians.

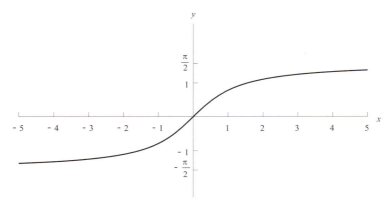

*Figure 4.16*: The arctangent function

ATAN2

We define the positive **atan2**-function for this book as the extended arctangent function which operates in an orthonormal $(x,y)$−frame from where it digests any coordinate pair $(y,x)$ into the corresponding angle $\theta \in [0, 2\pi[$ according to its quadrant.

$$\theta = \text{atan2}(y,x) = \begin{cases} \arctan\left(\frac{y}{x}\right) & \text{if } x > 0 \text{ and } y \geqslant 0 \quad \text{(quadrant I)} \\ \frac{\pi}{2} \text{ rad} & \text{if } x = 0 \text{ and } y > 0 \\ \arctan\left(\frac{y}{x}\right) + \pi & \text{if } x < 0 \quad \text{(quadrant II and III)} \\ \frac{3\pi}{2} \text{ rad} & \text{if } x = 0 \text{ and } y < 0 \\ \arctan\left(\frac{y}{x}\right) + 2\pi & \text{if } x > 0 \text{ and } y < 0 \quad \text{(quadrant IV)} \\ \text{indeterminate} & \text{if } x = 0 \text{ and } y = 0. \end{cases}$$

atan2(4,-1)=1.816 rad

atan2(4,1)=1.326 rad

atan2(-4,-1)=4.467 rad

atan2(-4,1)=4.957 rad

## 4.8 Maclaurin expansions

We define a **series expansion** of a function $f$ as a representation of it by an ongoing polynomial in its variable. This series equals the function restricted within a so-called **radius of convergence**, for which we kindly refer you to the specialised literature.

*Example*: We show such a restricted representation through the rational function

$$\frac{1}{1-x} = 1 + x + x^2 + x^3 + x^4 + \ldots \text{ which only applies within the radius } -1 < x < 1$$

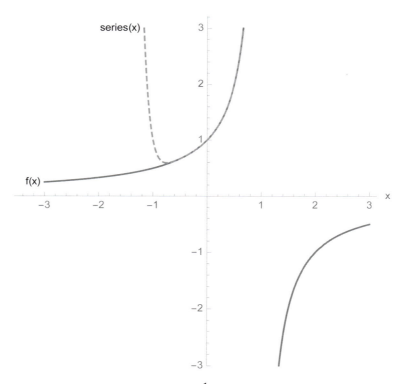

*Figure 4.17*: The rational function $f(x) = \dfrac{1}{1-x}$ portrayed along its series representation

We define the **maclaurin series** of a function $f$ as a type of series expanded around zero and valid within its radius of convergence around zero. The maclaurin series is named after the Scottish mathematician (and child prodigy) **Colin Maclaurin** (1698–1746), who attributed this series to his British contemporary **Brook Taylor** (1685–1731).

We define the **n$^{\text{th}}$ order maclaurin polynomial** of a function $f$ as the partial sum of the $n$ first terms of its maclaurin series, which approximates the function $f$ with a **residual error $\varepsilon$** corresponding to a zero-centred scope, for which we refer you to the specialised literature.

*Example*: We show such a 7$^{\text{th}}$ order maclaurin polynomial approximating the $\sin(x)$−function. Despite the radius of convergence of its corresponding maclaurin series is $\pm\infty$ this maclaurin polynomial is only accurate for a full sine period centred around zero. We illustrate this 7$^{\text{th}}$ order accuracy by estimating within the scope $-1 < x < 1$ the residual error $\varepsilon \leqslant 0.000\,003$.

$$\sin(x) \approx x - \frac{x^3}{3!} + \frac{x^5}{5!} - \frac{x^7}{7!} \quad \text{which is accurate within the scope } -\pi < x < \pi$$

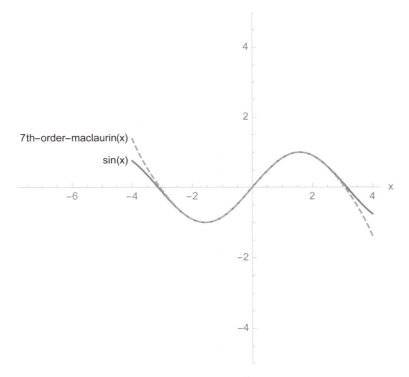

*Figure 4.18*: The sine function portrayed along its 7$^{\text{th}}$ order maclaurin polynomial

We see how the maclaurin polynomial can be applied as an approximation for functions $f$ and the more terms included, the better this approximation is. Maclaurin polynomials – amongst others – can be used by today's calculators to compute function return values

$f(x)$ when feeding function $f$ by arguments $x$. Often the latter requires recasting the polynomials into their chebyshev representation, named after the Russian mathematician **Pafnuty Lvovich Chebyshev** (1821–1894), fitting them to the clenshaw algorithm for which we once more refer you to the dedicated literature.

Moreover, we should keep in mind the linear and quadratic maclaurin polynomials can be used – within their restricted scopes – as approximations enabling solutions to otherwise unsolvable problems blocked by the actual function $f$.

*Examples*:  We show the maclaurin *series* of all functions covered in this chapter.

▷  Natural exponential function $\exp(x) = e^x$ given its natural base $e = 2.718281\ldots$

$$\exp(x) = 1 + x + \frac{x^2}{2!} + \frac{x^3}{3!} + \frac{x^4}{4!} + \frac{x^5}{5!} \ldots \quad \text{valid for all } -\infty < x < +\infty \quad (4.5)$$

▷  Sine function is confirmed as an odd function (see page 78) on arguments in radian

$$\sin(x) = x - \frac{x^3}{3!} + \frac{x^5}{5!} - \frac{x^7}{7!} + \frac{x^9}{9!} - \ldots \quad \text{valid for all } -\infty < x < +\infty \quad (4.6)$$

▷  Cosine function is shown to be an even function (see page 72) on radians

$$\cos(x) = 1 - \frac{x^2}{2!} + \frac{x^4}{4!} - \frac{x^6}{6!} + \frac{x^8}{8!} - \ldots \quad \text{valid for all } -\infty < x < +\infty \quad (4.7)$$

▷  Tangent function is confirmed as odd with its open domain $\left] \frac{-\pi}{2}, \frac{\pi}{2} \right[ \subset \mathbb{R}$ in radian

$$\tan(x) = x + \frac{x^3}{3} + \frac{2x^5}{15} + \ldots \quad \text{which only applies within the radius } -\frac{\pi}{2} < x < \frac{\pi}{2}$$

▷  Inverse sine function $\arcsin(x)$ confirms its domain $[-1, 1] \subset \mathbb{R}$ and range in radian

$$\arcsin(x) = x + \frac{x^3}{6} + \frac{3x^5}{40} + \ldots \quad \text{which only applies within the radius } -1 \leqslant x \leqslant 1$$

▷  Inverse cosine function $\arccos(x)$ confirms domain $[-1, 1] \subset \mathbb{R}$ and range in radian

$$\arccos(x) = \frac{\pi}{2} - x - \frac{x^3}{6} - \frac{3x^5}{40} - \ldots \quad \text{which only applies within the radius } -1 \leqslant x \leqslant 1$$

▷  Inverse tangent function $\arctan(x)$ requires the closed domain $[-1, 1] \subset \mathbb{R}$ and returns its range in radian

$$\arctan(x) = x - \frac{x^3}{3} + \frac{x^5}{5} - \ldots \quad \text{which only applies within the radius } -1 \leqslant x \leqslant 1$$

## 4.9 Exercises

**Exercise 32**  Prove that the slopes of perpendicular straight lines multiply to negative one. Hint: base your reasoning on the tangent of their inclination angles.

**Exercise 33**  The equation of a parabola is written as $y = ax^2 + bx + c$. Determine the sign of each coefficient $a, b$ and $c$ for each parabola type shown below.

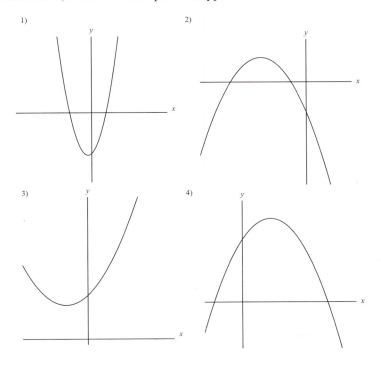

**Exercise 34**  During a computer game, an object is running on a straight line from the point $A$ with coordinates $(0, 20)$ to the point $B$ pinned by $(15, 30)$. Find the equation of this straight line $AB$. When the object is crossing the point $(30, 40)$, the gamer sends it 90° to the left on another straight line. Find the equation of this new straight line. Draw both lines, the former line $AB$ and the latter new one, in one $(x, y)$-frame.

**Exercise 35**  A car is driving on a straight line $3x + 5y = 8$. On another straight line $x + 3y = 4$ a brick wall is built. If the car keeps driving on its straight line, will it ever hit this wall? In case it does, at what point exactly? Draw both lines in one $(x, y)$-frame.

**Exercise 36**   Retrieve the intersection of the linear function $f(t) = 12t - 24$ and the quadratic function $g(t) = 3t^2 + 3t - 36$.

**Exercise 37**   Prove that the natural logarithmic $\log_e(x)$ and the natural exponential $\exp(x)$ make a pair of inverse functions. How do the graphs of such a pair of inverse functions relate in the same orthonormal $(x, y)$–plot frame?

Hint: base your reasoning on $y = \log_e(x)$ in which you formally interchange $y$ and $x$, and then solve the new expression for $y$ to retrieve its new recipe.

**Exercise 38**   Examine to which extent the composite functions $f(x) = \log_e(\exp(x))$ versus $g(x) = \exp(\log_e(x))$ differ by listing their:

▷   domains

▷   ranges

▷   roots

**Exercise 39**   Folding one sheet of newspaper works up to seven times, upon which we measure a final paper thickness of 1 cm. Retrieve the thickness of the initial sheet.

**Exercise 40**   Retrieve the recipes of both functions graphed below.

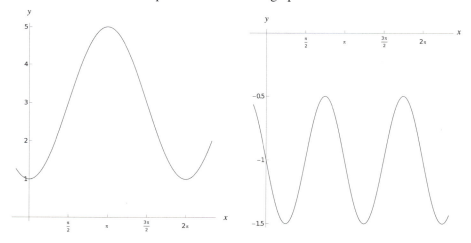

**Exercise 41**   Confirm the residual error of the $7^{\text{th}}$ order maclaurin polynomial to the elementary $\sin(x)$–function by recomputing its according $\varepsilon$ value (see page 87).

Hint: base your calculation on the scope of the approximation and the single $9^{\text{th}}$ maclaurin expansion term.

# Chapter 5 · The Golden Section

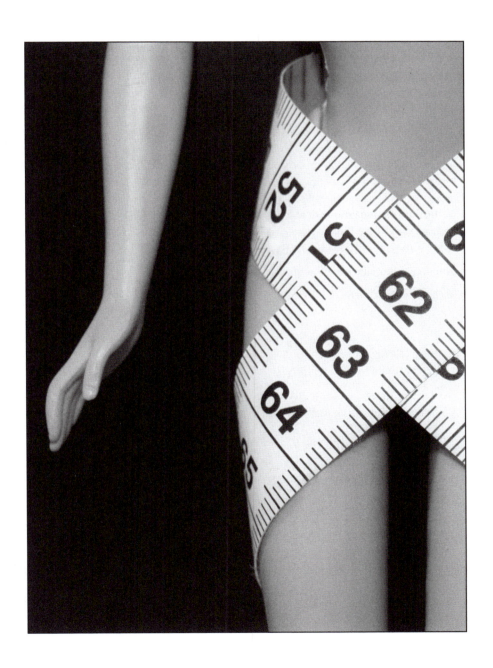

The golden number **golden number** $\frac{1+\sqrt{5}}{2}$ was constructed eras ago, presumably since antiquity, or at least 2 500 years back in time. Ever since, it has been kept in use as a famous irrational number. We easily find illustrations of it in nature as well as applications of it around the world. Throughout this chapter we promote the professional use of the golden section in graphical and digital arts. A pair of compasses, a ruler and a calculator are the required tools for this aesthetic chapter.

## 5.1 The golden number

It has been the Greek geometer **Euclid** (approx 300 – 250 Before Christ) who outlined a ruler-and-compass construction to cut a line segment into **mean proportional parts**, according to volume II (11th proposition of the Arithmetical Geometry) and to volume VI (30th proposition of the Theory of Ratio) of his thirteen-piece geometric standard 'The Elements'. Euclid stated we can cut a line segment into two parts in such a way that the ratio of the bigger part over the smaller part equals the ratio of the total segment over its bigger part. Ever since this true statement has been called the **golden section** for which we outline a modern ruler-and-compass construction.

*Construction*: the golden section.

Firstly, the construction of $\frac{1+\sqrt{5}}{2}$ implies the drawing of $\sqrt{5}$ in an exact way. We logically base the construction of square roots such as $\sqrt{5}$ on the Pythagorean Theorem (3.2).

1) We draw a right triangle $ABC$ having its right angle in $A$ and its lengths of sides $|AC| = 2|AB|$. Hence, the Pythagorean Theorem dictates the corresponding hypotenuse to be $|BC| = \sqrt{5}|AB|$.

2) We then extend the line segment $[AB]$ with the constructed segment $[BC]$ in such a way that the length of this new line segment becomes $|AD| = (1 + \sqrt{5})|AB|$.

3) We finally cut $[AD]$ through its midpoint $M$ which yields $|AM| = \frac{1+\sqrt{5}}{2}|AB|$.

So far, the ratio of the final segment $[AM]$ over the segment $[AB]$ results in $\frac{|AM|}{|AB|} = \frac{1+\sqrt{5}}{2}$. Secondly, we still need to prove that also the ratio of the bigger part $[AB]$ over the smaller part $[BM]$ simplifies to $\frac{1+\sqrt{5}}{2}$.

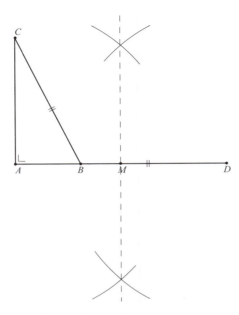

*Figure 5.1*: A construction of the golden section

*Proof*:

Setting $|AB| = a$ leads to

$$|BM| = |AM| - |AB| = \frac{1+\sqrt{5}}{2}a - a = \left(\frac{1+\sqrt{5}}{2} - 1\right)a = \frac{\sqrt{5}-1}{2}a.$$

We conclude

$$\frac{|AB|}{|BM|} = \frac{a}{\left(\frac{\sqrt{5}-1}{2}\right)a} = \frac{2}{\sqrt{5}-1},$$

since $a \neq 0$. We get rid of the square root in the denominator by multiplying the fraction with $1 = \frac{\sqrt{5}+1}{\sqrt{5}+1}$ and applying the Perfect Quotient (see formula (1.3)) in such a way that

$$\frac{|AB|}{|BM|} = \left(\frac{2}{\sqrt{5}-1}\right)\left(\frac{\sqrt{5}+1}{\sqrt{5}+1}\right) = \frac{2(\sqrt{5}+1)}{5-1} = \frac{1+\sqrt{5}}{2}.$$

∎

The segment $|AB|$ is called the mean proportional part between $|AM|$ and $|BM|$, also known as geometric mean of $|AM|$ and $|BM|$:

$$\frac{|AM|}{|AB|} = \frac{|AB|}{|BM|} \iff |AM| \cdot |BM| = |AB|^2.$$

We call this constant ratio the golden number which we typeset as

$$\Phi = \frac{1 + \sqrt{5}}{2} \approx 1.61803. \tag{5.1}$$

This constant proportion $\Phi$, named after the first character of the Greek sculptor **Phidias** (480 – 430 Before Christ), equals 1.618 only approximatively because of its infinite sequence of unpredictable fractional digits, just as all **irrational numbers** in $\mathbb{R}$ do.

## 5.2 The golden section

### THE GOLDEN TRIANGLE

We define the **golden triangle** as any isosceles triangle with an apex angle of $36°$.

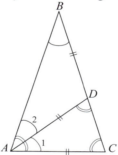

*Figure 5.2*: The golden triangle

The isosceles triangle $ABC$ with apex $B$ featuring angle $\hat{B} = 36°$ must have equal base angles $\hat{A} = 72° = \hat{C}$, because of its interior angles adding up to the straight angle. Measuring its base $[AC]$ as the radius of the circle $C(A, |AC|)$ up to the intersecting point $D$ on segment $[BC]$ creates a new isosceles triangle $CAD$. This latter triangle features base angles $\hat{C} = 72° = \hat{D}$ and must have an apex angle $\hat{A}_1 = 36°$, because of its interior angles adding up to the straight angle.

The similarity of the triangles $ABC$ and $CAD$ (see page 51) leads to the geometric ratios $\frac{|AB|}{|AC|} = \frac{|AC|}{|CD|} = \Phi$, hence their title of 'golden triangle'.

## THE GOLDEN RECTANGLE

Amongst the wide variety of rectangles, we tend to appreciate only one 'ideal rectangle'. It will not be a too stretched rectangle, nor will it be a square (which is also a valid example of a rectangle), but a typical shape somewhere in between. We define this unique 'ideal rectangle' in words: 'it is the rectangle which after cutting away a square is similar to the remaining rectangle'.

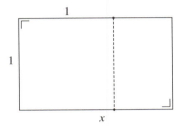

*Figure 5.3*: The golden rectangle

We formalise the above definition based on the similarity in its description:

$\frac{\text{length}}{\text{width}}$ of the golden rectangle $= \frac{\text{length}}{\text{width}}$ of its 'remaining rectangle'.

Setting the golden rectangle's width to 1 and its length to $x$, we can express this similarity as an equation $\frac{x}{1} = \frac{1}{x-1}$. Solving this equation for $x$ will lead us to the 'ideal ratio' we are searching for. To do so, we rewrite our expression to a standard quadratic equation

$$x^2 - x - 1 = 0, \tag{5.2}$$

subsequently applying the quadratic formula (see formula (1.6)) for it. We calculate its discriminant $D = 5$ and both roots as $x_1 = \frac{1+\sqrt{5}}{2}$ and $x_2 = \frac{1-\sqrt{5}}{2}$.

Given $\sqrt{5} \approx 2.23607$ and the fact that lengths have to be non-negative, we can only allow the $x_1$-value $\Phi = \frac{1+\sqrt{5}}{2} \approx 1.61803$ as the physical solution to the equation.

Alternatively we should not ignore the negative mathematical solution $x_2 = \frac{1-\sqrt{5}}{2}$ which we therefore distinguish with an accent mark as

$$\Phi' = \frac{1-\sqrt{5}}{2} \approx -0.61803. \tag{5.3}$$

In conclusion, we again rediscovered the ratio corresponding to the golden section as the only meaningful solution and thus rename the 'ideal rectangles' to **golden rectangles**.

THE GOLDEN SPIRAL

We start from a golden rectangle $ABCD$ for which $|AD| = \Phi \cdot |AB|$.

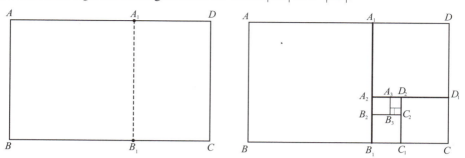

*Figure 5.4*: Construction of the golden Spiral within a golden rectangle

On segment $[AD]$ we choose the point $A_1$ and on segment $[BC]$ we choose the point $B_1$ in such a way that $|AA_1| = |AB| = |BB_1|$. This way we cut away a square from the golden rectangle. We therefore realise $|A_1D| = |AD| - |AB| = \Phi \cdot |AB| - |AB| = (\Phi - 1) \cdot |AB|$. Multiplying both sides of this equality with the factor $\Phi$ yields $\Phi \cdot |A_1D| = (\Phi^2 - \Phi) \cdot |AB|$. Due to the former quadratic equation $\Phi^2 - \Phi = 1$ for the golden number $\Phi$, we simplify the latter equality to $\Phi \cdot |A_1D| = |AB|$. Given the equal widths $|AB| = |CD|$ of the original rectangle, we find $|CD| = \Phi \cdot |A_1D|$. The latter proportion proves also the remaining rectangle $A_1B_1CD$ to be a golden rectangle.

We proceed by cutting away a square from this new golden rectangle $A_1B_1CD$, which will again lead to a smaller remaining golden rectangle. Iterating this process yields a sequence of progressively shrinking golden rectangles: $ABCD$, $DA_1B_1C$, $CD_1A_2B_1$, $B_1C_1D_2A_2$, $\ldots$.

We observe that the sequence of points $B, A_1, D_1, C_1, B_2, A_3, \ldots$ makes up a spiral. We can show this spiral in a non-exact way by drawing a quarter of a circle inside the progressively shrinking square cutaways, shifting the circle centre accordingly. Mathematically this so-called 'spiral' is just a chain of quarters of a circle.

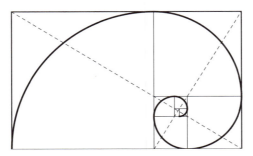

*Figure 5.5*: The golden spiral within a golden rectangle

In a similar way to our discovery of a **golden spiral** based on the sequence of progressively shrinking golden rectangles, we can sketch a golden spiral based on the golden triangle.

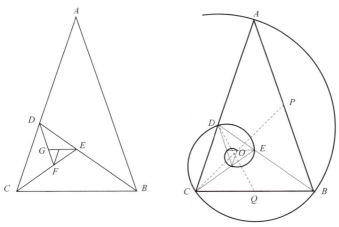

*Figure 5.6*: The golden spiral within the golden triangle

The isosceles triangles $ABC, BCD, CDE, DEF, EFG, \ldots$ are similar (see page 51) due to the constructed bisector of angles $\hat{B}$, $\hat{C}$, $\hat{D}$, $\hat{E}$, $\hat{F}$, $\hat{G}, \ldots$. The starting point $O$ for the golden spiral is the intersection point of the medians of the initial golden triangle and its remaining golden triangle. To a greater extent, the starting point is the intersection point of the medians $CP$ and $DQ$ of the golden triangles $ABC$ and $BCD$ respectively.

## THE GOLDEN PENTAGON

The diagonals of a pentagon intersect according to the golden section.

*Proof*:

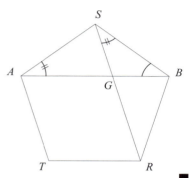

In this pentagon, the triangle *SBG* is similar to the triangle *ABS* because they have two equal interior angles (see page 51). The angle in vertex *B* is already shared and based on the symmetry of the pentagon we obtain $\hat{A} = \hat{S}$. Besides, $|AG| = |TR| = |SB|$ which instantly leads to

$$\frac{|AB|}{|BS|} = \frac{|SB|}{|BG|} \Rightarrow \frac{|AB|}{|AG|} = \frac{|AG|}{|BG|}.$$

In conclusion, the point *G* intersects the diagonal $[AB]$ according to the golden section.                                             ■

Due to the appearance of the golden section $\Phi$ on its diagonals, we give the pentagon its alias of **golden pentagon**.

## THE GOLDEN ELLIPSE

We consider two concentric circles around centre *O* with radii *a* and *b* in such a way that $a > b$. The area of the ring in between both circular discs $C(O,a)$ and $C(O,b)$ equals the difference of both circle areas $\pi a^2 - \pi b^2 = \pi \cdot (a^2 - b^2)$.

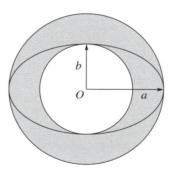

*Figure 5.7*: The golden ellipse

The inner ellipse centred on *O* featuring **semi-major axis** *a* and **semi-minor axis** *b* has an area equal to $\pi \cdot a \cdot b$.

Only in case the area of this ellipse equals the area of the ring, we conclude

$$\pi \cdot (a^2 - b^2) = \pi \cdot ab \Leftrightarrow a^2 - b^2 = ab \Leftrightarrow a^2 - a \cdot b - b^2 = 0.$$

Dividing both sides of the latter equation by $b^2$, since the existence of the ring assures $b \neq 0$, leads to

$$\frac{a^2}{b^2} - \frac{a \cdot b}{b^2} - \frac{b^2}{b^2} = 0 \Leftrightarrow \left(\frac{a}{b}\right)^2 - \frac{a}{b} - 1 = 0$$

or in other words, to the algebraic expression of the famous numbers $\Phi$ and $\Phi'$ (5.2).

Only in case of those equal areas, the ratio of the semi-axes $\frac{a}{b} = \Phi$ entitles it to be a **golden ellipse**.

## 5.3 Golden arithmetic

GOLDEN IDENTITIES

We prove the following golden identities easily by substituting $\Phi = \frac{1+\sqrt{5}}{2}$ and applying some elementary arithmetic.

**First Identity**

$$\Phi^2 = \Phi + 1. \tag{5.4}$$

*Proof*:

We substitute $\Phi = \frac{1+\sqrt{5}}{2}$ in both sides and then simplify them until we reach an equality.

For the left hand side we find

$$\Phi^2 = \left(\frac{1+\sqrt{5}}{2}\right)^2 = \left(\frac{1+\sqrt{5}}{2}\right)\left(\frac{1+\sqrt{5}}{2}\right) = \frac{1+2\sqrt{5}+5}{2 \cdot 2} = \frac{6+2\sqrt{5}}{4} = \frac{2 \cdot (3+\sqrt{5})}{4} = \frac{3+\sqrt{5}}{2},$$

while for the right hand side we reach $\Phi + 1 = \frac{1+\sqrt{5}}{2} + 1 = \frac{1+\sqrt{5}}{2} + \frac{2}{2} = \frac{3+\sqrt{5}}{2}$. ∎

**Second Identity**

$$-\frac{1}{\Phi} = \Phi'. \tag{5.5}$$

*Proof*:

$$-\frac{1}{\Phi} = \frac{-1}{\frac{1+\sqrt{5}}{2}}$$

$$= \frac{-2}{1+\sqrt{5}} = \left(\frac{-2}{1+\sqrt{5}}\right) \cdot \left(\frac{1-\sqrt{5}}{1-\sqrt{5}}\right) \qquad \text{via the Perfect Quotient (1.3)}$$

$$= \frac{-2(1-\sqrt{5})}{-4} = \frac{1-\sqrt{5}}{2}$$

$$= \Phi'. \qquad\qquad\qquad\qquad\qquad\qquad\qquad\qquad\qquad\qquad\qquad \blacksquare$$

## THE FIBONACCI NUMBERS

There is a famous number sequence which is prominently related to the golden section, called the **Fibonacci sequence** as mentioned in the novel '*The Da Vinci Code*'. We generate them as the sum of both previous numbers in their sequence by taking 1 and 1 as the two starting numbers: 1, 1, 2, 3, 5, 8, 13, 21, 34, 55, 89, .... Let us study those number sequences to a deeper extent.

In general, we define a number sequence $u_n$ for which every element equals the sum of both previous elements

$$u_{n+1} = u_n + u_{n-1} \qquad\qquad\qquad\qquad (5.6)$$

as **Lucas sequences**, named after the French mathematician **Edouard Lucas** (1842–1891). As soon as we initialise both starting numbers $u_1$ and $u_2$ for a Lucas sequence, we define it completely.

*Example 1*:  Setting $u_1 = 2, u_2 = 1$, we generate the Lucas sequence

$$2, 1, 3, 4, 7, 11, 18, 29, \ldots$$

and we call its elements **Lucas numbers**.

*Example 2*:  Setting $u_1 = 5, u_2 = 2$, we create the Lucas sequence

$$5, 2, 7, 9, 16, 25, 41, 66, 107, 173, 280, 453, 733, \ldots, 13153, 21282, \ldots.$$

Let us, for incrementing index values $n$, divide elements $u_n$ of this Lucas sequence by their previous element $u_{n-1}$ and display those fractions $\frac{u_n}{u_{n-1}}$ decimally.

| $\frac{u_n}{u_{n-1}}$ | ratio | decimally |
|---|---|---|
| $\frac{u_5}{u_4}$ | $\frac{16}{9}$ | 1.77778... |
| $\frac{u_{12}}{u_{11}}$ | $\frac{453}{280}$ | 1.61786... |
| $\frac{u_{13}}{u_{12}}$ | $\frac{733}{453}$ | 1.61810... |
| $\frac{u_{20}}{u_{19}}$ | $\frac{21282}{13153}$ | 1.61803... |
| ⋮ | | |

We notice how those ratios $\frac{u_n}{u_{n-1}}$ converge to $\Phi = 1.61803\ldots$ at incrementing indices.

*Example 3*:  Setting $u_1 = 1, u_2 = 1$, we build the Lucas sequence

$$1, 1, 2, 3, 5, 8, 13, 21, 34, 55, 89, 144, 233, \ldots, 17711, 28657, \ldots.$$

Lucas called the elements of this sequence **Fibonacci numbers**, named after the Italian **Leonardo of Pisa** (1170–1250, alias Fibonacci or Filius Bonacci meaning 'son of Bonacci'). This name was based on Fibonacci's popular 'rabbit population problem' published in his book 'Liber Abaci', as the above numbers were the solution to it.

The British author **Henry Dudeney** (1857–1930) described a more realistic version of it, known as the 'cow population problem' in his book '536 Puzzles and Curious Problems'. It also generates the Fibonacci sequence: *'If a cow in her second year produces her first calf (no bull) and continues like this annually, how many cows would this make in twelve years assuming no animal died?'*

| $\frac{f_n}{f_{n-1}}$ | ratio | decimally |
|---|---|---|
| $\frac{f_5}{f_4}$ | $\frac{5}{3}$ | 1.66667... |
| $\frac{f_{12}}{f_{11}}$ | $\frac{144}{89}$ | 1.61798... |
| $\frac{f_{13}}{f_{12}}$ | $\frac{233}{144}$ | 1.61806... |
| $\frac{f_{23}}{f_{22}}$ | $\frac{28657}{17711}$ | 1.61803... |
| ⋮ | | |

Again, for incrementing index values $n$, we divide elements $f_n$ of this Fibonacci sequence by their previous element $f_{n-1}$ and display those fractions $\frac{f_n}{f_{n-1}}$ decimally. Again we notice how the ratios $\frac{f_n}{f_{n-1}}$ converge to $\Phi = 1.61803\ldots$ at incrementing indices.

We assume the above property to be true for all Lucas sequences. Consequently we try to prove that *for all* Lucas sequences their corresponding sequence of fractions $\frac{u_n}{u_{n-1}}$ will converge to $\Phi$.

*Proof*:

1) Proving for a sequence of fractions to converge to one constant number is rather hard. This phenomenon is called **convergence**, for which we kindly refer you to the literature on real calculus, skipping this first part of the complete proof.

2) Referring to step 1 of this proof, we assume the sequence of fractions $\frac{u_n}{u_{n-1}}$ to tend to one constant limit $L \in \mathbb{R}$. We continue to use the definition of the Lucas sequence (5.6). Dividing both sides of it by $u_n$ yields

$$\frac{u_{n+1}}{u_n} = \frac{u_n}{u_n} + \frac{u_{n-1}}{u_n},$$

which brings the definition of the Lucas sequence to

$$\frac{u_{n+1}}{u_n} = 1 + \frac{u_{n-1}}{u_n}.$$

We assume in step 1 – omitting its proof – that in its limit we may replace the fraction $\frac{u_{n+1}}{u_n}$ by the constant $L$. Consequently we replace the inverse fraction $\frac{u_{n-1}}{u_n}$ by the inverse limit value $\frac{1}{L}$. After these replacements we encounter the equation

$$L = 1 + \frac{1}{L}$$

or after multiplying both sides with $L$ and rearranging terms, the standardised equation

$$L^2 - L - 1 = 0.$$

This means that in its limit we can rewrite the definition of Lucas sequences to the quadratic equation (5.2) valid for the golden number $\Phi$.

Conclusively, we either find $L = \Phi$ or $L = \Phi'$. Based on the non-negative fractions $\frac{u_{n+1}}{u_n} > 0$, we inevitably remain with the limit value $L = \Phi > 0$.  ■

## 5.4  The golden section worldwide

The golden section has probably been famous as a mathematically modelled shape, discovered in nature, for about two millennia. The golden section is an 'ideal proportion' featured during the growth and biological development of people, animals and plants. Hence, the golden section was and is often implemented by people as an 'ideal measure' for their buildings, paintings, sculptures, musical instruments and compositions.

Though we need to stay sceptical and cannot refer each proportion around us to Φ. A few prominent counterexamples are the A4 paper size, the Parthenon in Athens or the Great Pyramid of Giza. Despite our sound scepticism, we hereby draft a representative list of various golden section implementations.

About eighteen centuries beyond Euclid we credit **Luca Bartolomeo de Pacioli** (1447–1517) with rediscovering the 'divine' proportion Φ (1509, *De Divina Proportione*). Pacioli used to be a Franciscan munk who published revolutionary books on mathematics. The proportions of the human body were then successively related to the golden number by a famous pupil of Pacioli, **Leonardo da Vinci** (1490, *Vitruvian Man* as portrayed on the Italian 1€ coins), continued by both Germans **Heinrich Cornelius Agrippa** (1510, *Pentagram Man*) and **Adolf Zeising** (1854, *Neue Lehre von den Proportionen des menschlichen Körpers*).

*Figure 5.8*: Statue of the 'Vitruvian Man' in Vinci (Italy)

How do you think Barbie™ has tempted generations of girls? The golden section applies to the perfect human body in many unimaginable ways. Realising the average person is far from perfect, let us adopt the Barbie™ fashion doll as the ideal body. The length from her belly button to her feet is called the mean proportional part between her total length and her remaining length. We measure the same golden ratio on her arms, legs, hands and feet.

The golden section has also ruled architecture for ages, from the temples of antiquity and medieval cathedrals to the anthropometric standard called the 'modulor' as designed by the French-Swiss architect **Le Corbusier** (1887–1965). And we should not forget to add Islamic (Great Mosque of Kairouan), Buddhist (Borobudur, Java) and Spanish architecture being as eager to implement the golden section on many levels.

Many musical instruments are constructed according to $\Phi$; especially the traditional violin. Even musical compositions and performances can contain the golden number; it seems at least it seems to be the case for Beethoven, Haydn, Schubert, Bartók and Satie.

We should not forget painters: Seurat (*Bathers at Asnieres*), Juan Gris (cubism), Rembrandt (*The Night Watch*) and again Leonardo da Vinci, each of them applying the golden section in their art.

Finally, it goes without saying that graphic design makes use of the golden section. For instance, the logo of the Belgian railroad company, as designed by the Belgian **Henry Van de Velde** (1863–1957), is completely based on the golden section. The same applies to digital art, animation and web design. Talking about the latter, how about a nice two-column design for a webpage? Assuming a screen resolution of $1024 \times 768$, we can start from a webpage of width 1 000 pixels. Dividing its width by $\Phi$ yields 618 pixels for our main column, leaving 382 for the second column. Even this straightforward approach automatically guarantees an aesthetic result.

## 5.5 Exercises

**Exercise 42**   Given the construction

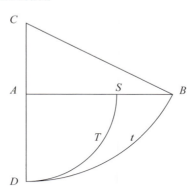

featuring $|AC| = \frac{1}{2} \cdot |AB|$ and the circular arc $\hat{t}$ drawn by the circle $C(C, |BC|)$ intersects the straight line $CA$ at the side of $A$ in point $D$. Next to this we have a second circular arc $\hat{T}$ drawn by the circle $C(A, |AD|)$.

Prove that the point $S$ intersects the segment $[AB]$ according to the golden section. In other words, setting length $|AB| = 1$ should imply length $|AS| = \frac{1}{\Phi} \approx 0.618$.

Hint: write the length of $[AD]$ in terms of the length of $[AB]$ and calculate their ratio $\frac{|AS|}{|AB|}$.

**Exercise 43**   Given the construction

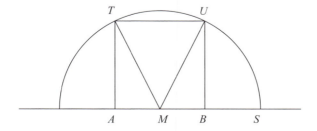

featuring the circular arc by the circle $C(M, |MT|)$, the square $ABUT$, a midpoint causing $|AM| = |MB|$ and $S$ the intersection point of the circular arc $C(M, |MT|)$ and the straight line $AB$.

Prove that the point $B$ intersects the segment $[AS]$ according to the golden section. In other words, setting length $|AB| = 1$ should imply length $|AS| = \Phi \approx 1.618$.

Hint:  write all lengths $|MU|$, $|MS|$, $|AS|$, $|AB|$ in terms of the length of $[AM]$ and then calculate the ratio $\frac{|AS|}{|AB|}$.

**Exercise 44**  Prove $\Phi^2 + \frac{1}{\Phi^2} = 3$ by substituting $\Phi = \frac{1+\sqrt{5}}{2}$.

**Exercise 45**  Prove $\frac{1}{\Phi} = \Phi - 1$ by substituting $\Phi = \frac{1+\sqrt{5}}{2}$.

**Exercise 46**  Prove $\Phi + \frac{1}{\Phi} = \sqrt{5}$ by substituting $\Phi = \frac{1+\sqrt{5}}{2}$.

**Exercise 47**  Prove by substituting, that $x_1 = -1, x_2 = -\Phi$ and $x_3 = -\Phi'$ are the solutions to the cubic equation $x^3 + 2 \cdot x^2 - 1 = 0$.

**Exercise 48**  The sequence $1, \Phi, \Phi^2, \Phi^3, \Phi^4, \Phi^5, \ldots$ is a Lucas sequence. Transform this Lucas sequence (up to 6 elements) to a number sequence linear in $\Phi$ (such as $1, \Phi, \Phi + 1$, ...) by means of the quadratic equation defined by the golden number (5.2).

**Exercise 49**  In Fibonacci's footsteps anno 1202, we go for a makeover of the solving of his famous rabbit population problem. How many couples of rabbits do we breed in one year, given the following conditions?

▷  Every rabbit couple is pubescent at the age of two months.

▷  Every rabbit couple that is pubescent produces one new rabbit couple monthly.

▷  No rabbit dies.

We hereby ignore their incest problem and their offsprings being constantly one male plus one female. Start in month 0 using one fresh rabbit couple.

**Exercise 50**  The **silver number** is the irrational number

$$\delta = 1 + \sqrt{2} \approx 2.41421$$

according to its original definition (128 Anno Domini, *Ostiahuis*).

1)  Find a ruler-and-compass construction based upon the Pythagorean Theorem to construct the **silver section**.

2)  We define the **silver rectangle** in words: 'it is the rectangle which after cutting away two squares, is similar to the remaining rectangle'. Find the meaningful solution $\delta$ to the quadratic equation corresponding to the above definition in words.

3)  We define the indexed elements $p_n \in \mathbb{N}$ as a number sequence of **John Pell** (1611–1685) by $p_{n+1} = 2p_n + p_{n-1}$. Their ratios $\frac{p_n}{p_{n-1}}$ converge to the limit $\delta = 2.41421\ldots$ for incrementing indices $n$. Prove this.

How does the computer keep track of moving objects? This chapter elaborates upon two different $2D$ coordinate systems in order to label each point (each pixel) efficiently. Subsequently it will also allow us to describe various lines and sets of points. Apart from the popular cartesian coordinate system, professional graphic design and computer animation involve the use of polar coordinates. For instance, rotation requires far less computing power (number of operations) and loss of precision (for which we refer you to the specialised literature) compared to the cartesian approach. Finally, using polar curves pushes digital design, computer art and animations to a higher level.

## 6.1 Cartesian coordinates

When we locate objects in a frame, we spontaneously make use of the cartesian coordinates. **Cartesian** means rectangular and was named after its inventor, the French mathematician **René Descartes** (1596–1650). So, cartesian coordinates are the popular rectangular $(x, y)$-coordinates and often orthonormal, which means their $x$-axis and $y$-axis are orthogonal and equally scaled. Logically, straight lines can be most efficiently expressed in this rectangular system.

As soon as the origin $O$ is defined as the intersection point of the $x$- and $y$-axis, we are able to label any point $P$ in $2D$ by two coordinate numbers: the horizontal $x$-coordinate called **abscissa** and the vertical $y$-coordinate called **ordinate**.

## 6.2 Parametric curves

We define a **parametric curve** as the locus of a moving point $P(x, y)$ with $x = x(t)$ and $y = y(t)$ both being functions of a parameter $t$, where the argument $t$ runs through a real domain.

We define the **parametric form** of such a curve as the coordinate functions

$$\begin{cases} x = x(t) \\ y = y(t) \end{cases}. \qquad (6.1)$$

*Example 1*: Whenever we eliminate the parameter $t$ of this 'curve'

$$\begin{cases} x = t \\ y = t \end{cases}$$

from abscissa $x$ and ordinate $y$, we obtain its cartesian form $y = x$.

*Example 2*:  When in the parametric form

$$\begin{cases} x = t \\ y = \sqrt{t^2} \end{cases}$$

we substitute abscissa $x$ in ordinate $y$ via the parameter $t$, we obtain the **absolute value**-function $y = \sqrt{x^2}$.  We typeset this absolute value-function either as $y = \mathrm{abs}(x)$ or as $y = |x|$.

$$\mathrm{abs}: \mathbb{R} \longrightarrow \mathbb{R}^+ : x \mapsto \sqrt{x^2} = \begin{cases} x & \text{if } x \geqslant 0 \\ -x & \text{if } x < 0 \end{cases} \tag{6.2}$$

We see on the graph its domain $\mathbb{R}$, range $\mathbb{R}^+$ and the solitary root $x_0 = 0$.  This graph shows very clearly that all return values are non-negative.

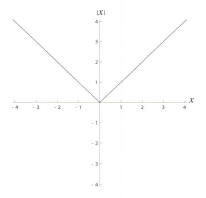

*Figure 6.1*: Absolute value-function $|x|$

*Example 3*:  The squaring and side to side addition of

$$\begin{cases} x = \cos t \\ y = \sin t \end{cases}$$

eliminates $t$ via the trigonometric Pythagorean Identity (3.6) to $(\cos t)^2 + (\sin t)^2 = 1^2$. This way we obtain the cartesian form $x^2 + y^2 = 1$ of the **unit circle** $C(O, 1)$ centred on the origin $O$ and having radius $r = 1$.

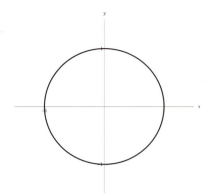

*Figure 6.2*: Unit circle in the cartesian coordinate frame

*Example 4*:  We eliminate the parameter $t$ from the parametric form

$$\begin{cases} x = a\cos t \\ y = b\sin t \end{cases}$$

via the trigonometric Pythagorean Identity as

$$\begin{cases} \dfrac{x}{a} = \cos t \\ \dfrac{y}{b} = \sin t \end{cases} \implies \begin{cases} \left(\dfrac{x}{a}\right)^2 = (\cos t)^2 \\ \left(\dfrac{y}{b}\right)^2 = (\sin t)^2 \end{cases} \implies \left(\dfrac{x}{a}\right)^2 + \left(\dfrac{y}{b}\right)^2 = 1.$$

This is the cartesian form of an **ellipse** centred on the origin $O$ and having semi-major axis $a > 0$ and semi-minor axis $b > 0$.

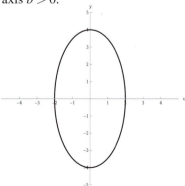

*Figure 6.3*: Ellipse centred on $O$, having semi-axes $a = 2$ and $b = 4$

*Example 5*:    Extending the squares of the previous ellipse to powers by any positive exponent $n \in \mathbb{R}^+$ we obtain a superellipse or so-called **Lamé curve**, named after the French mathematician **Gabriel Lamé** (1596–1650), satisfying the implicit equation

$$\left|\frac{x}{a}\right|^n + \left|\frac{y}{b}\right|^n = 1.$$

Notice the absolute value over both terms in this formula (see page 109).

We firstly define **convex** practically as spherical like a disc in 2D or a rugby ball in 3D. We define **concave** oppositely as hollow-shaped like a diamond in 2D or a diabolo in 3D.

The cartesian form of a Lamé superellipse centred on the origin is a closed curve resembling the ellipse keeping its semi-major axis $a > 0$ and semi-minor axis $b > 0$ and featuring the symmetry about these axes, but offering a wide class of square shapes running from concave stars, over diamonds to convex boxes.

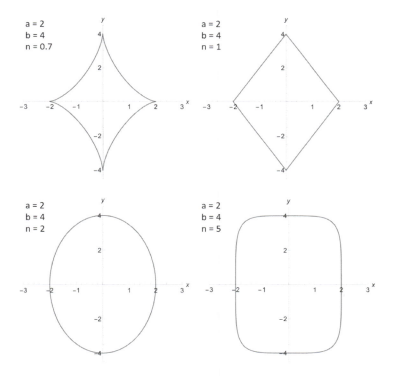

*Figure 6.4*: Lamé curves by $a = 2$ and $b = 4$ and exponents $n = 0.7$, 1, 2, up to 5

## 6.3 Polar coordinates

On the other hand, we can express curved and circular lines more efficiently in a non-rectangular system. Alternatively, we can locate a point $P$ in the plane by referring it to the pole $O$ and the horizontal axis. The **pole** $O$ is the central reference in the polar plane. The **polar axis** is the horizontal axis through the pole $O$.

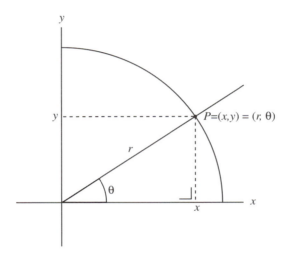

*Figure 6.5*: Cartesian and polar coordinates of a point $P$

Whenever we apply a coordinate system to the plane, we are able to locate a point $P$. This is the case for instance via its unique cartesian label $(x, y)$, but we can also locate it by using *two* alternative coordinate numbers. For polar coordinates we make use of the distance $r = d(O, P)$ from $P$ to the pole $O$, and the angle $\theta \in [0, 2\pi[$ made by the line $OP$ and the positive $x$-axis. We call them the **radial coordinate** $r$ and the **polar angle** $\theta$ respectively. We label a point $P$ in the polar plane uniquely by $(r, \theta)$. We note that the radial coordinate $r$ is non-negative and that the angular coordinate $\theta$ is described in radians measured counterclockwise.

For converting from cartesian to polar coordinates and vice versa, we refer to the previous chapter 3 on trigonometry. Converting is just a matter of applying the basic formulas of the right triangle (see paragraphs 3.2 and 3.3). Be careful in determining the polar angle $\theta$ based on the $x$- and $y$-values: a drawing of the location will confirm the correct positive value for the angle $\theta$.

We convert the polar coordinates of the point $P$ easily to cartesian coordinates by just applying the definitions of sine and cosine in the right triangle leading to

$$x = r\cos\theta \qquad \text{and} \qquad y = r\sin\theta. \qquad (6.3)$$

We convert the cartesian coordinates of the point $P$ straightforwardly to polar coordinates by implementing the Pythagorean Theorem and the definition of tangent in the right triangle (for its inversion see page 85) as

$$r = \sqrt{x^2 + y^2} \qquad \text{and} \qquad \tan\theta = \frac{y}{x} \Rightarrow \theta = \text{atan2}(y, x). \qquad (6.4)$$

*Example 1*:  Converting the cartesian labelled point $A(1, -1)_{\text{cc}}$ into its polar coordinates

$$r = \sqrt{1^2 + (-1)^2} = \sqrt{2}$$

$$\theta = \text{atan2}(-1, 1) = \frac{7\pi}{4}.$$

We conclude the polar coordinates of the point $A$ to become $\left(\sqrt{2}, \frac{7\pi}{4}\right)_{\text{pc}}$.

*Example 2*:  We convert the polar labelled point $B\left(1, \frac{\pi}{2}\right)_{\text{pc}}$ into its cartesian coordinates

$$x = 1\cos\frac{\pi}{2} = 0$$

$$y = 1\sin\frac{\pi}{2} = 1,$$

typeset as $B(0, 1)_{\text{cc}}$.

*Example 3*:
There are curves we can most efficiently describe by expressing their relation in $(r, \theta)$.

Amongst those curved and circular lines we can for instance express a 'default' spiral very efficiently in its polar form $r = \theta$.

We can rapidly plot its graph by sampling some polar angles $\theta$ to calculate their corresponding radial coordinate $r$.

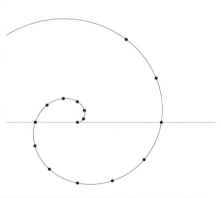

| $\theta$ (in rad) | 0 | $\frac{\pi}{6}$ | $\frac{\pi}{3}$ | $\frac{\pi}{2}$ | $\frac{2\pi}{3}$ | $\frac{5\pi}{6}$ | $\cdots$ |
|---|---|---|---|---|---|---|---|
| $r$ | 0 | $\frac{\pi}{6} \approx 0.52$ | $\frac{\pi}{3} \approx 1.05$ | $\frac{\pi}{2} \approx 1.57$ | $\frac{2\pi}{3} \approx 2.09$ | $\frac{5\pi}{6} \approx 2.62$ | $\cdots$ |

*Example 4*:  Convert the cartesian form of the circle $x^2 + y^2 + 2x = 0$ to its polar form.

We recall for $x^2 + y^2 = r^2$ and substitute the remaining $x$ by $r\cos\theta$. In this way we obtain

$$
\begin{aligned}
x^2 + y^2 + 2x = 0 &\Leftrightarrow r^2 + 2r\cos\theta = 0 \\
&\Leftrightarrow r(r + 2\cos\theta) = 0 \\
&\Leftrightarrow r = 0 \text{ or } r + 2\cos\theta = 0 \\
&\Leftrightarrow r = 0 \text{ or } r = -2\cos\theta.
\end{aligned}
$$

*Example 5*:  Convert the polar form $r = \frac{2}{1-\sin\theta}$ to its cartesian form.

Cancelling out the denominator leads to an equivalent form $r(1 - \sin\theta) = 2$. Expanding its left hand side we find $r - r\sin\theta = 2$. Finally we replace $r$ by $\sqrt{x^2 + y^2}$ and $r\sin\theta$ by $y$, which simplifies to

$$
\begin{aligned}
r - r\sin\theta = 2 &\Leftrightarrow \sqrt{x^2 + y^2} - y = 2 \\
&\Leftrightarrow \sqrt{x^2 + y^2} = y + 2 \\
&\Leftrightarrow x^2 + y^2 = (y + 2)^2 \\
&\Leftrightarrow x^2 = 4y + 4 \\
&\Leftrightarrow y = \frac{1}{4}x^2 - 1.
\end{aligned}
$$

We finally recognise the cartesian form as the equation of a parabola.

## 6.4 Polar curves

We define a **polar curve** as the locus of a moving point $P(r, \theta)$ with its radial coordinate $r = f(\theta)$ in terms of the polar angle $\theta$, where argument $\theta$ runs through a real domain.

We define the **polar equation** of a polar curve as the above mentioned relation $r = f(\theta)$ in radial coordinate $r$ and polar angle $\theta$.

*Example 1*: We define the **crosier curve** of opening $a > 0$ as the polar equation $r = \dfrac{a}{\sqrt{\theta}}$ given its polar angle $\theta$ runs through an interval of $\mathbb{R}^+ \backslash \{0\}$.

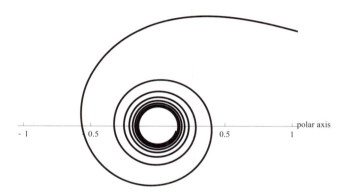

*Figure 6.6*: A crosier curve of opening $a = 1$

*Example 2*: We define the odd-petalled **rose** of size $a \neq 0$ as the polar equation $r = a\cos(m\theta)$ for each odd $m$, given its polar angle $\theta$ runs through $[0, 2\pi[$.

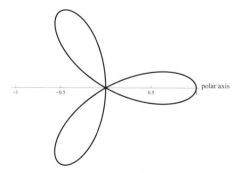

*Figure 6.7*: A three-petalled rose described by $r = 1\cos(3\theta)$

*Example 3*: We define the **lemniscate** with semi-axis $a > 0$ as the polar equation $r = a(\cos(2\theta))^{\frac{1}{2}}$ given its polar angle $\theta$ runs through $[0, 2\pi[$. Notice how in this case the coefficient of the angle $\theta$ stands for the two lobes of the lemniscate, just like it stood for the three petals in the case of a rose. Consequently we may interpret the lemniscate as the *even-petalled* rose $r = a(\cos(n\theta))^{\frac{1}{2}}$ featuring the smallest petal number $n = 2$.

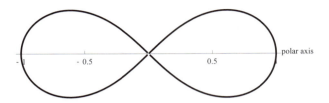

*Figure 6.8*: Lemniscate with semi-axis $a = 1$

A polar superformula

Initially seeking a mathematical model to fit square bamboo, in 1997 the Belgian engineer and botanist **Johan Gielis** (1962) developed in a powerful polar equation capable of plotting almost every polar curve. By implementing six **shaping parameters**, Gielis's **superformula** draws circles, polygons, roses, stars, spirals and all kinds of mixtures of them. Due to its unseen combination of computing efficiency (less source code and less computing overhead) and graphical capacity, the superformula used to be plugged in into Adobe Illustrator®.

We define Gielis's superformula as the polar equation

$$r = \left( \left| \frac{\cos\left(\frac{m}{4}\theta\right)}{a} \right|^{n_2} + \left| \frac{\sin\left(\frac{m}{4}\theta\right)}{b} \right|^{n_3} \right)^{-\frac{1}{n_1}}$$

given its polar angle $\theta$ runs through the interval $[0, 2\pi[$. Notice the absolute value over both terms in this formula (see page 109). The superformula implements six shaping parameters:

 ▷  the semi-major axis $a > 0$,

 ▷  the semi-minor axis $b > 0$,

 ▷  the coefficient $m$ of the angle $\theta$ implies the number of petals,

 ▷  the global shaping exponent $n_1$,

 ▷  the major or horizontal shaping exponent $n_2$ and

 ▷  the minor or vertical shaping exponent $n_3$.

Varying each of the above shaping parameters $a, b, m, n_1, n_2$ and $n_3$ reveals an impressive gallery of exotic polar curves.

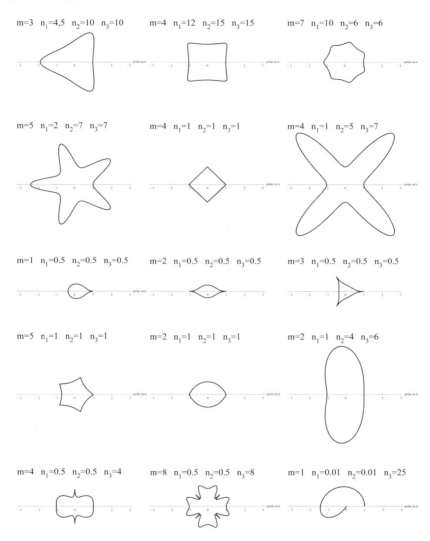

*Figure 6.9*: Superformula of $a = b = 1$ and varying $m, \ n_1, n_2, n_3$

## 6.5 Exercises

**Exercise 51**  Convert from polar to cartesian coordinates. Draw the points in a cartesian frame.

1)  $\left(1, \frac{\pi}{2}\right)_{pc}$

2)  $\left(3, \frac{7\pi}{4}\right)_{pc}$

3)  $\left(4, \frac{3\pi}{2}\right)_{pc}$

4)  $\left(2, \frac{5\pi}{4}\right)_{pc}$

**Exercise 52**  Convert from cartesian to polar coordinates. Draw the points in the polar plane.

1)  $\left(2, 2\right)_{cc}$

2)  $\left(0, -1\right)_{cc}$

3)  $\left(\frac{1}{2}, \frac{\sqrt{3}}{2}\right)_{cc}$

4)  $\left(\sqrt{2}, -\sqrt{2}\right)_{cc}$

**Exercise 53**  Rotate the point $\left(\frac{-10}{4}, \frac{5\sqrt{5}}{2}\right)_{cc}$ counterclockwise $90°$ around the origin $O$. Hint: first label the point in polar coordinates.

**Exercise 54**  Convert the following equations to their polar form.

1)  $y = 3x$

2)  $y = -2x + 3$

3)  $x^2 + y^2 + 2x + 2y = 0$

**Exercise 55**  Convert the following equations to their cartesian form.

1)  $r = \frac{1}{1 - \cos\theta}$

2)  $r = 5\theta$

**Exercise 56**  Draw with pen and paper the **spiral of Theodorus of Cyrene** (5th century Before Christ), taking an isosceles right triangle with sides of length 1 for a start. Put again a side of length 1 perpendicular on its hypotenuse, and keep repeating this for some steps. We hereby picture the initial triangle $ABC$ and the first step.

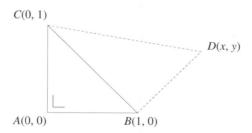

Determine the coordinates of the point $D$.

Determine the irrational number sequence corresponding to the growing hypotenuses.

**Exercise 57** The parametric form of the crosier curve with opening $a > 0$ is

$$\begin{cases} x = \dfrac{a\cos t}{\sqrt{t}} \\ y = \dfrac{a\sin t}{\sqrt{t}} \end{cases} \quad . \tag{6.5}$$

Convert the parametric form to the polar equation of the crosier curve by the subsequent use of the Pythagorean Identity and Theorem (see page 56). Interpret the remaining parameter $t$ as polar angle $\theta$.

**Exercise 58** We define the **scarabaeus curve** with coefficients $a \geqslant 0$ and $b \geqslant 0$ as the polar equation $r = b\cos(2\theta) - a\cos(\theta)$ given its polar angle $\theta$ runs through the interval $[0, 2\pi[$.

Determine with pen and paper the scarabaeus curve coefficients $a$ and $b$ describing

1) a circle with radius 1.

2) a rose of size 3.

Draw on computer the scarabaeus curves with coefficients

1) $a = 2$ and $b = 3$ and vice versa,

2) $a = 3$ and $b = 2$.

**Exercise 59** Determine the shaping parameter(s) $a, b, m, n_1, n_2$ and $n_3$ for the superformula to describe

1) the unit circle (see figure 6.2) and try this with pen and paper.

2) the vertical ellipse centred on the origin $O$ on semi-axes of length 2 and 4 (see figure 6.4) using a computer for this.

We all know numbers. But numbers as such fail to describe motion, because they are unable to express direction. We model the motions in kinematics and forces in dynamics by using vectors. Vectors became increasingly popular through their massive usefulness in physics. During the last few decades, vectors became even more important as they are applied in robotics, computer animation, computer games, pattern matching and security. In this chapter, we introduce vectors before studying all main operations on them: vector addition, scalar multiplication, dot product and cross product. We hereby already hint at the creation of free vectors via the subtraction of location vectors as a *key competence* to every professional. We finally explain some required techniques for 3D animation such as producing a normal vector on polygon surfaces.

## 7.1 The concept of a vector

Distance and displacement are different measures. For instance, if we drive from Brussels to Amsterdam to Brussels, then we travel quite a distance but eventually cause no displacement. Distance is just a non-negative number and differs a lot from displacement, which is a vector.

### VECTORS AS ARROWS

We define **scalar** measures as being completely determined by a numerical value, and often expressed in a certain unit. The definition of a scalar implies it can not determine a direction. Illustrations of scalar measures are *distance* (500 m), *speed* (90 km per hour), temperature (23° C), the height of a person (182.5 cm), energy (123.4 Joule) or the solutions to an algebraic equation ($2x - 6 = 0 \Leftrightarrow x = 3$), ....

We define a **vector** as a measure determined by a magnitude (or absolute value) and a direction. The definition of a vector implies it can be represented by an arrow when it is in 2D or 3D space. Illustrations of vectors are *displacement*, *velocity*, acceleration, force, pressure, ....

Usually we represent a vector by an arrow with its length proportional to the vector's magnitude, pointing in a certain direction and with its initial point meaningfully located.

In this figure the origin $O$ of the plane and two
points in it $A$ and $B$, are shown. The location of
the points $A$ and $B$ is fully determined by the arrow
from $O$ to $A$ and by the arrow from $O$ to $B$ respectively. Such arrows are called vectors. We typeset
the vector ending in $A$ using its lowercase character
as $\vec{a}$ and the one ending in $B$ likewise as $\vec{b}$. Then

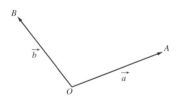

there is one exceptional vector which we cannot draw since it starts and ends in the origin
$O$. This special vector is called the **null vector** or **zero vector**: we typeset it accordingly
as $\vec{o}$.

Vectors can even be applied apart from their geometrical use in 2D and 3D space, for
holding ordered lists of numbers called arrays. As for this (programming) purpose the
number of components often exceeds three, we are no longer able to draw these vectors
as arrows. A multicomponent vector $\vec{x}$ can for instance hold four variables $x_1$ till $x_4$ as
$\vec{x} = (x_1, x_2, x_3, x_4)$ makes an array (see Quaternions, paragraph 17.5). Multimedia programming performs operations on vectors, which again requires us to define vectors as
ordered lists of numbers.

## VECTORS AS ARRAYS

We introduced vectors as arrows in the plane or in 3D space. This approach offers a good
geometrical insight but is less helpful when it comes to performing operations on vectors.
For this reason, let us apply the cartesian coordinate system on the plane and on 3D space,
in order to label points and consequently determine vectors as well.

In this figure, the origin $O$ and two points $A$ and $B$
in the plane are shown again. Applying the cartesian coordinate system on this plane, its two orthonormal axes intersecting in $O$ are shown. Consequently point $A$ gets coordinates $(5,2)$ and point
$B$ coordinates $(-3,4)$. Subsequently, these coordinates are as labels inherited by their corresponding
vectors $\vec{a}$ and $\vec{b}$, and typeset in a column of **vector
components**:

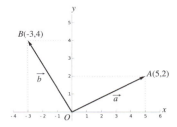

$$\vec{a} = \begin{pmatrix} 5 \\ 2 \end{pmatrix}, \quad \vec{b} = \begin{pmatrix} -3 \\ 4 \end{pmatrix}, \quad \vec{o} = \begin{pmatrix} 0 \\ 0 \end{pmatrix}.$$

We can extend cartesian coordinates to 3D space in which we label a point $P$ by three coordinates $x, y$ and $z$ as located on three perpendicular axes. Here we need to decide whether the $z$-axis 'disappears in' our sheet or 'appears outwards' from it. In other words, whether ascending $z$-values escape from or approach us. Both options are valid and accepted. The former is known as a left-handed frame, the latter as a right-handed frame.

We define a **right-handed frame** by aligning all four right-hand fingers in the direction of the positive $x$-axis and then turn them to the positive $y$-axis. Doing so, our thumb will point in the direction of the positive $z$-axis. We call this practical definition the **right-hand grip rule**. We similarly define a **left-handed frame** by equivalently using our left hand. Alternatively, we can define a right-handed frame by applying the **corkscrew rule**. We rotate the corkscrew from the $x$-axis to the $y$-axis. If the resulting effect is in the direction of the positive $z$-axis, we are in a right-handed frame.

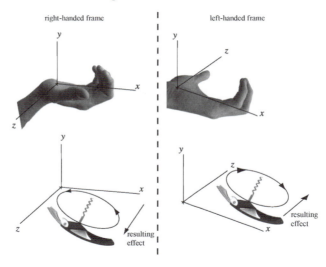

*Figure 7.1*: Right-handed and left-handed frame

For instance Adobe® Flash® holds a right-handed frame for which the $x$-axis runs to the right, the $y$-axis points downwards and the $z$-axis disappears in their plane. In this book we work in the right-handed frames only.

Applying the cartesian coordinate system to 3D space will accordingly label its vectors from arrows to arrays. We typeset the $x$-coordinate of a point $A$ in 3D space as $a_1$, its $y$-coordinate as $a_2$ and its $z$-coordinate as $a_3$. This point $A(a_1, a_2, a_3)$ as well as its corresponding vector $\vec{a}$ are now uniquely labelled, by $\vec{a} = \begin{pmatrix} a_1 \\ a_2 \\ a_3 \end{pmatrix}$ in components.

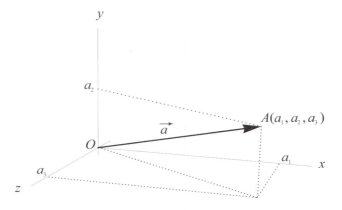

*Figure 7.2*: Vector in 3D space

A **vector** $\vec{a}$ is as an arrow uniquely determined by its **length** (or **norm** or **magnitude**) and its **direction** (holding an orientation and a sense).

To be more precise, a vector $\vec{a}$ features

▷  a non-negative **length** $\|\vec{a}\| \in \mathbb{R}^+$;

▷  an **orientation** parallel to its arrow;

▷  a **sense** given by its arrow-head.

We call the given combination of the vector's orientation and sense the vector's direction.

We typeset a **vector** $\vec{a} \in \mathbb{R}^n$ as

$$\vec{a} = \begin{pmatrix} a_1 \\ a_2 \\ \vdots \\ a_n \end{pmatrix}. \tag{7.1}$$

We call the numbers $a_1, a_2, \ldots, a_n$ the **components** of the vector $\vec{a}$. We typeset the **length** of this vector as $\|\vec{a}\| \in \mathbb{R}^+$.

We calculate the length of a plane vector $\vec{a} = \begin{pmatrix} a_1 \\ a_2 \end{pmatrix}$ via the Pythagorean Theorem (3.2),

$$\|\vec{a}\| = \sqrt{a_1^2 + a_2^2}.$$

We easily extend this calculation for multidimensional vectors. We calculate the length of a 3D vector by applying the Pythagorean Theorem twice, leading to: $\|\vec{a}\| = \sqrt{a_1^2 + a_2^2 + a_3^2}$.

Logically, we calculate the length or **norm** of an $n$-dimensional vector $\vec{a} \in \mathbb{R}^n$ via

$$\|\vec{a}\| = \sqrt{a_1^2 + a_2^2 + a_3^2 + \cdots + a_n^2}. \tag{7.2}$$

We define **unit vectors** as vectors of length 1. We may distinguish unit vectors from all others by typesetting these by overhead hat $\hat{a}$ (instead of arrowed). For instance, vector $\hat{a} = \begin{pmatrix} \frac{3}{5} \\ \frac{4}{5} \end{pmatrix}$ is a unit vector because $\|\hat{a}\| = \sqrt{\frac{9}{25} + \frac{16}{25}} = 1$.

## FREE VECTORS

We define **location vectors** as vectors having the origin $O$ as initial point or **tail**. We typeset location vectors by a lowercase character referring to their terminal point or **head** $A$, like $\vec{a} = \overrightarrow{OA}$. We define **free vectors** as vectors whose initial point is not the origin $O$, but a free floating tail. We typeset a free vector from tail $A$ to head $B$ by an overhead arrow as $\overrightarrow{AB}$ or in physics by overhead arrowed single capital letters such as for force $\overrightarrow{F}$.

Two vectors $\overrightarrow{AB}$ and $\overrightarrow{CD}$ are equal ($\overrightarrow{AB} = \overrightarrow{CD}$) only if their length and direction are equal. Two vectors $\overrightarrow{AB}$ and $\overrightarrow{CD}$ are **parallel** ($\overrightarrow{AB} \uparrow\uparrow \overrightarrow{CD}$) if their direction is equal. Practically we create equal vectors by parallel displacing a given vector. Two vectors $\overrightarrow{AB}$ and $\overrightarrow{CD}$ are **antiparallel** ($\overrightarrow{AB} \uparrow\downarrow \overrightarrow{CD}$) if their orientation is equal and their sense is opposite.

## BASE VECTORS

The vectors $\begin{pmatrix} 1 \\ 0 \\ 0 \end{pmatrix}, \begin{pmatrix} 0 \\ 1 \\ 0 \end{pmatrix}$ and $\begin{pmatrix} 0 \\ 0 \\ 1 \end{pmatrix}$ are the **base vectors** of 3D space. We typeset these base vectors as $\vec{e}_1, \vec{e}_2$ and $\vec{e}_3$ as they align the $x-, y-$ and $z-$axis respectively. Due to these alignments they are perpendicular. Similarly the vectors $\begin{pmatrix} 1 \\ 0 \end{pmatrix}$ and $\begin{pmatrix} 0 \\ 1 \end{pmatrix}$ are the base vectors of $\mathbb{R}^2$. We define base vectors more elaborately later on in this chapter (see page 134).

## 7.2 Addition of vectors

We already introduced a vector $\vec{a}$ as an arrow from the origin $O$ to a terminal point $A$. Somehow, such a location vector $\vec{a}$ and its head $A$ are holding equivalent information, so we could wonder why we ever introduced vectors. One main reason for it is the operations we perform on vectors, which don't apply to points. For instance, the sum $A + B$ of two points $A$ and $B$ is meaningless, as is the product of a point $A$ and a number $\lambda$. On the contrary, the sum of two vectors $\vec{a}$ and $\vec{b}$ and even the product of a vector $\vec{a}$ and a number $\lambda$ are meaningfully defined.

### VECTORS AS ARROWS

The figure shows how to add two vectors geometrically. Vector $\vec{c}$ heading the point $C$ is the sum of the vectors $\vec{a}$ and $\vec{b}$, briefly $\vec{c} = \vec{a} + \vec{b}$. We construct the sum $\vec{c}$ by parallel displacing vector $\vec{b}$, putting its tail at the head of the vector $\vec{a}$. We call this the **head-to-tail rule** for vector addition and typeset it as $\vec{a} + \vec{b}$. We could as well have been parallel displacing vector $\vec{a}$, putting its tail at the head of vector $\vec{b}$ to reach $\vec{b} + \vec{a}$. Apparently the order in which we perform vector addition is indifferently leading to

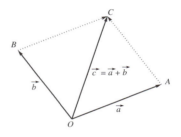

the same sum $\vec{c} = \vec{a} + \vec{b} = \vec{b} + \vec{a}$, which we define as the **commutative property** of vector addition.

Alternatively, the result vector $\vec{c}$ can be recognised as the diagonal of the parallelogram $AOBC$. This *parallelogram method* applied to two vectors $\vec{a}$ and $\vec{b}$, states that their sum $\vec{a} + \vec{b}$ is the diagonal vector $\vec{c}$ in their spanned parallelogram.

### VECTORS AS ARRAYS

Vectors are most easy to add in cartesian coordinate systems. We add vectors by simply adding their corresponding cartesian components.

$$\vec{a} + \vec{b} = \begin{pmatrix} 5 \\ 2 \end{pmatrix} + \begin{pmatrix} -3 \\ 4 \end{pmatrix} = \begin{pmatrix} 5 + (-3) \\ 2 + 4 \end{pmatrix} = \begin{pmatrix} 2 \\ 6 \end{pmatrix}.$$

VECTOR ADDITION SUMMARISED

▷ geometric interpretation: head-to-tail rule or parallelogram method

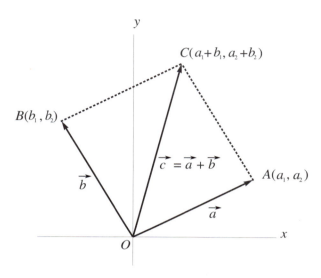

*Figure 7.3*: Vector addition

▷ in words: The addition of vectors is defined as the addition of their corresponding cartesian components.

▷ in symbols: Given $\vec{a} = \begin{pmatrix} a_1 \\ a_2 \\ \vdots \\ a_n \end{pmatrix}$ and $\vec{b} = \begin{pmatrix} b_1 \\ b_2 \\ \vdots \\ b_n \end{pmatrix}$ are vectors in $\mathbb{R}^n$, then their sum is

$$\vec{a} + \vec{b} = \begin{pmatrix} a_1 + b_1 \\ a_2 + b_2 \\ \vdots \\ a_n + b_n \end{pmatrix} \in \mathbb{R}^n. \tag{7.3}$$

## 7.3 Scalar multiplication of vectors

We define the **scalar multiplication** of vectors as a number $\lambda$ times a vector $\vec{a}$, resulting in a scaled vector $\lambda \vec{a}$.

### VECTORS AS ARROWS

The figure shows some examples of vectors $\vec{a}$ and $\vec{b}$ multiplied by real numbers. We obtain the scaled vector $3\vec{b}$ by adding $\vec{b} + \vec{b} + \vec{b}$. Geometrically we therefore extend $\vec{b}$ via the head-to-tail rule by vector $\vec{b}$ itself, which results in the doubled vector $2\vec{b}$. Extending $2\vec{b}$ again by parallel displacing the tail of the vector $\vec{b}$ to the head of $2\vec{b}$, results in $3\vec{b}$. Hence, the vector $3\vec{b}$ keeps the direction of $\vec{b}$ and has a length of 3 times $\|\vec{b}\|$.

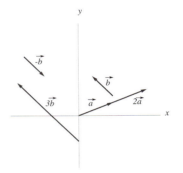

Given the vector $\vec{a}$ and a real number $\lambda$, then the scaled vector $\lambda \vec{a}$ is obtained from the vector $\vec{a}$ by

▷ multiplying its **length** by the absolute value $|\lambda|$,

▷ keeping its **orientation**,

▷ adjusting its **sense** according to $\lambda$:

  ▶ in case $\lambda > 0$ then $\lambda \vec{a}$ keeps the sense of $\vec{a}$;

  ▶ in case $\lambda < 0$ then $\lambda \vec{a}$ gets the opposite sense of $\vec{a}$;

  ▶ in case $\lambda = 0$ then $\lambda \vec{a} = 0\vec{a} = \vec{o}$. (The zero vector $\vec{o}$ has no direction at all.)

Consequently, the **opposite vector** $-\vec{a} = (-1)\vec{a}$ has the same length and orientation of the vector $\vec{a}$, but an opposite sense. We say the vectors $-\vec{a}$ and $\vec{a}$ are antiparallel. We meanwhile also define the subtraction of vectors as $\vec{a} - \vec{b} = \vec{a} + (-\vec{b})$.

### VECTORS AS ARRAYS

The scalar multiplication of a vector is the scalar multiplication of each of its cartesian components, for instance, by the same number $\lambda = 4$.

$$4\vec{a} = \vec{a} + \vec{a} + \vec{a} + \vec{a} = \begin{pmatrix} 3 \\ 1 \end{pmatrix} + \begin{pmatrix} 3 \\ 1 \end{pmatrix} + \begin{pmatrix} 3 \\ 1 \end{pmatrix} + \begin{pmatrix} 3 \\ 1 \end{pmatrix}$$

$$4\vec{a} = \begin{pmatrix} 3+3+3+3 \\ 1+1+1+1 \end{pmatrix} = \begin{pmatrix} 4\cdot 3 \\ 4\cdot 1 \end{pmatrix} = \begin{pmatrix} 12 \\ 4 \end{pmatrix}.$$

SCALAR MULTIPLICATION SUMMARISED

▷ in words: We multiply a vector by a scalar by multiplying each component by that real number.

▷ in symbols: Given $\vec{a} = \begin{pmatrix} a_1 \\ a_2 \\ \vdots \\ a_n \end{pmatrix} \in \mathbb{R}^n$ and $\lambda \in \mathbb{R}$, we obtain

$$\lambda\vec{a} = \begin{pmatrix} \lambda a_1 \\ \lambda a_2 \\ \vdots \\ \lambda a_n \end{pmatrix}. \tag{7.4}$$

NORMALISATION

We define the **normalisation** of a vector $\vec{a}$ as the division of $\vec{a}$ by its length $\|\vec{a}\|$. Therefore, $\vec{a}$ turns into a unit vector along the same direction as the original vector, which we may distinguish by typesetting it overhead hat $\hat{a}$ (instead of arrowed).

For instance, normalising the vector $\vec{a} = \begin{pmatrix} 3 \\ 4 \end{pmatrix}$ by its length $\|\vec{a}\| = \sqrt{9+16} = \sqrt{25} = 5$, creates its corresponding unit vector $\hat{a} = \begin{pmatrix} \frac{3}{5} \\ \frac{4}{5} \end{pmatrix}$.

PROPERTIES

Given $\vec{a}, \vec{b}, \vec{c} \in \mathbb{R}^n$ and $\lambda, \mu \in \mathbb{R}$ we summarise the addition, scalar multiplication and subtraction of vectors arithmetically.

closure $\qquad\qquad \vec{a}+\vec{b} \in \mathbb{R}^n$

associative property $\quad (\vec{a}+\vec{b})+\vec{c} = \vec{a}+(\vec{b}+\vec{c})$

zero vector                      $\vec{a} + \vec{o} = \vec{a} = \vec{o} + \vec{a}$

opposite vector              $\vec{a} + (-\vec{a}) = \vec{o} = (-\vec{a}) + \vec{a}$

commutative property    $\vec{a} + \vec{b} = \vec{b} + \vec{a}$

We define an **algebraic structure** as a set that takes a meaningful operation on its elements. A structure featuring the above properties is called a commutative **group**. Hence, we conclude the vector addition is a commutative group.

mixed distributive law     $\lambda(\vec{a} + \vec{b}) = \lambda\vec{a} + \lambda\vec{b}$

$(\lambda + \mu)\vec{a} = \lambda\vec{a} + \mu\vec{a}$

mixed associative law      $(\lambda \cdot \mu)\vec{a} = \lambda(\mu\vec{a})$

## 7.4 Vector subtraction

CREATING FREE VECTORS

For the multimedia, animation and game professional professional, creating a free vector by subtracting its underlying location vectors is a key competence. It is an inevitable step in problem solving and programming of various vector related topics such as kinematics, dynamics, collision handling and linear transformations.

▷ **Point-to-Vector formula**

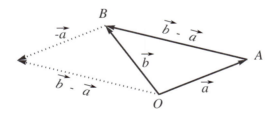

Aiming for the components of a free vector $\overrightarrow{AB}$, we recall

$$\vec{a} + \overrightarrow{AB} = \vec{b} \iff \overrightarrow{AB} = \vec{b} - \vec{a}. \tag{7.5}$$

Hence we calculate the *components* of the vector $\overrightarrow{AB}$ by subtracting the *coordinates* of its initial point $A$ from those of its terminal point $B$. We advise you to memorise this important rule for instance as 'head minus tail' since it is of key importance for all subsequent chapters in this book.

*Example*:  We calculate the components of the free vector $\overrightarrow{AB}$, pointing from tail $A(4,1,3)$ to head $B(-1,4,5)$.

$$\overrightarrow{AB} = \overrightarrow{OB} - \overrightarrow{OA} = \begin{pmatrix} -1 \\ 4 \\ 5 \end{pmatrix} - \begin{pmatrix} 4 \\ 1 \\ 3 \end{pmatrix} = \begin{pmatrix} -5 \\ 3 \\ 2 \end{pmatrix}.$$

▷ **Distance between two points**

We are now able to calculate the distance between two given points $A$ and $B$ straightforwardly as the length of their created free vector $\left\| \overrightarrow{AB} \right\|$ (see formula (7.2)).

EULER'S METHOD FOR TRAJECTORIES

Especially in game programming, the **trajectory** by a moving location vector $\vec{s}(t)$ due to an acceleration $\vec{a}(t)$ is often realised via a method named after the Swiss mathematician **Leonhard Euler** (1707–1783). Euler estimates the moving point's trajectory, given its acceleration function $\vec{a}(t)$, as an initial value problem. For such a problem, we know at time zero $t_0$ what we nowadays call the **frame data** of point $S$:

▷   the initial location vector $\vec{s}_0$,

▷   its initial velocity $\vec{v}_0$,

▷   and the scalar **delta time** $\Delta t \in \mathbb{R}^+$ between two successive frames.

Taking off with our initial location $\vec{s}_0$ and velocity $\vec{v}_0$, we now try to determine $\vec{s}_1$ and $\vec{v}_1$ straightforwardly for the next frame. In case of sufficiently small time intervals $\Delta t$, we may assume the instantaneous velocity and acceleration equal to the average velocity and acceleration:

$$\begin{cases} \vec{v} & \approx & \dfrac{\Delta \vec{s}}{\Delta t} \\[2mm] \vec{a} & \approx & \dfrac{\Delta \vec{v}}{\Delta t}. \end{cases}$$

Accepting this approximation yields a **recursive** formula for $\vec{s}$ and $\vec{v}$ to determine them for a frame $i+1$, based upon their former values in frame $i$.

$$\begin{cases} \vec{v}_i & = & \dfrac{\vec{s}_{i+1} - \vec{s}_i}{\Delta t} \\[2mm] \vec{a}_i & = & \dfrac{\vec{v}_{i+1} - \vec{v}_i}{\Delta t} \end{cases} \Longleftrightarrow \begin{cases} \vec{v}_i \Delta t & = & \vec{s}_{i+1} - \vec{s}_i \\[2mm] \vec{a}_i \Delta t & = & \vec{v}_{i+1} - \vec{v}_i \end{cases} \Longleftrightarrow \begin{cases} \vec{s}_i + \vec{v}_i \Delta t & = & \vec{s}_{i+1} \\[2mm] \vec{v}_i + \vec{a}_i \Delta t & = & \vec{v}_{i+1} \end{cases}$$

In this way the initial values $\vec{s}_0$ and $\vec{v}_0$ lead to $\vec{s}_1$ and $\vec{v}_1$ for the next frame, given the acceleration function $\vec{a}(t)$ can either be described within the source, event-driven by the user or a combination of both and takes off at $\vec{a}_0 = \vec{a}(0)$.

$$\vec{s}_{i+1} \;=\; \vec{s}_i + \vec{v}_i \Delta t$$

$$\vec{v}_{i+1} \;=\; \vec{v}_i + \vec{a}_i \Delta t \tag{7.6}$$

Euler's method to determine parametric trajectories is far from accurate unless the slope of the *velocity curve* remains more or less constant during each time step $\Delta t$. If this slope changes significantly between two frames, then usually after a number of successive frames $0, 1, 2, 3, \ldots$ the difference with respect to the actual velocities $\vec{v}(t)$ and corresponding locations $\vec{s}(t)$ may grow out of bounds. Since this technique is economical and easy to implement, a lot of game programmers still like to use it.

## 7.5 Decomposition of vectors

DECOMPOSITION OF A PLANE VECTOR

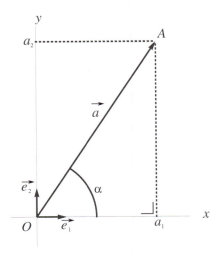

*Figure 7.4*: Decomposition of a 2D vector

We are able to calculate the $x$- and $y$-component of a vector $\vec{a} \in \mathbb{R}^2$ in 2D space whenever its length $\|\vec{a}\|$ and its direction $\alpha$ (the angle made by the vector $\vec{a}$ and the positive $x$-axis) are given, which we define as the **decomposition of a plane vector**. Notice the equivalence of this to the description of a point $A$ in polar coordinates (see chapter 6). We recall that for the right triangle holds

$$\cos \alpha = \frac{a_1}{||\vec{a}||} \Rightarrow a_1 = ||\vec{a}|| \cos \alpha,$$

$$\sin \alpha = \frac{a_2}{||\vec{a}||} \Rightarrow a_2 = ||\vec{a}|| \sin \alpha,$$

leading to the trigonometric components of the location vector

$$\vec{a} = \begin{pmatrix} ||\vec{a}|| \cos \alpha \\ ||\vec{a}|| \sin \alpha \end{pmatrix}. \tag{7.7}$$

BASE VECTORS DEFINED

▷ Base vectors are spanning

We obtain any vector as a sum of scalar multiplied base vectors $\vec{e}_i$.

$$\begin{pmatrix} 3 \\ 2 \end{pmatrix} = \begin{pmatrix} 3 \\ 0 \end{pmatrix} + \begin{pmatrix} 0 \\ 2 \end{pmatrix} = 3 \begin{pmatrix} 1 \\ 0 \end{pmatrix} + 2 \begin{pmatrix} 0 \\ 1 \end{pmatrix}$$

$$= 3\,\vec{e}_1 + 2\,\vec{e}_2$$

We call such a sum of scalar multiplied vectors a **linear combination**. Any vector in $\mathbb{R}^n$ can be written as a linear combination of its $n$ base vectors. For this reason, we call base vectors a **spanning set**.

$$\vec{a} = \begin{pmatrix} a_1 \\ a_2 \\ \vdots \\ a_n \end{pmatrix} = a_1 \begin{pmatrix} 1 \\ 0 \\ \vdots \\ 0 \end{pmatrix} + a_2 \begin{pmatrix} 0 \\ 1 \\ \vdots \\ 0 \end{pmatrix} + \cdots + a_n \begin{pmatrix} 0 \\ 0 \\ \vdots \\ 1 \end{pmatrix}$$

$$= a_1 \vec{e}_1 + a_2 \vec{e}_2 + \cdots + a_n \vec{e}_n = \sum_{i=1}^{n} a_i \vec{e}_i$$

▷ Base vectors are independent

Base vectors are said to be **linearly independent**: a given base vector can never be a linear combination of the remaining base vectors.

▷ Dimension

We define the **dimension** of a space as its number of base vectors.

## 7.6 Dot product

The **dot product** or 'inner product' multiplies two compatible vectors and returns a *scalar*. We consistently typeset this product operation by a dot symbol.

DEFINITION

▷ in words: The dot (product) of two vectors is the projection of one vector onto the other, and therefore returning a real *scalar* (see page 136).

▷ in symbols: Given $\vec{a} = \begin{pmatrix} a_1 \\ a_2 \\ a_3 \end{pmatrix}$ and $\vec{b} = \begin{pmatrix} b_1 \\ b_2 \\ b_3 \end{pmatrix}$ two vectors in $\mathbb{R}^3$, we define

their dot product as the real number

$$\vec{a} \cdot \vec{b} = a_1 b_1 + a_2 b_2 + a_3 b_3 \tag{7.8}$$

$$= \sum_{i=1}^{3} a_i b_i.$$

As a constraint, given $\vec{a} \in \mathbb{R}^n$ and $\vec{b} \in \mathbb{R}^m$ given $m \neq n$, then their dot product is not defined.

*Examples*:

$$\vec{a} = \begin{pmatrix} 2 \\ 3 \end{pmatrix} \in \mathbb{R}^2 \qquad\qquad \vec{b} = \begin{pmatrix} 6 \\ -1 \end{pmatrix} \in \mathbb{R}^2$$

$$\vec{c} = \begin{pmatrix} 1 \\ 2 \\ -3 \end{pmatrix} \in \mathbb{R}^3 \qquad\qquad \vec{d} = \begin{pmatrix} 2 \\ -1 \\ 0 \end{pmatrix} \in \mathbb{R}^3$$

$$\vec{a} \cdot \vec{b} = 2 \cdot 6 + 3 \cdot (-1) = 9 \in \mathbb{R}$$
$$\vec{b} \cdot \vec{a} = 6 \cdot 2 + (-1) \cdot 3 = 9 \in \mathbb{R}$$
$$\vec{c} \cdot \vec{d} = 1 \cdot 2 + 2 \cdot (-1) + (-3) \cdot 0 = 0 \in \mathbb{R}$$
$$\vec{c} \cdot \vec{a} \text{ violates compatibility and is therefore not defined}$$

*Properties*:  Given $\vec{a}, \vec{b} \in \mathbb{R}^3$ and $\lambda \in \mathbb{R}$.

independence of a scalar    $\lambda \vec{a} \cdot \vec{b} = \lambda \, (\vec{a} \cdot \vec{b})$

commutative property    $\vec{a} \cdot \vec{b} = \vec{b} \cdot \vec{a}$

square of a vector    $(\vec{a})^2 = \vec{a} \cdot \vec{a} = a_1^2 + a_2^2 + a_3^2 = \|\vec{a}\|^2$

*Base vectors*:

We dot product the default 3D base vectors $\vec{e_1} = \begin{pmatrix} 1 \\ 0 \\ 0 \end{pmatrix}, \vec{e_2} = \begin{pmatrix} 0 \\ 1 \\ 0 \end{pmatrix}$ and $\vec{e_3} = \begin{pmatrix} 0 \\ 0 \\ 1 \end{pmatrix}$,

to discover:
$$\vec{e_1} \cdot \vec{e_1} = \vec{e_2} \cdot \vec{e_2} = \vec{e_3} \cdot \vec{e_3} = 1,$$
$$\vec{e_1} \cdot \vec{e_2} = \vec{e_2} \cdot \vec{e_1} = \vec{e_2} \cdot \vec{e_3} = \vec{e_3} \cdot \vec{e_2} = \vec{e_3} \cdot \vec{e_1} = \vec{e_1} \cdot \vec{e_3} = 0.$$

### GEOMETRIC INTERPRETATION

Animation programming may imply calculating angles between two vectors. Applying the dot product on two vectors leads us to their interior angle, expressed via the components of both vectors.

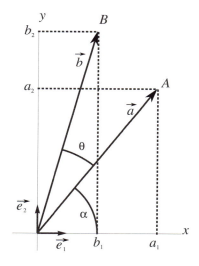

*Figure 7.5*: The interior angle subtended between two vectors

We draw two location vectors $\vec{a}$ and $\vec{b}$ and both cartesian base vectors $\vec{e_1}$ and $\vec{e_2}$ in the plane spanned by these vectors $\vec{a}$ and $\vec{b}$. Vector $\vec{b}$ subtends an interior angle $\theta$ with vector $\vec{a}$. Vector $\vec{a}$ subtends an angle $\alpha$ with base vector $\vec{e_1}$ carrying the $x$-axis. This allows us to decompose both location vectors $\vec{a}$ and $\vec{b}$ to their base (see formula (7.7)) as

$$\vec{a} = \begin{pmatrix} \|\vec{a}\| \cos\alpha \\ \|\vec{a}\| \sin\alpha \end{pmatrix} \text{ and } \vec{b} = \begin{pmatrix} \|\vec{b}\| \cos(\alpha+\theta) \\ \|\vec{b}\| \sin(\alpha+\theta) \end{pmatrix}$$

and calculate their dot product as

$$
\begin{aligned}
\vec{a} \cdot \vec{b} &= a_1 b_1 + a_2 b_2 \\
&= \|\vec{a}\| \, \|\vec{b}\| \, \cos\alpha \, \cos(\alpha + \theta) + \|\vec{a}\| \, \|\vec{b}\| \, \sin\alpha \, \sin(\alpha + \theta) \\
&= \|\vec{a}\| \, \|\vec{b}\| \, \big(\cos\alpha \, \cos(\alpha + \theta) + \sin\alpha \, \sin(\alpha + \theta)\big) \\
&= \|\vec{a}\| \, \|\vec{b}\| \, \cos\big(\alpha - (\alpha + \theta)\big) \\
&= \|\vec{a}\| \, \|\vec{b}\| \, \cos(-\theta) = \|\vec{a}\| \, \|\vec{b}\| \, \cos\theta.
\end{aligned}
$$

The interior angle $\theta$ subtended between two vectors $\vec{a}$ and $\vec{b}$ in $\mathbb{R}^3$ is the angle which satisfies

$$
\vec{a} \cdot \vec{b} = \|\vec{a}\| \, \|\vec{b}\| \cos\theta \tag{7.9}
$$

and for which equivalently

$$
\theta = \arccos\left( \frac{a_1 b_1 + a_2 b_2 + a_3 b_3}{\sqrt{a_1^2 + a_2^2 + a_3^2} \, \sqrt{b_1^2 + b_2^2 + b_3^2}} \right) \tag{7.10}
$$

allows to determine $0 \le \theta \le \pi$ rad (see figure 3.15).

*Example 1*:

$$
\vec{a} = \begin{pmatrix} 3 \\ 4 \\ 2 \end{pmatrix} \quad \text{and} \quad \vec{b} = \begin{pmatrix} 6 \\ -1 \\ 1 \end{pmatrix}
$$

$$
\begin{aligned}
\cos\theta &= \frac{3 \cdot 6 + 4 \cdot (-1) + 2 \cdot 1}{\sqrt{3^2 + 4^2 + 2^2} \, \sqrt{6^2 + (-1)^2 + 1^2}} \\
&= \frac{18 - 4 + 2}{\sqrt{9 + 16 + 4} \, \sqrt{36 + 1 + 1}} \\
&= 0.4819 \\
\theta &= \arccos(0.4819) \\
&= 61.1851^\circ
\end{aligned}
$$

*Example 2*:

$$\vec{c} = \begin{pmatrix} 3 \\ \sqrt{3} \end{pmatrix} \quad \text{and} \quad \vec{d} = \begin{pmatrix} \sqrt{3} \\ -3 \end{pmatrix}$$

$$\cos\theta = \frac{3\sqrt{3} - 3\sqrt{3}}{\sqrt{3^2 + 3}\,\sqrt{3 + (-3)^2}} = 0$$

$$\theta = \arccos(0)$$

$$= 90°$$

*Example 3*:  Applying the length of a vector proves the Law of Cosines straightforwardly.

Choosing for the perspective of vertex $A$ within its cartesian frame yields the default version of the Law of Cosines for scalene triangles (see formula (3.8)).

$$\overrightarrow{AC} + \overrightarrow{CB} = \overrightarrow{AB} \quad \Leftrightarrow$$

$$\overrightarrow{CB} = \overrightarrow{AB} - \overrightarrow{AC} \quad \Rightarrow$$

$$\left(\overrightarrow{CB}\right)^2 = \left(\overrightarrow{AB} - \overrightarrow{AC}\right)^2 \Leftrightarrow$$

$$\left(\overrightarrow{CB}\right)^2 = \left(\overrightarrow{AB}\right)^2 - 2\overrightarrow{AB} \cdot \overrightarrow{AC} + \left(\overrightarrow{AC}\right)^2 \Leftrightarrow$$

$$\left\|\overrightarrow{CB}\right\|^2 = \left\|\overrightarrow{AB}\right\|^2 - 2\left\|\overrightarrow{AB}\right\|\left\|\overrightarrow{AC}\right\|\cos\alpha + \left\|\overrightarrow{AC}\right\|^2 \Leftrightarrow$$

$$a^2 = c^2 - 2cb\cos\alpha + b^2 \Leftrightarrow$$

$$a^2 = b^2 + c^2 - 2bc\cos\alpha$$

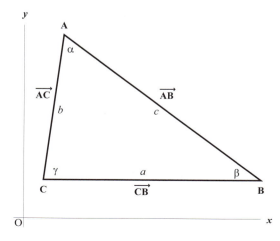

ORTHOGONALITY

Two vectors $\vec{a}$ and $\vec{b}$ are orthogonal in $\mathbb{R}^3$ if and only if their dot product equals zero.

$$\vec{a} \perp \vec{b} \Longleftrightarrow \vec{a} \cdot \vec{b} = 0 \qquad (7.11)$$

We note that the zero vector $\vec{o}$ is orthogonal (or perpendicular) to any vector.

*Proof:*

| $\Rightarrow$ | $\Leftarrow$ |
|---|---|
| given: $\vec{a} \perp \vec{b}$ <br> to prove: $\vec{a} \cdot \vec{b} = 0$ | given: $\vec{a} \cdot \vec{b} = 0$ <br> to prove: $\vec{a} \perp \vec{b}$ |
| partial proof: | partial proof: |
| 1. $\vec{a} \neq \vec{o} \neq \vec{b}$ <br><br> $\vec{a} \perp \vec{b} \Rightarrow \theta = \frac{\pi}{2}$ <br><br> $\vec{a} \cdot \vec{b} = \|\vec{a}\|\ \|\vec{b}\| \cos \frac{\pi}{2}$ <br><br> $\vec{a} \cdot \vec{b} = \|\vec{a}\|\ \|\vec{b}\| \cdot 0$ <br><br> $\vec{a} \cdot \vec{b} = 0$ <br><br> 2. $\vec{a} = \vec{o}$ or $\vec{b} = \vec{o}$ <br> $\vec{a} \cdot \vec{b} = 0$ | $\vec{a} \cdot \vec{b} = 0$ <br><br> $\Downarrow$ <br><br> $\|\vec{a}\|\ \|\vec{b}\| \cos \theta = 0$ <br><br> $\Downarrow$ <br><br> $\|\vec{a}\| = 0$ or $\|\vec{b}\| = 0$ or $\cos \theta = 0$ <br><br> $\Downarrow$ <br><br> $\vec{a} = \vec{o}$ or $\vec{b} = \vec{o}$ or $\theta = \frac{\pi}{2}$ <br> $\Downarrow$ <br> $\vec{a} \perp \vec{b}$ ∎ |

Applying this criterion for orthogonality on a computer, we need to compare the dot product to a tiny threshold value $\varepsilon \approx 10^{-5}$ instead of to zero itself: $\mathrm{abs}(\vec{a} \cdot \vec{b}) < \varepsilon \Leftrightarrow \vec{a} \perp \vec{b}$.

In case of a dot product significantly different from zero, we are able to interpret its sign (positive or negative). Given $\theta$ as the interior angle subtended between vectors $\vec{a}$ and $\vec{b}$, then for

▷ $\vec{a} \cdot \vec{b} < 0$ we conclude $90° < \theta < 180°$,

▷ $\vec{a} \cdot \vec{b} > 0$ we deduct $0° < \theta < 90°$.

## 7.7 Cross product

This powerful operation exists in the 3D vector space only. That is to say, it is sadly restricted to 3D within the conventional approach – compared to its superior **geometric algebra** version which allows equivalents of the cross products as the so-called *outer products* in nD (and for which we kindly refer you to the specialised literature [19], [50]). The **cross product** or 'vector product' multiplies two 3D vectors and returns another 3D vector. We consistently typeset this product operation by a cross symbol.

D EFINITION

▷ in words: The cross product $\vec{a} \times \vec{b}$ of two vectors $\vec{a} \neq \vec{b}$ returns a vector featuring these three vector aspects (see page 142):

1) $\vec{a} \times \vec{b}$ is perpendicular to the plane spanned by $\vec{a}$ and $\vec{b}$ (orientation of $\vec{a} \times \vec{b}$),

2) the threesome $\vec{a}, \vec{b}$ and $\vec{a} \times \vec{b}$ creates a right-handed frame (sense of $\vec{a} \times \vec{b}$),

3) $\|\vec{a} \times \vec{b}\|$ is the area of the parallelogram subtended by $\vec{a}$ and $\vec{b}$ (length of $\vec{a} \times \vec{b}$).

▷ in symbols: Given $\vec{a} = \begin{pmatrix} a_1 \\ a_2 \\ a_3 \end{pmatrix}$ and $\vec{b} = \begin{pmatrix} b_1 \\ b_2 \\ b_3 \end{pmatrix}$ two vectors in $\mathbb{R}^3$, we define their cross product as the returned vector:

$$\vec{a} \times \vec{b} = \begin{pmatrix} a_2 b_3 - b_2 a_3 \\ -a_1 b_3 + b_1 a_3 \\ a_1 b_2 - b_1 a_2 \end{pmatrix}. \tag{7.12}$$

As advise, this component wise definition should be known by heart, which is much easier via a determinant (see formula (11.3)). As a condition, the cross product requires both $\vec{a} \in \mathbb{R}^3$ and $\vec{b} \in \mathbb{R}^3$, then yields $\vec{a} \times \vec{b} \in \mathbb{R}^3$. For our better understanding, we elaborate on each of the above defining aspects.

1) The first aspect describes the direction of $\vec{a} \times \vec{b}$, which is orthogonal to the vector $\vec{a}$ and orthogonal to the vector $\vec{b}$.

2) The second aspect can be realised using the right-hand grip rule or corkscrew rule (see figure 7.1). Applying the right-hand grip rule, we align our right-hand fingers to vector $\vec{a}$ and turning them to vector $\vec{b}$ makes the thumb points in the direction of the vector $\vec{a} \times \vec{b}$. Applying the corkscrew rule, turning the corkscrew from the vector $\vec{a}$ to the vector $\vec{b}$ effects its movement in the direction of the vector $\vec{a} \times \vec{b}$.

3) The third aspect says that the length of the cross product is given by the area of the parallelogram edged by the vectors $\vec{a}$ and $\vec{b}$.

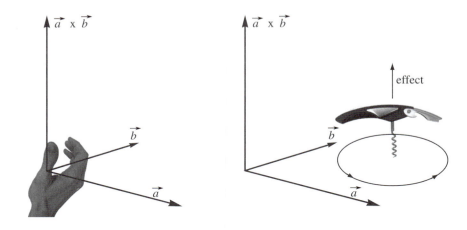

*Figure 7.6*: The right-hand grip rule or corkscrew rule directing the cross product

*Example*:  Given $\vec{a} = \begin{pmatrix} 5 \\ -6 \\ 0 \end{pmatrix} \in \mathbb{R}^3$ and $\vec{b} = \begin{pmatrix} 1 \\ 2 \\ 3 \end{pmatrix} \in \mathbb{R}^3$.

$$\vec{a} \times \vec{b} = \begin{pmatrix} -6 \cdot 3 - 0 \cdot 2 \\ -5 \cdot 3 + 0 \cdot 1 \\ 5 \cdot 2 - (-6) \cdot 1 \end{pmatrix}$$

$$= \begin{pmatrix} -18 \\ -15 \\ 16 \end{pmatrix} \in \mathbb{R}^3$$

We verify the cross product $\vec{a} \times \vec{b}$ to be perpendicular to the vector $\vec{a}$ by $(\vec{a} \times \vec{b}) \cdot \vec{a} = -18 \cdot 5 + (-15) \cdot (-6) + 16 \cdot 0 = 0$. Similarly we confirm the cross product $\vec{a} \times \vec{b}$ to be orthogonal to the vector $\vec{b}$.

*Properties*: Given $\vec{a}, \vec{b} \in \mathbb{R}^3$ and $\lambda \in \mathbb{R}$.

independence of a scalar $\qquad \lambda \vec{a} \times \vec{b} = \lambda \ (\vec{a} \times \vec{b})$

**anticommutative** property $\qquad \vec{a} \times \vec{b} = -(\vec{b} \times \vec{a})$

any vector crossed to itself $\qquad \vec{a} \times \vec{a} = \begin{pmatrix} a_2 a_3 - a_2 a_3 \\ -a_1 a_3 + a_1 a_3 \\ a_1 a_2 - a_1 a_2 \end{pmatrix} = \begin{pmatrix} 0 \\ 0 \\ 0 \end{pmatrix} = \vec{o}$

*Base vectors*:

We cross product the 3D base vectors $\vec{e_1} = \begin{pmatrix} 1 \\ 0 \\ 0 \end{pmatrix}$, $\vec{e_2} = \begin{pmatrix} 0 \\ 1 \\ 0 \end{pmatrix}$ and $\vec{e_3} = \begin{pmatrix} 0 \\ 0 \\ 1 \end{pmatrix}$,

to discover how they relate:

$$\vec{e_1} \times \vec{e_1} = \vec{e_2} \times \vec{e_2} = \vec{e_3} \times \vec{e_3} = \vec{o}$$

The threesome default 3D base vectors cyclically create the right-handed cartesian frame:

$$\vec{e_1} \times \vec{e_2} = \vec{e_3},$$
$$\vec{e_2} \times \vec{e_3} = \vec{e_1},$$
$$\vec{e_3} \times \vec{e_1} = \vec{e_2}.$$

Alternatively, a left-handed version of it would be defined similarly via:

$$\vec{e_1} \times \vec{e_2} = -\vec{e_3},$$
$$\vec{e_2} \times \vec{e_3} = -\vec{e_1},$$
$$\vec{e_3} \times \vec{e_1} = -\vec{e_2}.$$

## GEOMETRIC INTERPRETATION

Applying the component wise definition of the cross product (see formula (7.12)), we obtain for any two different nonzero location vectors $\vec{a}$ and $\vec{b}$ decomposed within their own cartesian plane (see formula (7.7) applied on figure 7.5) as

$$\vec{a} = \begin{pmatrix} \|\vec{a}\| \cos \alpha \\ \|\vec{a}\| \sin \alpha \\ 0 \end{pmatrix} \text{ and } \vec{b} = \begin{pmatrix} \|\vec{b}\| \cos(\alpha + \theta) \\ \|\vec{b}\| \sin(\alpha + \theta) \\ 0 \end{pmatrix},$$

their returned vector as

$$\vec{a} \times \vec{b} = \begin{pmatrix} a_2 b_3 - b_2 a_3 \\ -a_1 b_3 + b_1 a_3 \\ a_1 b_2 - b_1 a_2 \end{pmatrix}$$

$$= \begin{pmatrix} \|\vec{a}\| \sin \alpha \cdot 0 - \|\vec{b}\| \sin(\alpha + \theta) \cdot 0 \\ -\|\vec{a}\| \cos \alpha \cdot 0 + \|\vec{b}\| \cos(\alpha + \theta) \cdot 0 \\ \|\vec{a}\| \cos \alpha \|\vec{b}\| \sin(\alpha + \theta) - \|\vec{b}\| \cos(\alpha + \theta) \|\vec{a}\| \sin \alpha \end{pmatrix}$$

$$= \begin{pmatrix} 0 \\ 0 \\ \|\vec{a}\| \|\vec{b}\| (\sin(\alpha + \theta) \cos \alpha - \cos(\alpha + \theta) \sin \alpha) \end{pmatrix} = \begin{pmatrix} 0 \\ 0 \\ \|\vec{a}\| \|\vec{b}\| \sin((\alpha + \theta) - \alpha) \end{pmatrix}$$

$$= \begin{pmatrix} 0 \\ 0 \\ \|\vec{a}\| \|\vec{b}\| \sin \theta \end{pmatrix}.$$

The interior angle $\theta$ subtended between two vectors $\vec{a}$ and $\vec{b}$ in $\mathbb{R}^3$ satisfies

$$\|\vec{a} \times \vec{b}\| = \|\vec{a}\| \, \|\vec{b}\| \sin \theta \qquad (7.13)$$

$$\text{with } \theta = \arcsin \left( \frac{\|\vec{a} \times \vec{b}\|}{\|\vec{a}\| \, \|\vec{b}\|} \right)$$

fails to determine all $0 \le \theta \le \pi$ rad, given arcsine's returns restricted to $\left[ -\frac{\pi}{2}, \frac{\pi}{2} \right]$ (see figure 3.14).

Furthermore, we verify the triplex definition of the cross product:

1)  $\vec{a} \times \vec{b}$ is indeed perpendicular to the plane spanned by $\vec{a} \ne \vec{b}$ since we calculate $(\vec{a} \times \vec{b}) \cdot \vec{a} = 0$ and $(\vec{a} \times \vec{b}) \cdot \vec{b} = 0$ for the above result,

2)  $\vec{a}, \vec{b}$ and $\vec{a} \times \vec{b}$ indeed create a right-handed frame since for all subtended angles $0 \le \theta \le \pi$ the cross product's $z$-component $\|\vec{a}\| \|\vec{b}\| \sin \theta$ proves $\vec{a} \times \vec{b} \uparrow\uparrow \vec{e}_3$,

3)  $\|\vec{a} \times \vec{b}\| = \|\vec{a}\| \|\vec{b}\| \sin \theta$ indeed yields the area of the parallelogram edged by the vectors $\vec{a}$ and $\vec{b}$.

*Proof:*

$$\text{area}_{parallelogram} = 2 \times \text{area}_{scalene\ triangle}$$

$$= 2 \left( \frac{1}{2} a \, b \sin \theta \right) \text{ (see formula (3.9 ))}$$

$$= \|\vec{a}\| \|\vec{b}\| \sin \theta \qquad \blacksquare$$

*Example*: Relying on the equal length of cross products proves the general Law of Sines.

Based on the perspectives of vertices $A$ and $B$ (see figure 3.10), we prove the first shackle of the Law of Sines for scalene triangles (see formula (3.7)).

$$\left\|\overrightarrow{AB} \times \overrightarrow{AC}\right\| = \left\|\overrightarrow{BA} \times \overrightarrow{BC}\right\| \quad \Leftrightarrow \quad \left\|\overrightarrow{AB}\right\| \left\|\overrightarrow{AC}\right\| \sin\alpha = \left\|\overrightarrow{BA}\right\| \left\|\overrightarrow{BC}\right\| \sin\beta$$

$$\Leftrightarrow \quad cb\sin\alpha = ca\sin\beta$$

$$\Leftrightarrow \quad b\sin\alpha = a\sin\beta$$

$$\Leftrightarrow \quad \frac{b}{\sin\beta} = \frac{a}{\sin\alpha} \qquad\blacksquare$$

## PARALLELISM

Two vectors $\vec{a}$ and $\vec{b}$ are parallel or antiparallel in $\mathbb{R}^3$ if and only if their cross product equals the zero vector.

$$\vec{a} \uparrow\uparrow \vec{b} \text{ or } \vec{a} \uparrow\downarrow \vec{b} \Longleftrightarrow \vec{a} \times \vec{b} = \vec{o}$$

We note that the zero vector $\vec{o}$ is parallel (and antiparallel) to any vector. *Proof:*

$\Rightarrow$

given: $\vec{a} \uparrow\uparrow \vec{b}$ or $\vec{a} \uparrow\downarrow \vec{b}$
to prove: $\vec{a} \times \vec{b} = \vec{o}$

partial proof:

1. $\vec{a} \neq \vec{o} \neq \vec{b}$

$\left(\vec{a} \uparrow\uparrow \vec{b} \Rightarrow \theta = 0\right)$ or $\left(\vec{a} \uparrow\downarrow \vec{b} \Rightarrow \theta = \pi\right)$

$\left\|\vec{a} \times \vec{b}\right\| = \|\vec{a}\| \|\vec{b}\| \sin\theta$

$\left\|\vec{a} \times \vec{b}\right\| = \|\vec{a}\| \|\vec{b}\| 0$

$\left\|\vec{a} \times \vec{b}\right\| = 0$

2. $\vec{a} = \vec{o}$ or $\vec{b} = \vec{o}$

$\left\|\vec{a} \times \vec{b}\right\| = 0$

$\Downarrow$

$\vec{a} \times \vec{b} = \vec{o}$

$\Leftarrow$

given: $\vec{a} \times \vec{b} = \vec{o}$
to prove: $\vec{a} \uparrow\uparrow \vec{b}$ or $\vec{a} \uparrow\downarrow \vec{b}$

partial proof:

$\vec{a} \times \vec{b} = \vec{o}$

$\Downarrow$

$\left\|\vec{a} \times \vec{b}\right\| = \|\vec{a}\| \|\vec{b}\| \sin\theta = 0$

$\Downarrow$

$\|\vec{a}\| = 0$ or $\|\vec{b}\| = 0$ or $\sin\theta = 0$

$\Downarrow$

$\vec{a} = \vec{o}$ or $\vec{b} = \vec{o}$ or $\theta = 0$ or $\theta = \pi$

$\Downarrow$

$\vec{a} \uparrow\uparrow \vec{b}$ or $\vec{a} \uparrow\downarrow \vec{b}$

Applying this criterion for (anti)parallelism on a computer, we need to compare the length of the cross product to a tiny threshold value $\varepsilon \approx 10^{-5}$ instead of to zero itself:

$$\left\| \vec{a} \times \vec{b} \right\| < \varepsilon \Leftrightarrow \vec{a} \uparrow\uparrow \vec{b} \quad ( \text{or } \vec{a} \uparrow\downarrow \vec{b}).$$

In case a cross product is significantly different from the zero vector $\vec{o}$, then this cross product $\vec{a} \times \vec{b}$ is perpendicular to the triangle edged by both vectors $\vec{a} \neq \vec{b}$. We define such a cross product (perpendicular to its triangle) as a **normal vector** of this triangle, which leads us straight into the subsequent paragraph.

## 7.8 Normal vectors

All 3D objects are modelled by **polygons**. Polygons are flat surfaces determined by vertices and bordered by straight edges. Triangles are very popular polygons since most render engines require an input sequence of polygons featuring three edges. One single 3D object is often described by thousands of polygons.

Apart from the essential polygon data (the cartesian $x$-, $y$- and $z$-coordinates of its vertices) it is as useful to store the components of the polygon's chosen **normal vector**. The normal vector is perpendicular to the polygon's surface. Normal vectors determine whether the surface should be visible on screen and calculate its shading and reflection, depending on the type of ambient light. The separate sequence of all normal vectors of one single 3D object is called a **normal map**.

We construct for a triangle $ABC$ a normal vector using the 3D cartesian coordinates of its three vertices $A$, $B$ and $C$. Each polygon in 3D space has a vector (and its opposite vector) which is perpendicular to it. By calculating for instance $\overrightarrow{BA} \times \overrightarrow{BC}$ we immediately gain a vector which is orthogonal to the free vector $\overrightarrow{BA}$ and to its neighbouring vector $\overrightarrow{BC}$, and hence is orthogonal to the polygon $ABC$ they edge. In order to be able to calculate *percentages* of reflection later on, it is a common practice to divide a normal vector by its own length $\|\overrightarrow{BA} \times \overrightarrow{BC}\|$ which scales it into a unit vector:

$$\hat{n} = \frac{\overrightarrow{BA} \times \overrightarrow{BC}}{\|\overrightarrow{BA} \times \overrightarrow{BC}\|}. \tag{7.14}$$

We have already defined normalising a vector as scaling it into a unit vector by dividing it by its own length. We call such a normalised normal vector $\hat{n}$ a **unit normal vector**.

*Example*: Imagine we need to construct a unit normal vector on the polygon defined by the three vertices: $A(-2,0,1)$, $B(1,5,1)$ and $C(0,7,2)$. The plane containing these three points $A, B$ and $C$ can for our purpose be defined by two vectors each aligning one of the straight edges $[AB], [BC]$ or $[CA]$. We are free to choose for instance the vectors $\overrightarrow{BA}$ and $\overrightarrow{BC}$ for it. We emphasise that other pairs of nonparallel vectors would also work.

Firstly, we find the components of our chosen free vectors $\overrightarrow{BA}$ and $\overrightarrow{BC}$.

$$\overrightarrow{BA} = \overrightarrow{OA} - \overrightarrow{OB} = \begin{pmatrix} -2 \\ 0 \\ 1 \end{pmatrix} - \begin{pmatrix} 1 \\ 5 \\ 1 \end{pmatrix} = \begin{pmatrix} -3 \\ -5 \\ 0 \end{pmatrix}$$

$$\overrightarrow{BC} = \overrightarrow{OC} - \overrightarrow{OB} = \begin{pmatrix} 0 \\ 7 \\ 2 \end{pmatrix} - \begin{pmatrix} 1 \\ 5 \\ 1 \end{pmatrix} = \begin{pmatrix} -1 \\ 2 \\ 1 \end{pmatrix}$$

Secondly, we calculate the cross product of $\overrightarrow{BA}$ and $\overrightarrow{BC}$ because this yields a normal vector to the polygon defined by $A, B$ and $C$.

$$\overrightarrow{BA} \times \overrightarrow{BC} = -5\vec{e_1} + 3\vec{e_2} - 11\vec{e_3} = \begin{pmatrix} -5 \\ 3 \\ -11 \end{pmatrix}$$

Finally, to obtain its corresponding unit normal vector, we need to divide this normal vector by its own length $\left\| \overrightarrow{BA} \times \overrightarrow{BC} \right\| = \sqrt{155}$, which yields

$$\hat{n} = \frac{1}{\sqrt{155}} \begin{pmatrix} -5 \\ 3 \\ -11 \end{pmatrix}$$

$$\approx \begin{pmatrix} -0.402 \\ 0.241 \\ -0.884 \end{pmatrix}.$$

We can easily verify that $\|\hat{n}\| \approx 1$, for $\hat{n}$ indeed is a unit vector.

Verifying that $\hat{n} \cdot \overrightarrow{BA} = 0$ and that $\hat{n} \cdot \overrightarrow{BC} = 0$, proves $\hat{n}$ to be orthogonal to both free vectors $\overrightarrow{BA}$ and $\overrightarrow{BC}$, hence $\hat{n}$ is a normal vector of the polygon these free vectors subtend.

## 7.9 Exercises

**Exercise 60** Calculate the length and the direction of the net force $\vec{F}_1 + \vec{F}_2 + \vec{F}_3 + \vec{F}_4$ caused by the four forces $\vec{F}_1, \vec{F}_2, \vec{F}_3$ and $\vec{F}_4$, given their magnitudes $||\vec{F}_1|| = 120N$, $||\vec{F}_2|| = 100N$, $||\vec{F}_3|| = 80N$ and $||\vec{F}_4|| = 40N$.

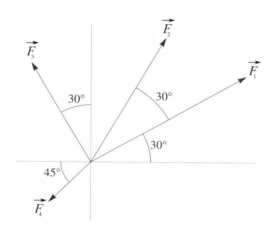

**Exercise 61** An aeroplane is flying at a speed of 800 kilometres per hour and in the direction of 210°. At the same time, there is a wind blowing at a speed of 90 kilometres per hour and from the direction of 165° (from the SE to the NW). Find the 'ground speed' and the resulting direction of this aeroplane. Hint: start finding all vector components for a straightforward solution. Please consider that in navigation directions are measured clockwise, starting from N, which is labelled as 0°.

**Exercise 62** Given the location vectors

$$\vec{a} = \begin{pmatrix} 3 \\ 2 \\ -4 \end{pmatrix} \qquad \vec{b} = \begin{pmatrix} -2 \\ 0 \\ 4 \end{pmatrix} \qquad \vec{c} = \begin{pmatrix} -5 \\ 1 \\ 4 \end{pmatrix}$$

Find

1) $-2(\vec{b} + 5\vec{c}) + 5(\vec{a} - 3\vec{b})$

2) $3(\vec{a} \cdot \vec{b})\vec{c} - 5(\vec{b} \cdot \vec{c})\vec{a}$

3) $(\vec{a} - 3\vec{b}) \cdot (4\vec{c})$

4) $(\vec{a} \cdot \vec{b})\vec{c} - \vec{b}$

5) $(-\vec{a} + 2\vec{c}) \times (-\vec{b})$

6) $(2\vec{a}) \times (-\vec{b} + 5\vec{c})$

**Exercise 63** A camera is positioned in cartesian coordinate $C(1,4)$ and has a 'focus vector' $\vec{f} = \begin{pmatrix} 5 \\ 3 \end{pmatrix}$ describing its line of sight. Meanwhile we spot an animal roaming in $A(7,2)$. Given that our camera swings over an angle of $90°$ to both sides, gaining an eyeshot of $180°$, can it film this animal?

**Exercise 64** Some fancy sports cars feature a tiny spoiler at the back of their roof. After simplifying its bodywork, we can model such a sports car frame in 3D cartesian coordinates, as shown in the figure. Assume the spoiler $\overrightarrow{FG}$ has a length $0.3$ and is perpendicular to the rear window $ABCD$. Find the coordinates of the point $G$, making use of vectors.

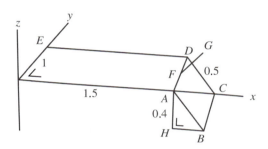

Hints:

▷ Calculate the distance $|HB|$ to determine the coordinates of several useful points.

▷ Given the spoiler $\overrightarrow{FG}$ is perpendicular to the rear window, it must be a normal vector of the polygon $ABCD$. Hence, the direction of $\overrightarrow{FG}$ can be obtained via the cross product of two vectors which align along the nonparallel edges of $ABCD$. Calculate an appropriate cross product.

▷ Given the length of the spoiler, determine the coordinates of the point $G$.

**Exercise 65** Find the area of the parallelogram spanned by the location vectors $\vec{a} = (4, -10, 5)$ and $\vec{b} = (-3, -1, -3)$.

**Exercise 66** Find the interior angle in the point $B$, subtended by the straight lines connecting the points $A(1,1,1)$, $B(2,3,4)$ and $C(-2,5,2)$.

Hint: choose two free vectors to start from.

# Chapter 8 · Parameters

Lines and planes are of major importance in setting up digital landscapes. In paragraph 4.2 we outlined how to obtain the equation of a straight line, in order to border a landscape or to describe an object's trajectory. Such an equation suits only 2D space (see formula (4.3)). Because 3D applications are taking over, it becomes inevitable for us to know how to describe straight lines in 3D space. Additionally, we also explain the equation of a plane in 3D coordinates.

## 8.1 Parametric equations

It is straightforward to describe a moving object by cartesian coordinates $(x, y)$ which are themselves functions of the running time:

$$x = x(t) \qquad \text{and} \qquad y = y(t).$$

We call such an approach, by means of a running parameter $t$, the **parametric equation** of a function (see formula (6.1)). We typeset its parameter as $\lambda$ or $t$ as it often refers to a running time. To each value of the parameter $\lambda$, there corresponds exactly one point on the curve drawn by $(x(\lambda), y(\lambda))$.

*Example:* Consider the parametric equation

$$x = \lambda^2 \qquad \text{and} \qquad y = \lambda.$$

Calculate the $x$- and $y$-coordinates for a sample of parameter values $\lambda$.

| $\lambda$ | $-3$ | $-2$ | $-1$ | 0 | 1 | 2 | 3 |
|---|---|---|---|---|---|---|---|
| $x$ | 9 | 4 | 1 | 0 | 1 | 4 | 9 |
| $y$ | $-3$ | $-2$ | $-1$ | 0 | 1 | 2 | 3 |

Plotting points on these locations, we obtain the horizontal parabola (see page 72). Eliminating parameter $\lambda$ via $x = \lambda^2 = y^2$ yields $y^2 = x$, which is indeed the cartesian equation of the horizontal parabola.

## 8.2 Vector equation of a line

This figure shows several scalar multiples of a location vector $\vec{a} = \overrightarrow{OA}$. The heads of these scalar multiples are all on the same straight line $s$. The other way around, each point of this straight line running through the origin $O$ and the given point $A$ can be reached by a scalar multiplication of the vector $\vec{a}$ and a real number $\lambda$. In other words, for each point $X$ we can express its location vector $\vec{x}$ on the line $s$ as $\vec{x} = \lambda \vec{a}$ given $\lambda$ a real number. For this reason we call $\vec{x} = \lambda \vec{a}$ a **vector equation** of the line $s$. In this equation we call $\vec{a}$ a **direction vector** of the straight line $s$. And we call the running number $\lambda$ a free parameter.

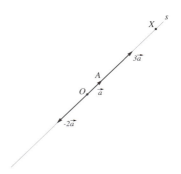

We take a similar approach for straight lines running through any arbitrary **anchor point** outside of the origin, along the direction of $\overrightarrow{BA}$. We can express the vector with head $C$ on line $r$ as $\vec{b} + 3\overrightarrow{BA}$. Repeating this for a vector with arbitrary head $X$ on line $r$ we get $\vec{x} = \vec{b} + \lambda \overrightarrow{BA}$ for an arbitrary value of $\lambda$. The vector expression $\vec{x} = \vec{b} + \lambda \overrightarrow{BA}$ means that each point on the line $r$ can be reached by taking an appropriate distance from the anchor point $B$ along the direction vector $\overrightarrow{BA}$. In other words, for each parameter value $\lambda$, the location vector $\vec{x}$ finds his head on the line $r$. For this reason, we call $\vec{x} = \vec{b} + \lambda \overrightarrow{BA}$ a vector equation of the line $r$.

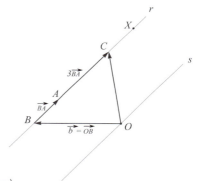

We may call the location vector $\vec{b} = \overrightarrow{OB}$ a **position vector** of the straight line $r$. We require the position vector $\vec{b}$ to aim at a chosen anchor point $B$ on the line $r$ and require the **direction vector** $\overrightarrow{BA}$ to be different from the null vector.

We realise that only the *orientation* of the direction vector is essential, not its length nor sense. For instance, the direction vectors $\begin{pmatrix} 4 \\ 2 \end{pmatrix}$, $\begin{pmatrix} 2 \\ 1 \end{pmatrix}$ or $\begin{pmatrix} -1 \\ -0.5 \end{pmatrix}$ describe the same line. Secondly, we realise that *each* vector which takes the origin as tail and any point at the line as head makes a valid position vector. Given the above, we understand we have several ways to describe the same line. We conclude that a straight line has no unique vector equation.

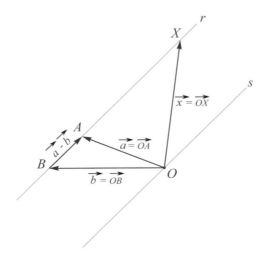

*Figure 8.1*: A straight line $r$ through two points $A$ and $B$

In general, we consider a straight line $r$ through two points $A$ and $B$, pinned by their location vectors $\vec{a} = \overrightarrow{OA}$ and $\vec{b} = \overrightarrow{OB}$. To find its vector equation, we just need to choose an appropriate position vector and a direction vector. We again can express the location vector $\vec{x}$ of a variable point $X$ on the line $r$ as the sum vector

$$\vec{x} = \vec{b} + \overrightarrow{BX}.$$

Since $\overrightarrow{BX}$ and $\overrightarrow{BA}$ are parallel vectors, we are able to express $\overrightarrow{BX} = \lambda \overrightarrow{BA} = \lambda(\vec{a} - \vec{b})$. As for a vector equation of a straight line connecting two given points $A$ and $B$ we conclude:

$$\vec{x} = \vec{b} + \lambda(\vec{a} - \vec{b}).$$

We memorise such a vector equation for a straight line more easily in words:

$$\vec{x} = \text{position vector} + \lambda \, \text{direction vector}. \tag{8.1}$$

Rewriting a vector equation $\vec{x} = \vec{b} + \lambda \vec{c}$ of a straight line on position vector $\vec{b}$ along direction vector $\vec{a}$ by its vector components $\vec{b} = \begin{pmatrix} b_1 \\ b_2 \\ b_3 \end{pmatrix}$ and $\vec{c} = \begin{pmatrix} c_1 \\ c_2 \\ c_3 \end{pmatrix}$ in 3D space, we obtain

its **parametric equation**:

$$\begin{cases} x = b_1 + \lambda c_1 \\ y = b_2 + \lambda c_2 \\ z = b_3 + \lambda c_3 \end{cases}.$$

Eliminating the parameter $\lambda$ from its three equations, we obtain its **cartesian equation**:

$$\frac{x - b_1}{c_1} = \frac{y - b_2}{c_2} = \frac{z - b_3}{c_3} (= \lambda).$$

*Example*: We aim for an equation of the straight line connecting the points $A(8, 2, 4)$ and $B(-2, -2, -2)$.

We determine the orientation of this line by

$$\overrightarrow{AB} = \overrightarrow{OB} - \overrightarrow{OA} = \begin{pmatrix} -2 - 8 \\ -2 - 2 \\ -2 - 4 \end{pmatrix} = \begin{pmatrix} -10 \\ -4 \\ -6 \end{pmatrix}.$$

We may take $\begin{pmatrix} -10 \\ -4 \\ -6 \end{pmatrix}$ for direction vector, but since $\begin{pmatrix} -10 \\ -4 \\ -6 \end{pmatrix} = -2 \begin{pmatrix} 5 \\ 2 \\ 3 \end{pmatrix}$, the

vector $\begin{pmatrix} 5 \\ 2 \\ 3 \end{pmatrix}$ is also a valid direction vector for the straight line through the points $A$

and $B$. Any location vector to an arbitrary point of the line makes a valid position vector

for the line $AB$. We may for instance choose $\overrightarrow{OA} = \begin{pmatrix} 8 \\ 2 \\ 4 \end{pmatrix}$ for it. The above decisions

yield this vector equation for the line $AB$:

$$\vec{x} = \begin{pmatrix} 8 \\ 2 \\ 4 \end{pmatrix} + \lambda \begin{pmatrix} 5 \\ 2 \\ 3 \end{pmatrix},$$

or rewritten to its parametric equation:

$$\begin{cases} x = 8 + 5\lambda \\ y = 2 + 2\lambda \\ z = 4 + 3\lambda \end{cases}.$$

For instance, at parameter value $\lambda = -1$ we generate its point $P(3, 0, 1)$. After elimination of this free parameter $\lambda$, we find the corresponding cartesian equation for the line $AB$ in 3D space:

$$\frac{x - 8}{5} = \frac{y - 2}{2} = \frac{z - 4}{3}.$$

Alternatively, we could have chosen the direction vector $\overrightarrow{AB}$ and for instance $\overrightarrow{OB}$ as the position vector. Consequently, another valid vector equation describing the same straight line $AB$ would be

$$\vec{x} = \begin{pmatrix} -2 \\ -2 \\ -2 \end{pmatrix} + \mu \begin{pmatrix} -10 \\ -4 \\ -6 \end{pmatrix}.$$

Referring to its point $P(3,0,1)$ again, in this equation it is generated at $\mu = \frac{-1}{2}$.

*Example*: To further illustrate that a vector equation of a line is not unique, we hereby show some alternatives for the straight line $AB$ through the points $A(2,7)$ and $B(-3,-3)$ in 2D space.

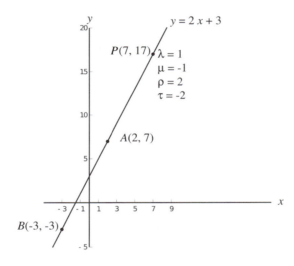

| position vector | direction vector | vector equation | cartesian equation |
|---|---|---|---|
| $\vec{a} = \begin{pmatrix} 2 \\ 7 \end{pmatrix}$ | $\overrightarrow{BA} = \begin{pmatrix} 5 \\ 10 \end{pmatrix}$ | $\begin{pmatrix} x \\ y \end{pmatrix} = \begin{pmatrix} 2 \\ 7 \end{pmatrix} + \lambda \begin{pmatrix} 5 \\ 10 \end{pmatrix}$ | $\frac{x-2}{5} = \frac{y-7}{10}$ $\Leftrightarrow y = 2x + 3$ |
| $\vec{a} = \begin{pmatrix} 2 \\ 7 \end{pmatrix}$ | $\overrightarrow{AB} = \begin{pmatrix} -5 \\ -10 \end{pmatrix}$ | $\begin{pmatrix} x \\ y \end{pmatrix} = \begin{pmatrix} 2 \\ 7 \end{pmatrix} + \mu \begin{pmatrix} -5 \\ -10 \end{pmatrix}$ | $\frac{x-2}{-5} = \frac{y-7}{-10}$ $\Leftrightarrow y = 2x + 3$ |
| $\vec{b} = \begin{pmatrix} -3 \\ -3 \end{pmatrix}$ | $\overrightarrow{BA} = \begin{pmatrix} 5 \\ 10 \end{pmatrix}$ | $\begin{pmatrix} x \\ y \end{pmatrix} = \begin{pmatrix} -3 \\ -3 \end{pmatrix} + \rho \begin{pmatrix} 5 \\ 10 \end{pmatrix}$ | $\frac{x+3}{5} = \frac{y+3}{10}$ $\Leftrightarrow y = 2x + 3$ |
| $\vec{b} = \begin{pmatrix} -3 \\ -3 \end{pmatrix}$ | $\overrightarrow{AB} = \begin{pmatrix} -5 \\ -10 \end{pmatrix}$ | $\begin{pmatrix} x \\ y \end{pmatrix} = \begin{pmatrix} -3 \\ -3 \end{pmatrix} + \tau \begin{pmatrix} -5 \\ -10 \end{pmatrix}$ | $\frac{x+3}{-5} = \frac{y+3}{-10}$ $\Leftrightarrow y = 2x + 3$ |

We can reach any point on the line $AB$ using one of these four alternative vector equations. For instance, its point $P(7, 17) \in AB$ corresponds alternatively to $\lambda = 1$, $\mu = -1$, $\rho = 2$ or $\tau = -2$.

## 8.3 Intersecting straight lines

We consider the two straight lines $r$ given by $\vec{x} = \vec{b} + \lambda \vec{c}$ and $s$ given by $\vec{x} = \vec{q} + \mu \vec{p}$. To find their possible intersection point, we have to compare their recipes in one equation which we solve for the parameters $\lambda$ and $\mu$ (see paragraph 4.3).

In case of a solution, we will find the intersection point of both lines. In case of an inconsistent system, both lines have no intersection point. This latter case can be caused either by parallel lines (which do not coincide) or by skew lines. We define **intersecting** lines (in 2D and 3D) in case they have exactly one point in common. We define **parallel** lines (in 2D and 3D) in case they are not intersecting and lying in a common plane. We define **skew** lines (in 3D) in case they are not intersecting and *not* lying in a common plane. Notice that coinciding lines are also considered as parallel lines.

*Example 1*: We aim for the possible intersection point of the straight lines

$$r : \begin{pmatrix} x \\ y \\ z \end{pmatrix} = \begin{pmatrix} 1 \\ 1 \\ 0 \end{pmatrix} + \lambda \begin{pmatrix} 1 \\ 1 \\ 1 \end{pmatrix} \text{ and } s : \begin{pmatrix} x \\ y \\ z \end{pmatrix} = \begin{pmatrix} 2 \\ 0 \\ 2 \end{pmatrix} + \mu \begin{pmatrix} 1 \\ -1 \\ 2 \end{pmatrix}.$$

Comparing their recipes in one equation and solving it by elimination for the parameters $\lambda$ and $\mu$ yields a solution to their underdetermined system (see paragraph 2.2).

$$\begin{cases} 1+\lambda &=& 2+\mu \\ 1+\lambda &=& -\mu \\ \lambda &=& 2+2\mu \end{cases} \Leftrightarrow \begin{cases} \lambda - \mu &=& 1 \quad |1 \qquad |1 \\ \lambda + \mu &=& -1 \quad |-1 \\ \lambda - 2\mu &=& 2 \qquad\qquad |-1 \end{cases}$$

$$\Leftrightarrow \begin{cases} \lambda - \mu &=& 1 \\ -2\mu &=& 2 \\ \mu &=& -1 \end{cases}$$

$$\Leftrightarrow \begin{cases} \lambda - \mu &=& 1 \\ \mu &=& -1 \\ \mu &=& -1 \end{cases}$$

$$\Leftrightarrow \begin{cases} \lambda &=& 0 \\ \mu &=& -1 \\ \mu &=& -1 \end{cases}$$

Given this solution, we generate the intersection point $S$ of the straight lines $r$ and $s$. Replacing $\lambda = 0$ in the parametric equation of $r$ leads to

$$\begin{pmatrix} x \\ y \\ z \end{pmatrix} = \begin{pmatrix} 1 \\ 1 \\ 0 \end{pmatrix} + 0 \begin{pmatrix} 1 \\ 1 \\ 1 \end{pmatrix} = \begin{pmatrix} 1 \\ 1 \\ 0 \end{pmatrix} = S.$$

Evaluating the parametric equation of $s$ for $\mu = -1$ obviously leads to the same intersection point $S$:

$$\begin{pmatrix} x \\ y \\ z \end{pmatrix} = \begin{pmatrix} 2 \\ 0 \\ 2 \end{pmatrix} - 1 \begin{pmatrix} 1 \\ -1 \\ 2 \end{pmatrix} = \begin{pmatrix} 1 \\ 1 \\ 0 \end{pmatrix} = S.$$

*Example 2*: We try to determine the possible intersection point of the straight lines

$$\begin{cases} x &=& \lambda \\ y &=& -1+\lambda \\ z &=& 1+\lambda \end{cases} \text{ and } \begin{cases} x &=& -1+2\mu \\ y &=& 4-\mu \\ z &=& 0. \end{cases}$$

After comparing both recipes in one equation, we again apply the elimination method for solving it.

$$\begin{cases} \lambda = -1 + 2\mu \\ -1 + \lambda = 4 - \mu \\ 1 + \lambda = 0 \end{cases} \Leftrightarrow \begin{cases} \lambda - 2\mu = -1 & |1 \\ \lambda + \mu = 5 & |2 \\ \lambda = -1 \end{cases}$$

$$\Leftrightarrow \begin{cases} \lambda - 2\mu = -1 \\ 3\lambda = 9 \\ \lambda = -1 \end{cases}$$

$$\Leftrightarrow \begin{cases} \lambda - 2\mu = -1 \\ \lambda = 3 \\ \lambda = -1 \end{cases}$$

Because the parameter $\lambda$ can impossibly hold simultaneously values 3 and $-1$, we classify the above system as inconsistent. For this reason, we conclude that both straight lines do not intersect at all.

## 8.4 Vector equation of a plane

From the previous chapter, we understand there can be many vector equations describing the same straight line. As a constraint to these equations, we can only choose from direction vectors which are equal apart from a nonzero number factor. Every direction vector we can choose is a scalar multiple of another direction vector for that straight line. We call such vectors **linearly dependent**. Two vectors $\vec{a}$ and $\vec{b}$ are linearly dependent when there is a real number $k$ relating both vectors by scalar multiplication $k\vec{a} = \vec{b}$. In case there does not exist such a scalar factor $k \in \mathbb{R}$ to relate them, then we have two **linearly independent** vectors.

Therefore, two linearly independent vectors $\vec{a}$ and $\vec{b}$ cannot lie on the same straight line. This allows us to decompose any arbitrary location vector $\vec{x}$ lying in the plane spanned by $\vec{a}$ and $\vec{b}$ in two directions: one along vector $\vec{a}$ and one along vector $\vec{b}$. In other words, we are able to write location vector $\vec{x}$ as a linear combination of the independent vectors $\vec{a}$ and $\vec{b}$, briefly $\vec{x} = \lambda \vec{a} + \mu \vec{b}$.

In 3D space, all heads $X$ of these linear combinations make up the plane $v_O$ through the origin $O$. In a similar way to that of straight lines, we can express such a plane through the origin $O$ by means of a vector equation: $\vec{x} = \lambda \vec{a} + \mu \vec{b}$.

We call those vectors $\vec{a}$ and $\vec{b}$ **direction vectors** of the plane $v_O$. The plane $v_O$ is spanned by both direction vectors $\vec{a}$ and $\vec{b}$. Any two linearly independent vectors lying in the plane

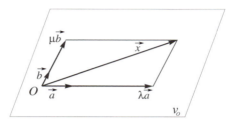

*Figure 8.2*: The plane $v_O$ through the origin $O$

$v_O$ can be chosen as its direction vectors. Consequently, we have several ways to describe the same plane. We conclude a plane has no unique vector equation.

We take a similar approach for a plane $v_C$ through an arbitrary **anchor point** different from the origin $C$. Such a plane $v_C$ has been shifted over the location vector $\vec{c} = \overrightarrow{OC} \ (\neq \vec{o})$ with respect to the plane $v_O$. This allows us to express any vector $\vec{x}$ of such a plane as a sum of a vector lying in $v_O$ and a position vector heading from the origin $O$ to the plane $v_C$.

Summarised, the vector equation of a general plane $v_C$ is $\vec{x} = \vec{c} + \lambda \overrightarrow{CA} + \mu \overrightarrow{CB}$. Vector $\vec{c} = \overrightarrow{OC}$ is the **position vector** and vectors $\overrightarrow{CA}$ and $\overrightarrow{CB}$ are both **direction vectors** of the plane $v_C$. Just as in case of the plane $v_O$, any two linearly independent vectors lying in the plane $v_C$ can be chosen as direction vectors for it. We are also free to choose an appropriate position vector $\vec{c}$, as long as it is heading from the origin $O$ to an anchor point $C$ lying in the plane $v_C$.

We memorise such a vector equation for a plane more easily in words:

$$\vec{x} = \text{position vector} + \lambda \, \text{direction vector}_1 + \mu \, \text{direction vector}_2, \qquad (8.2)$$

given its position vector running from the origin to an arbitrary anchor point in the plane and both direction vectors linearly independent.

Eliminating parameters $\lambda$ and $\mu$ from the plane's vector equation, we obtain its cartesian equation

$$v_C : ax + by + cz + d = 0.$$

*Example*: We aim for the vector equation of the plane $v_A$ determined by the points $A(1,2,7), B(-1,0,-3)$ and $C(0,3,8)$. The anchor point of the plane $v_A$ is, as hinted by its notation, chosen to be $A(1,2,7)$.

We need a position vector and two independent direction vectors for this.

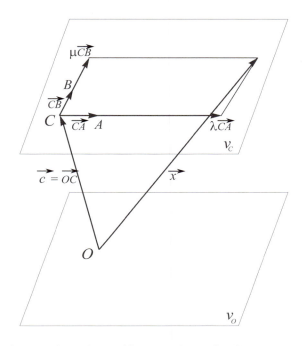

*Figure 8.3*: The plane $v_C$ through an arbitrary anchor point $C$

▷  position vector: $\overrightarrow{OA} = \begin{pmatrix} 1 \\ 2 \\ 7 \end{pmatrix}$

▷  direction vectors: $\overrightarrow{AB} = \begin{pmatrix} -2 \\ -2 \\ -10 \end{pmatrix}$ and $\overrightarrow{AC} = \begin{pmatrix} -1 \\ 1 \\ 1 \end{pmatrix}$

Vectors $\overrightarrow{AB}$ and $\overrightarrow{AC}$ are indeed linearly independent, because there is no $k \in \mathbb{R}$ relating $\overrightarrow{AB} = k\overrightarrow{AC}$. In conclusion, the desired vector equation is

$$\begin{pmatrix} x \\ y \\ z \end{pmatrix} = \begin{pmatrix} 1 \\ 2 \\ 7 \end{pmatrix} + \lambda \begin{pmatrix} -2 \\ -2 \\ -10 \end{pmatrix} + \mu \begin{pmatrix} -1 \\ 1 \\ 1 \end{pmatrix}.$$

or separated into its parametric shape

$$\begin{cases} x = 1 - 2\lambda - \mu \\ y = 2 - 2\lambda + \mu \\ z = 7 - 10\lambda + \mu. \end{cases}$$

To obtain the cartesian equation from this parametric equation, we need to eliminate its parameters. We can for instance solve the first equation to $\mu$, which results in: $\mu = 1 - 2\lambda - x$. Replacing this in the second equation yields

$$y = 2 - 2\lambda + \mu = 2 - 2\lambda + (1 - 2\lambda - x) = 3 - 4\lambda - x.$$

Subsequently solving the second equation to $\lambda$, results in: $\lambda = \frac{1}{4}(3 - x - y)$. Finally we replace the expressions for $\lambda$ and $\mu$ in the third equation.

$$\begin{aligned}
z &= 7 - 10\lambda + \mu \\
&= 7 - \frac{10}{4}(3 - x - y) + (1 - 2\lambda - x) \\
&= 7 - \frac{30}{4} + \frac{10}{4}x + \frac{10}{4}y + 1 - \frac{1}{2}(3 - x - y) - x \\
&= 7 - \frac{30}{4} + \frac{10}{4}x + \frac{10}{4}y + 1 - \frac{3}{2} + \frac{1}{2}x + \frac{1}{2}y - x \\
&= 2x + 3y - 1
\end{aligned}$$

We conclude that the cartesian equation of the plane $v_A$ is $2x + 3y - z - 1 = 0$.

## 8.5 Exercises

**Exercise 67**   Referring to chapter 6, find the parametric equation of a circle $C(O,r)$.

**Exercise 68**

1)  Find the parametric equation of the straight line running through the point $P(1,0,4)$ and parallel to the straight line $AB$, given its points $A(6,3,7)$ and $B(2,-1,-1)$.

2)  Find the vector equation of the straight line running through the point $P(1,-2,7)$ and parallel to the straight line

$$\frac{x+3}{4} = \frac{y+1}{3} = z-1.$$

**Exercise 69**

1)  Find the vector equation of the straight line $r$ through the point $P(4,2,8)$ and parallel to the vector $\vec{a} = \begin{pmatrix} -1 \\ 4 \\ 3 \end{pmatrix}$. Continue aiming for its parametric and cartesian equation.

2)  Find the parametric equation of the straight line $r$ through the points $M(5,8,21)$ and $N(7,10,31)$, culminating in its cartesian equation.

3)  Find the possible intersection point of the above straight lines $r$ and $s$.

**Exercise 70**

1)  Find the cartesian equation of the straight line through the point $P(2,2,3)$ and perpendicular to the plane $v_C$ spanned by the direction vectors $\vec{u} = \begin{pmatrix} 1 \\ 1 \\ 0 \end{pmatrix}$, $\vec{e} = \begin{pmatrix} 0 \\ 0 \\ 1 \end{pmatrix}$ and anchored by the point $C(5,5,5)$.

2)  Find a parametric equation of this plane. Prove that the point $P(8,8,4)$ belongs to this plane.

**Exercise 71**   The three points $P(3,1,0), Q(-4,1,1)$ and $R(5,9,3)$ determine the plane $v_P$.

1)  Find a vector equation of this plane $v_P$.

2)  Find a normal vector $\vec{n}$ on this plane $v_P$.

**Exercise 72**

1) Find a parametric equation of the plane $v_P$ anchored in $P(3,6,2)$ and parallel to the plane $v_O$ through the origin $O$ and the points $A(1,2,3)$ and $B(4,2,1)$.

2) Find the cartesian equation of this plane $v_P$.

**Exercise 73**   Vectors $\vec{v} = \begin{pmatrix} 1 \\ 2 \\ 2 \end{pmatrix}$ and $\vec{w} = \begin{pmatrix} 5 \\ -1 \\ 1 \end{pmatrix}$ are given. A polygon of a 3D object is pinned by the origin $O$ and the terminal points of vectors $\vec{v}$ and $\vec{w}$. Find a parametric equation and the cartesian equation of this polygon's plane. Find a normal vector $\vec{n}$ on this polygon.

**Exercise 74**   Are the straight lines

$$l : \frac{x+8}{5} = \frac{y-2}{2} = \frac{z+4}{3}$$

and

$$r : \frac{x-4}{9} = \frac{y+1}{1} = \frac{z-2}{6}$$

1) parallel (be it coinciding),

2) skew, or

3) intersecting?

In case of the latter situation, find their intersection.

**Exercise 75**   Find the intersection of the straight line $l = AB$ given $A(6,3,7)$ and $B(2,-1,-1)$ and the plane $v_P$ containing the points $P(3,1,0)$, $Q(-4,1,1)$ and $R(5,9,3)$.

**Exercise 76**   Find and explain the intersection of the three planes

$$u_A : 8x - 7y + 2z = 3,$$

$$v_A : 11x + 5y - 7z = 9,$$

$$w_A : x - 2y + z = 0.$$

**Exercise 77**   Find and explain the intersection of the straight line

$$l : \frac{x-1}{4} = \frac{y-2}{-2} = \frac{z-3}{3},$$

and the sphere

$$B : (x-1)^2 + (y-2)^2 + (z-3)^2 = 58.$$

# Chapter 9 · Kinematics

Kinematics – as the study of motion – provides us with the basic *equations* which govern all motion. In this way, we mathematically *describe how* objects move. Of course this is critically important to game programming as nearly every action game involves objects flying, driving, sailing or moving in any other way. Applying even basic kinematics will give our games a touch of realism. In this chapter, we discuss delta time via frame rate, to base as well translational and circular motion on it, and to eventually outline the popular projectile motion.

## 9.1 Measures

A **measure** is a quality or aspect from reality that records a directly observable or computable value.

Precision
We define numerical **precision** as the number of correct **significant digits** of a real value. In physics, precision indicates how accurately we capture nature in digits. Throughout this chapter, we apply various numerical precisions appropriate to the situation. As **trailing zeros** matter in physics, they reflect the accuracy of a measurement. For instance, a length $l$ of one metre ought to be typeset as $1.00\ m$ when measured up to centimetres.

Units
In our modern world, we quantify measures standardised by the **SI** (the International System of Units) for which we refer to Annex C (see page 398). Nature's fundamental quantities length $l$, mass $m$ and time $t$ are measured in metres $m$, kilograms $kg$ and seconds $s$ respectively. Each of these fundamental **units** is physically based on an operational definition, for which we refer you to the expert literature. We may typeset units by putting *square brackets* around their corresponding measures.

| measure | symbol | SI-unit | |
|---|---|---|---|
| length | $l$ | $[l] = m$ | metre |
| mass | $m$ | $[m] = kg$ | kilogram |
| time | $t$ | $[t] = s$ | second |

## 9.2 Deltatime

Games can be understood as fancy flipbooks. We picture discrete frames and hook them together to create the impression of motion whilst scrolling through our flipbook. Physics engines behave similarly for they execute discrete time steps which march the simulation forward in time. We may view this propagation as extrapolation: given an initial position and velocity, it predicts a next position and velocity.

We define the measure **delta time** $\Delta t$ as the time elapsed between two successive frames $n$ and $n+1$. Therefore, delta time is measured in seconds: $[\Delta t] = s$.

We define **frame rate** $frt$ as the refresh rate of our runtime screen, which creates the impression of motion. Therefore, frame rate is measured in frames per second: $[frt] = fps$.

These game programmers' twin measures are reciprocal: $\Delta t \times frt = 1$ frame. Completely controlling delta time instead of accepting it as a default global constant allows

   ▷   to keep our action game pace to be kept steady,

   ▷   object wise time scaling (for which we refer you to 'frame rate independence' in the expert literature),

   ▷   pausing actions (e.g. by runtime setting $\Delta t = 10^6 s$).

It is good practice to constraint delta time by a frame rate of 30 *fps* or in other words to limit $\Delta t \leqslant \frac{1}{30}s \approx 33ms$.

Per time step $t_n$ we define for frame $n$ its according **frame data** per moving point $P$ as:

   ▷   its actual location vector $\vec{s}_n$,

   ▷   its actual velocity $\vec{v}_n$,

   ▷   and its (object wise) delta time $\Delta t$ to the former frame $n-1$.

## 9.3 Translational motion

Translational motions describe the linear movement of an object through space without any side effects: the flight of a golf ball (without spinning), of an aeroplane (without its rigid body motions) or a bird (without its propelling wings) are just some straightforward examples. In this chapter, we outline the constant cases of rectilinear and circular motion. Finally we also discuss the various plane motions in general, approached either via polar coordinates or via the Independence of Motion Principle.

To avoid any further confusion, we need to differentiate some specific terms from their day to day use. Firstly, we distinguish between the object's **location vector** $\vec{s}$ which heads its corresponding **position** at point $P$, creating its **trajectory** as the locus of all its successive positions over time $t$.

Secondly, we need to distinguish between the **displacement** vector and the scalar travel distance.

*Example:* Imagine some movements of a train engine along a stretch of its rail track.

Given that the engine drives along its track from location $A$ to location $B$, where it halts for some seconds before a reverse motion in the opposite direction from $B$ to $C$, we may graph its history over time. Therefore we describe its successive locations in one dimension by the location vectors $s_A$, $s_B$ and $s_C$ with letter $s$ referring to *space*. Note that we do not overarrow 1D vectors in this physics chapter, where 2D and 3D vectors are overarrowed. We may display the train's history via a space-over-time graph (see figure 9.1): its engine initially waits for three seconds in $A$, then speeds to $B$ in one second where it halts for another three seconds, to finally drive in reverse to $C$ in ten seconds.

For our deeper insight, we calculate all the above readings essentially applying the famous Point-To-Vector formula (7.5).

$$\Delta s_{AB} = \quad s_B - s_A = \quad 5 - (-2) = 7\ m$$
$$\Delta s_{BB} = \quad s_B - s_B = \quad 5 - 5 = 0\ m$$
$$\Delta s_{BC} = \quad s_C - s_B = \quad 1 - 5 = -4\ m$$
$$\Delta s_{AC} = \quad s_C - s_A = \quad 1 - (-2) = 3\ m$$

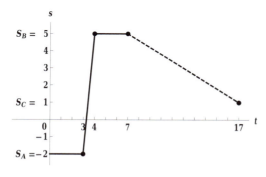

| | displacement in $m$ | distance in $m$ |
|---|---|---|
| | $\Delta s_{AB} = 7$ | $d_{AB} = 7$ |
| | $\Delta s_{BB} = 0$ | $d_{BB} = 0$ |
| | $\Delta s_{BC} = -4$ | $d_{BC} = 4$ |
| | $\Delta s_{AC} = 3$ | $d_{AC} = 11$ |
| | $\Delta s$ implies direction | $d \geqslant 0$ |

*Figure 9.1*: Exemplary $(s,t)$-graph

Thirdly, we distinguish between the vector velocity and the scalar speed. We define

$$\textbf{velocity} = \frac{\text{displacement}}{\text{delta time}} \quad \text{or in symbols as} \quad \vec{v} = \frac{\Delta \vec{s}}{\Delta t}, \tag{9.1}$$

whereas the **speed** is but its magnitude $v = \|\vec{v}\|$. Therefore, $\vec{v}$ implies direction as inherited by its numerator $\Delta \vec{s}$, and with letter $v$ referring to *velocity*. When we drive a car instead of a train, for instance, velocity $\vec{v}$ is the combined information of our GPS's direction and dashboard's speed $v$. In our one dimensional example we do not overarrow its 1D vectors, but typeset the above

$$\text{in 1D simply as} \quad v = \frac{\Delta s}{\Delta t}, \tag{9.2}$$

whereas the speed is but its absolute value $|v|$ (see formula (6.2)).

For our deeper insight, we calculate the above readings (see figure 9.2), including their unit conversion from metre per second to kilometre per hour.

$$v_{AA} = \frac{\Delta s_{AA}}{\Delta t} = \frac{s_A - s_A}{t_A - t_0} = \frac{-2 - (-2)}{3 - 0} = \frac{0}{3} = 0 \, \frac{m}{s} = 0 \, \frac{km}{h}$$

$$v_{AB} = \frac{\Delta s_{AB}}{\Delta t} = \frac{s_B - s_A}{t_B - t_A} = \frac{5 - (-2)}{4 - 3} = \frac{7}{1} \, \frac{m}{s} = \frac{7}{1} \left( \frac{3600}{1000} \right) \frac{1000 \, m}{3600 \, s} = 7(3.6) \, \frac{km}{h} = 25.2 \, \frac{km}{h}$$

$$v_{BB} = \frac{\Delta s_{BB}}{\Delta t} = \frac{s_B - s_B}{t_B' - t_B} = \frac{5 - 5}{7 - 4} = \frac{0}{3} = 0 \, \frac{m}{s} = 0 \, \frac{km}{h}$$

$$v_{BC} = \frac{\Delta s_{BC}}{\Delta t} = \frac{s_C - s_B}{t_C - t_B'} = \frac{1 - 5}{17 - 7} = \frac{-4}{10} \, \frac{m}{s} = -0.4(3.6) \, \frac{km}{h} = -1.44 \, \frac{km}{h}$$

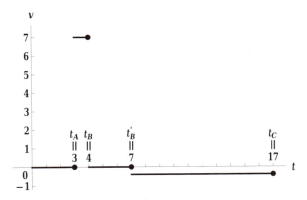

| | velocity in $\frac{m}{s}$ | speed in $\frac{m}{s}$ |
|---|---|---|
| | $v_{AA} = 0$ | $|v_{AA}| = 0$ |
| | $v_{AB} = 7$ | $|v_{AB}| = 7$ |
| | $v_{BB} = 0$ | $|v_{BB}| = 0$ |
| | $v_{BC} = -0.4$ | $|v_{BC}| = 0.4$ |
| | $v$ implies direction | $|v| \geqslant 0$ |

*Figure 9.2*: Exemplary $(v,t)$-graph

## RECTILINEAR MOTION WITH CONSTANT VELOCITY (RMCV)

For an overview of the three types of rectilinear motion with constant velocity, we stick to the former example of our train engine moving along its track, for which we restart each of its motions at $t_0 = 0$ to let them continue up to an arbitrary time step $t$. Location equations of rectilinear motion at constant velocity are linear functions (see formula (4.2)).

For a complete insight, we retrieve each of the location equations based on our previous results, given the velocities are constant.

$$v_{AB} = \frac{s - s_0}{t - t_0} \implies 7 = \frac{s - (-2)}{t - 0} \implies 7t = s + 2 \Leftrightarrow s = 7t - 2$$

$$v_{BB} = \frac{s - s_0}{t - t_0} \implies 0 = \frac{s - 5}{t - 0} \implies 0 \cdot t = s - 5 \Leftrightarrow s = 0 + 5 \Leftrightarrow s = 5$$

$$v_{BC} = \frac{s - s_0}{t - t_0} \implies -0.4 = \frac{s - 5}{t - 0} \implies -0.4t = s - 5 \Leftrightarrow s = -0.4t + 5$$

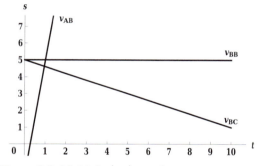

| RMCV | location equation |
|---|---|
| from $A$ to $B$ | $s = 7t - 2$ |
| from $B$ to $B$ | $s = 5$ |
| from $B$ to $C$ | $s = -0.4t + 5$ |
| initiated at $t_0 = 0$ | $s = vt + s_0$ |

*Figure 9.3*: Multiple $(s,t)$-graph

## RECTILINEAR MOTION WITH CONSTANT ACCELERATION (RMCA)

In order to learn about rectilinear motion with constant acceleration, we revisit the former example of a train engine moving along its track. We therefore define

$$\textbf{acceleration} = \frac{\text{change of velocity}}{\text{delta time}} \quad \text{or in symbols as} \quad \vec{a} = \frac{\Delta \vec{v}}{\Delta t}, \quad (9.3)$$

whereas for its unit we obtain $[\vec{a}] = \frac{[\Delta \vec{v}]}{[\Delta t]} = \frac{\frac{m}{s}}{s} = \frac{m}{s^2}$. The vector $\vec{a}$ implies direction as inherited by its numerator $\Delta \vec{v}$, with letter $a$ referring to *acceleration*. In our one dimensional example, we do not overarrow its 1D vectors, but typeset the above

$$\text{in 1D simply as} \quad a = \frac{\Delta v}{\Delta t}. \quad (9.4)$$

We realise that the train's former $(v,t)$-graph was far from realistic, since in nature it takes at least a little time to change velocities (see figure 9.2). We may overcome its displayed discontinuities by underneath successive adaptions.

Firstly, we could halt the at 25.2 $\frac{km}{h}$ speeding engine either abruptly in only one tenth of a second, or smoothly in nearly two minutes. Stopping a moving train requires an external force, for which we refer to the physics literature and which at the same time reveals a new motion type. We calculate each of both corresponding decelerations $a_h$ and $a_s$ answered in $\frac{m}{s^2}$ to finally compare them graphically via their successive $(v,t)$-graphs (see figure 9.4).

$$a_h = \frac{0 - 25.2 \frac{km}{h}}{0.1 - 0\ s} = \frac{-25.2 \left(\frac{1000}{3600}\right) \frac{m}{s}}{0.1\ s} = \frac{-7 \frac{m}{s}}{0.1\ s} = -70 \frac{m}{s^2}$$

$$a_s = \frac{0 - 25.2 \frac{km}{h}}{100 - 0\ s} = \frac{-25.2 (0.28) \frac{m}{s}}{100\ s} = \frac{-7 \frac{m}{s}}{100\ s} = -0.07 \frac{m}{s^2}$$

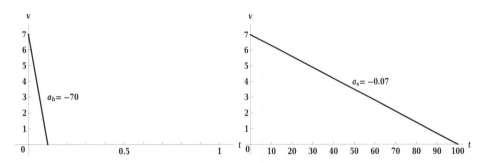

*Figure 9.4*: Comparative $(v,t)$-graphs of a deceleration

Secondly, we could also ... keep the engine at rest. Keeping a train's zero velocity constant means no external force acts upon it, which we extend to cruise controlled nonzero velocities as well. Calculating either $a_k$ to keep at rest or alternatively $a_v$ to keep any velocity constant yields a zero acceleration. Graphically this draws a horizontal line in the corresponding $(v,t)$-graph (see figure 9.5).

$$a_k = \frac{0 - 0 \frac{m}{s}}{3 - 0\ s} = \frac{0 \frac{m}{s}}{3\ s} = 0 \frac{m}{s^2}$$

$$a_v = \frac{7 - 7 \frac{m}{s}}{4 - 3\ s} = \frac{0 \frac{m}{s}}{1\ s} = 0 \frac{m}{s^2}$$

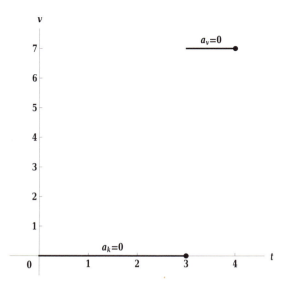

*Figure 9.5*: The $(v, t)$-graphs of a zero acceleration

Eventually, we reversed the halted engine to the left at velocity $-1.44 \frac{km}{h}$, but we could have restarted it to the right at $+1.44 \frac{km}{h}$ as well. We apply a delta time of $5\ s$ to calculate both required accelerations in $\frac{m}{s^2}$ and to compare them via their successive $(v, t)$-graphs (see figure 9.6).

$$a_l = \frac{-1.44 - 0 \frac{km}{h}}{5 - 0\ s} = \frac{-1.44 \left( \frac{1000}{3600} \right) \frac{m}{s}}{5\ s} = \frac{-0.4 \frac{m}{s}}{5\ s} = \frac{-0.8 \frac{m}{s}}{10\ s} = -0.08 \frac{m}{s^2}$$

$$a_r = \frac{+1.44 - 0 \frac{km}{h}}{5 - 0\ s} = \frac{+1.44 (0.28) \frac{m}{s}}{5\ s} = \frac{+0.4 \frac{m}{s}}{5\ s} = \frac{+0.8 \frac{m}{s}}{10\ s} = +0.08 \frac{m}{s^2}$$

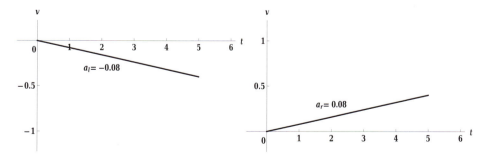

*Figure 9.6*: Comparative $(v, t)$-graphs of accelerations

As both vectors velocity $\vec{v}$ and acceleration $\vec{a}$ imply a direction, we define

$$\textbf{deceleration} = \frac{\text{decrease of velocity}}{\text{delta time}} \quad \text{via their dot product} \quad \vec{a} \cdot \vec{v} < 0, \quad (9.5)$$

whereas for velocity increasing accelerations $\vec{a} \cdot \vec{v} > 0$ applies. As 1D vectors $v$ and $a$ are signed positively in the direction of their axis (and negatively otherwise), we define a deceleration $a$ in one dimension

when multiplied with its 1D velocity $v$, by the sign of their product $\quad a \cdot v < 0,$

whereas for speed increasing 1D accelerations $a \cdot v > 0$ holds. In this respect, we leave the former example of a moving train engine by verifying all the latter nonzero accelerations applied to it.

$$a_h = -70 \, \frac{m}{s^2} \quad \text{is a deceleration} \quad \Leftrightarrow a_h \cdot v = (-70)(7) = -490 < 0$$

$$a_s = -0.07 \, \frac{m}{s^2} \quad \text{is a deceleration} \quad \Leftrightarrow a_s \cdot v = (-0.07)(7) = -0.49 < 0$$

$$a_l = -0.08 \, \frac{m}{s^2} \quad \text{is an acceleration} \quad \Leftrightarrow a_l \cdot v = (-0.08)(-0.4) = 0.032 > 0$$

$$a_r = +0.08 \, \frac{m}{s^2} \quad \text{is an acceleration} \quad \Leftrightarrow a_r \cdot v = (+0.08)(+0.4) = 0.032 > 0$$

FREE FALL

To more straightforwardly illustrate the RMCA, we opt for Earth's constant **acceleration due to gravity**. We study an object in **free fall** as dropped at a height of only 1 *km* during only 10 *s*, which allows us to ignore **friction** and to consider the acceleration due to gravity vector $\vec{g}$ indeed having a constant magnitude of on average $9.81 \frac{m}{s^2}$. Just like the previous examples, free fall is a one dimensional motion.

## Constant acceleration

It is common practice to orient the vertical axis $s$ upwards, whilst Earth's acceleration due to gravity acts downwards and therefore adopts a negative sign. In this one dimensional example we do not overarrow its 1D vectors, but typeset the above as $g = -9.81 \frac{m}{s^2}$. Since this acceleration due to gravity $g$ is a constant, it draws a horizontal line in its $(a,t)$-graph (see figure 9.7).

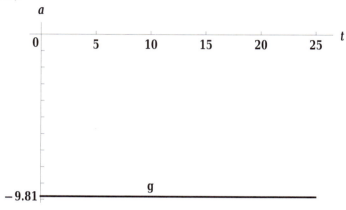

*Figure 9.7*: Constant $(a,t)$-graph of free fall

## Velocity equation

Given an arbitrary acceleration $a$ is constant, we retrieve the velocity equation of the RMCA via its general definition.

*Proof:*

$$a = \frac{\Delta v}{\Delta t} \quad \Leftrightarrow \quad a = \frac{v - v_0}{t - t_0}$$

$$\Rightarrow \quad \text{if we set } t_0 = 0 \text{ then } a = \frac{v - v_0}{t - 0}$$

$$\Leftrightarrow \quad at = v - v_0$$

$$\Leftrightarrow \quad v = v_0 + at \qquad\qquad \blacksquare$$

We notice this velocity equation to be linear in $t$ and therefore may interprete it as the linear velocity function at any meaningful time $t \in \mathbb{R}^+$, given the initial velocity $v_0$:

$$v(t) = v_0 + at. \tag{9.6}$$

We apply it to our free fall case to draw its corresponding $(v,t)$-graph (see figure 9.8).

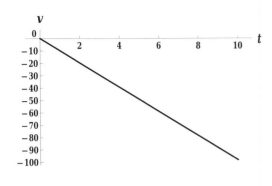

The RMCA velocity function

$v(t) = v_0 + at$

applied to free fall yields

$v(t) = v_0 + gt$

$v(t) = 0 + (-9.81)t$

$v(t) = -9.81t$

e.g. $v(10) = -9.81 \frac{m}{s^2}(10\ s)$

$= -98.1 \frac{m}{s} = -350 \frac{km}{h}$

*Figure 9.8*: Linear $(v,t)$-graph of free fall

## Location equation

Given its velocity $v = v_0 + at$ is linear, we prove the location equation of the RMCA is quadratic (see formula (1.6)).

*Proof:* Despite being restricted to 1D vectors, the proven result applies to 2D and 3D vectors as well. We interpret 1D displacements as areas in $(v,t)$-graphs.

Firstly, in case of an RMCV the generally time-dependent velocity equation simplifies to a constant. In its elementary $(v,t)$-graph we interpret the rectangular area $v_0 \cdot t$ as the displacement $\Delta s$ caused by the constant velocity $v_0$ during a variable time $t$.

Equation $v = v_0 + at$

simplifies for the RMCV to

$v = v_0 + 0t \Rightarrow$

$v = v_0$ a constant.

Since area $v_0 \cdot t = \Delta s$

$\Leftrightarrow s - s_0 = v_0 \cdot t$

$\Leftrightarrow s = s_0 + v_0 \cdot t$

Secondly, in case of an RMCA, the time-dependent velocity equation draws an inclined line. In its elementary $(v,t)$-graph we interpret the trapezoidal area as the displacement $\Delta s$ caused by the linear velocity $v$ during a variable time $t$.

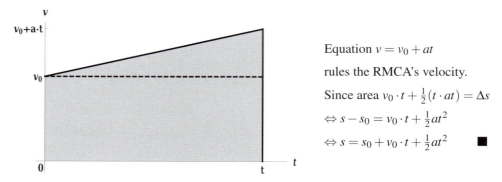

Equation $v = v_0 + at$

rules the RMCA's velocity.

Since area $v_0 \cdot t + \frac{1}{2}(t \cdot at) = \Delta s$

$\Leftrightarrow s - s_0 = v_0 \cdot t + \frac{1}{2}at^2$

$\Leftrightarrow s = s_0 + v_0 \cdot t + \frac{1}{2}at^2$  ∎

We notice this location equation to be quadratic in $t$ and therefore interprete it as the quadratic location function at any meaningful time $t \in \mathbb{R}^+$, given the initial location $s_0$ and initial velocity $v_0$:

$$s(t) = s_0 + v_0 t + \frac{1}{2}at^2. \qquad (9.7)$$

We apply it to our free fall case to draw its corresponding $(s,t)$-graph (see figure 9.9). We hereby emphasise the $(s,t)$-graph is *not* the trajectory of the falling object, which is a vertical line segment.

SUMMARY

We advise you to know both the velocity and the location equation for the RMCA by heart, because they imply those for the RMCV if we simply set $a = 0$:

$$v = v_0 + at,$$

$$s = s_0 + v_0 t + \frac{1}{2}at^2.$$

Finally, we prove the **time-independent formula** for the RMCA

$$v^2 = v_0^2 + 2\vec{a} \cdot \Delta \vec{s}, \qquad (9.8)$$

by eliminating the scalar variable $t$ between the velocity and the location equation.

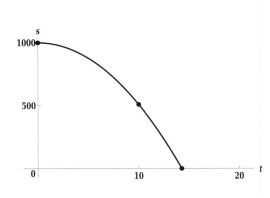

The RMCA location function

$$s(t) = s_0 + v_0 t + \tfrac{1}{2} a t^2$$

applied to free fall yields

$$s(t) = 1000 + 0t + \tfrac{1}{2}(-9.81)t^2$$

Retrieving its roots at

$$t_{1,2} = \pm \sqrt{\frac{2(-1000)}{-9.81}}$$

$t_1 \approx +14\ s$ (physical) or $t_2 \approx -14\ s$

Determining its vertex at

$$t_{vertex} = \frac{-v_0}{2g} = \frac{-0}{2(-9.81)} = 0\ s$$

e.g. $s(10) = 1000 + \tfrac{1}{2}(-9.81)10^2$
$$= 509.5\ m$$

*Figure 9.9*: Quadratic $(s,t)$-graph of free fall

For this occasion, we extend both former equations to general 2D and 3D vectors

$$\vec{v} = \vec{v}_0 + \vec{a}t \ \text{ and }$$
$$\vec{s} = \vec{s}_0 + \vec{v}_0 t + \frac{1}{2}\vec{a}t^2.$$

*Proof:* Therefore we square the vector velocity equation, and substitute for the location equation.

$$\vec{v} = \vec{v}_0 + \vec{a}t \ \Rightarrow \ \vec{v}^2 = (\vec{v}_0 + \vec{a}t)^2$$
$$\Leftrightarrow \ \vec{v}^2 = \vec{v}_0^2 + 2\vec{v}_0 \cdot \vec{a}t + (\vec{a}t)^2 \qquad \text{see formula (1.2),}$$
$$\Leftrightarrow \ v^2 = v_0^2 + 2\vec{a} \cdot \vec{v}_0 t + \vec{a}^2 t^2 \qquad \text{see page 135,}$$
$$\Leftrightarrow \ v^2 = v_0^2 + 2\vec{a} \cdot \left( \vec{v}_0 t + \frac{1}{2}\vec{a}t^2 \right) \qquad \text{factoring out } 2\vec{a}$$

$$\vec{s} = \vec{s}_0 + \vec{v}_0 t + \frac{1}{2}\vec{a}t^2 \ \Leftrightarrow \ \vec{s} - \vec{s}_0 = \vec{v}_0 t + \frac{1}{2}\vec{a}t^2$$
$$\Rightarrow \ v^2 = v_0^2 + 2\vec{a} \cdot (\vec{s} - \vec{s}_0) \qquad \text{displacement}$$
$$\Leftrightarrow \ v^2 = v_0^2 + 2\vec{a} \cdot \Delta\vec{s}$$

## 9.4 Circular motion

As the previous part of this chapter dealt with translational motions only, circular motion describes the motion of spinning objects as well. A golfer may use spin to cause effect on the golf ball, it takes four spinning wheels to drive a car and an aeroplane makes use of different 3D-rotations during flight (see page 282). These are just some straightforward illustrations of where you will need to model circular motion in your programming.

UNIFORM CIRCULAR MOTION (UCM)

We define a **uniform motion** as having a constant speed. Suppose an object is moving at a constant speed in a circular orbit of radius $r$. How do you describe this object's position $P$ at any given time $t$? We describe such a circular motion most conveniently in polar coordinates $P(r, \theta)$, of which we define $\theta$ as the **angular location** of an object (see figure 6.5). Similarly to rectilinear motion, we consequently define

$$\textbf{angular speed} = \frac{\textbf{angular displacement}}{\text{delta time}} \quad \text{or in symbols as} \quad \omega = \frac{\Delta\theta}{\Delta t}. \quad (9.9)$$

When we usually set $t_0 = 0$ seconds, the angular speed leads to

$$\omega = \frac{\Delta\theta}{\Delta t} = \frac{\theta - \theta_0}{t - t_0} = \frac{\theta - \theta_0}{t - 0} = \frac{\theta - \theta_0}{t}. \quad (9.10)$$

The angular speed $\omega$ comes out in the unit $[\omega] = \frac{[\Delta\theta]}{[\Delta t]} = \frac{\text{rad}}{s}$. Firstly, we consider the initial angular location usually set as $\theta_0 = 0$ radians. This further simplifies the angular speed to

$$\omega = \frac{\theta - \theta_0}{t} = \frac{\theta - 0}{t} = \frac{\theta}{t}. \quad (9.11)$$

| $\theta$ in rad | $t$ in $s$ | $\omega$ in $\frac{\text{rad}}{s}$ |
| --- | --- | --- |
| 0 | 0 | $\frac{0}{0} =?$ indeterminate |
| $\omega$ | 1 | $\frac{\omega}{1}$ |
| $2\pi$ | $T$ | $\frac{2\pi}{T}$ |

The angular speed $\omega$ is the constant with which to complete the full angle of $2\pi$ radians in one **general period** of time $T$ seconds:

$$\omega = \frac{2\pi}{T}. \quad (9.12)$$

Vertical projection of the uniform circular motion

Retrieving only the vertical projection of this uniform circular motion, we rely on the definition of sine in its right triangle as $\sin\theta = \frac{y}{r}$ which solves for $y = r\sin\theta$. Since the zero based formula of a constant angular speed $\omega = \frac{\theta}{t}$ solves for $\theta = \omega t$, its vertical projection leads to the function

$$y(t) = r\sin(\omega t). \tag{9.13}$$

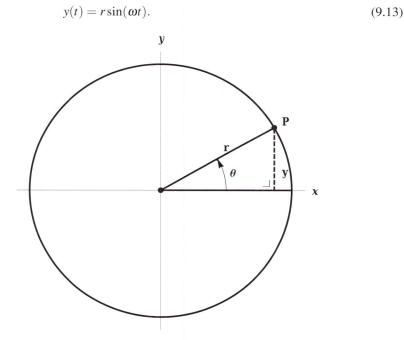

Secondly, let us consider an object $P$ in a uniform circular motion taking off at an initial angular location $\theta_0$ at initial time $t_0 = 0$. Since the according definition of the constant angular speed $\omega = \frac{\theta - \theta_0}{t}$ in this case solves for $\theta = \omega t + \theta_0$, our former function in $t$ generalises to

$$f(t) = r\sin(\omega t + \theta_0). \tag{9.14}$$

For a complete outline of this vertically projected motion, we refer to the general sine function in $x$ (see formula (4.4)). Meanwhile notice the similarity between the angular location $\theta = \omega t + \theta_0$ of the UCM and its linear equivalent $s = vt + s_0$ for the RMCV (see figure 9.3).

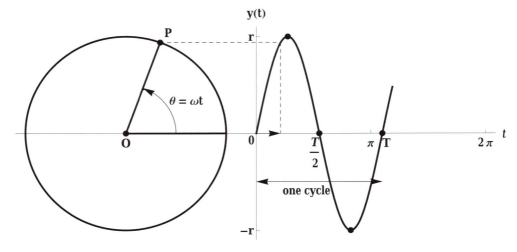

*Figure 9.10*: Vertical $(y,t)$-graph of a UCM given $\theta_0 = 0$

Planar uniform circular motion (PUCM)

In case we prefer a linear speed in units $\frac{m}{s}$ or $\frac{km}{h}$, we use a pair of compasses to recall that the ratio of circular arc length $s$ in $m$ to its radius $r$ in $m$, yields the angle

$$\theta = \frac{s}{r} \quad \text{in rad}.$$

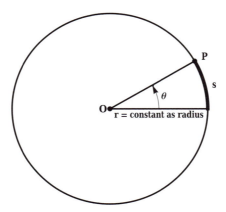

*Figure 9.11*: From angular to linear measures for circular motion

Rewriting the former definition for radians to the circular arc length $s = r\theta$ reveals the UCM's linear speed via substitution as

$$
\begin{aligned}
v &= \frac{\Delta s}{\Delta t} = \frac{\Delta(r\theta)}{\Delta t} \quad \text{in} \quad \frac{m}{s} \\
&= \frac{r\Delta\theta}{\Delta t}, && \text{given the constant radius} \quad r \geqslant 0 \\
&= r\frac{\Delta\theta}{\Delta t} = r\omega, && \text{given the angular speed (see formula (9.9)).} \\
\Downarrow & \\
v &= r\omega \quad \text{in} \quad \frac{m}{s}. && (9.15)
\end{aligned}
$$

This linear speed is (in absolute value) equal to the magnitude of the continuously varying **tangential velocity** vector $\vec{v}$. To be more precise, this free vector $\vec{v}$ features

▷   a constant magnitude commonly typeset as $v = r\omega$ (since mostly $v \geqslant 0$);

▷   a *varying* orientation perpendicular to its location vector $\vec{r}$;

▷   the sense in the direction of the angular displacement $\Delta\theta$.

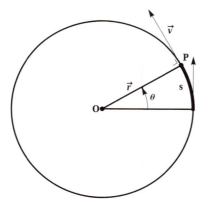

*Figure 9.12*: Tangential velocity vector of uniform circular motion

Since this velocity vector $\vec{v}$ varies, we define its vectorial rate of change with time by proving its properties (see formula (9.3)). To be more precise, we prove the according acceleration vector $\vec{a}_c$ to feature

▷   a constant magnitude $a_c = r\omega^2$ in $\frac{m}{s^2}$;

▷   a *varying* orientation antiparallel to its location vector $\vec{r}$;

▷   the sense inwards to the centre, therefore $\vec{a}_c$ is called **centripetal acceleration**.

*Proof:*

Proving $\vec{a}_c = \dfrac{\Delta \vec{v}}{\Delta t}$ implies the need to determine vector $\Delta \vec{v}$. To be more precise, we need to determine

▷   which magnitude $\|\Delta \vec{v}\|$;

▷   which orientation (with respect to its location vector $\vec{r}$); and

▷   either sense along its orientation.

Since this tangential velocity $\vec{v}$ is *continuously* varying, only tiny angular displacements $\Delta \theta \approx 0$ are meaningful to describe the corresponding changes $\Delta \vec{v}$.

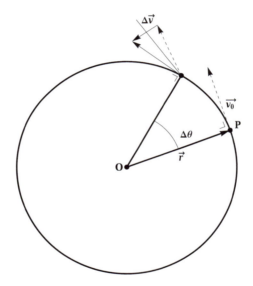

Picturing such a tiny positive angular displacement between times $t$ and $t_0$ shows its corresponding difference $\Delta \vec{v} = \vec{v} - \vec{v}_0$, aiming at the circle centre and therefore already proves for tiny $\Delta \theta \approx 0$

▷   the orientation of $\Delta \vec{v}$ to be antiparallel to its location vector $\vec{r}$;

▷   which implies its sense as inwards to the motion centre.

Picturing such a tiny positive angular displacement also reveals an isosceles triangle since $\|\vec{v}\| = \|\vec{v}_0\|$ with its apex in the revolving point $P$ and its base given by $\|\Delta \vec{v}\|$.

Halving this isosceles triangle along its altitude into two right triangles allows us to apply the definition of sine.

$$\sin\left(\frac{\Delta\theta}{2}\right) = \frac{\frac{\|\Delta\vec{v}\|}{2}}{\|\vec{v}\|} \implies \|\Delta\vec{v}\| = 2\|\vec{v}\|\sin\left(\frac{\Delta\theta}{2}\right) \qquad \text{as rewritten for } \|\Delta\vec{v}\|$$

$$\implies \|\Delta\vec{v}\| = 2\|\vec{v}\|\frac{\Delta\theta}{2} \qquad \text{in case of tiny } \Delta\theta \approx 0$$

$$\implies \|\Delta\vec{v}\| = \|\vec{v}\|\Delta\theta = (r\omega)\Delta\theta \qquad \text{substituting formula (9.15)}$$

To be more precise, this completes for tiny $\Delta\theta \approx 0$ the difference vector $\Delta\vec{v}$ proved with

▷  magnitude $\|\Delta\vec{v}\| = r\omega\Delta\theta$;

▷  an orientation antiparallel to its location vector $\vec{r}$;

▷  and therefore its sense as inwards to the motion centre.

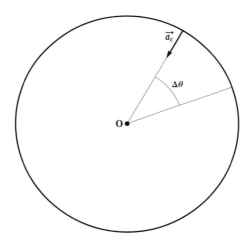

Returning to our initial rate of change with time of the tangential velocity vector, we now calculate the centripetal acceleration's magnitude as

$$\|\vec{a}_c\| = \frac{\|\Delta\vec{v}\|}{\Delta t} = \frac{r\omega\Delta\theta}{\Delta t} = r\omega\frac{\Delta\theta}{\Delta t} = r\omega\omega, \text{ in } m\frac{\text{rad}}{s}\frac{\text{rad}}{s} \qquad \text{see formula (9.9)}$$

$$\Downarrow$$

$$a_c = r\omega^2 \text{ in } \frac{m}{s^2}. \qquad (9.16)$$

Acceleration $\vec{a}_c$ inherits the direction of its numerator $\Delta\vec{v}$, completing this proof.  ∎
We recall rad is not a physical dimension, but a dimensionless mathematical tag (see formula (3.1)).

Referring to the definition of vectorial deceleration, we discover the centripetal acceleration $\vec{a}_c$ to be neither a deceleration nor a 'true' acceleration, in the sense of changing the
*magnitude v*. This means a pure centripetal acceleration $\vec{a}_c$ cannot affect the linear *speed*,
which is proved by the dot product $\vec{a}_c \cdot \vec{v} = 0$ due to their mutual orientation $\vec{a}_c \perp \vec{v}$ (see
formula (9.5)).

Extending the former discovery, the uniform circular motion clearly offers us the insight
that in kinematics the velocity vector $\vec{v}(t)$ is tangent to the object's trajectory in the direction of its moving location vector $\vec{s}(t)$, whilst the acceleration vector $\vec{a}(t)$ can take any
direction.

Spatial uniform circular motion (SUCM)

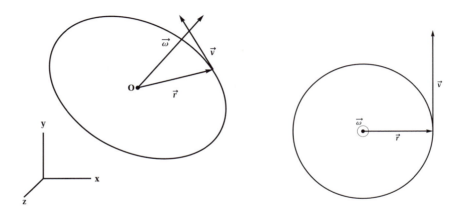

*Figure 9.13*: Spatial uniform circular motion

Based on all the above, we finally define in 3D the **angular velocity** vector $\vec{\omega}$ as the first
factor in the cross product of its corresponding tangential velocity

$$\vec{v} = \vec{\omega} \times \vec{r}. \tag{9.17}$$

Therefore, this free vector $\vec{v}$ more precisely confirms its

▷  magnitude recalculated as $v = \left\| \vec{\omega} \times \vec{r} \right\| = \left\| \vec{\omega} \right\| \left\| \vec{r} \right\| \sin \frac{\pi}{2} = r\omega$;

▷  orientation as orthogonal to both factor $\vec{\omega}$ and factor $\vec{r}$;

▷  sense according to the right-hand grip rule or corkscrew rule (see figure 7.1).

## Nonuniform circular motion (NCM)

How do you describe an object's position when it is moving at a variable angular velocity $\vec{\omega}$ along a spatial circle of radius $r$? Similarly to the previous, we consequently define

$$\textbf{angular acceleration} = \frac{\textbf{change of angular velocity}}{\text{delta time}} \quad \text{or} \quad \vec{\alpha} = \frac{\Delta \vec{\omega}}{\Delta t}. \quad (9.18)$$

The vector $\vec{\alpha}$ implies direction as inherited by its numerator $\Delta \vec{\omega}$, with the Greek character $\alpha$ referring to *acceleration*. This angular acceleration comes out in the unit $\left[ \| \vec{\alpha} \| \right] = \frac{\left[ \| \Delta \vec{\omega} \| \right]}{[\Delta t]} = \frac{\frac{rad}{s}}{s} = \frac{rad}{s^2}$.

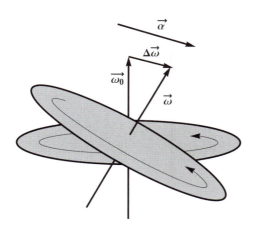

*Figure 9.14*: Spatial nonuniform circular motion

## Constant angular acceleration

Limiting our case to constant angular accelerations only, we prove its corresponding angular velocity equation.

*Proof:*

$$\vec{\alpha} = \frac{\Delta \vec{\omega}}{\Delta t} \quad \Leftrightarrow \quad \vec{\alpha} = \frac{\vec{\omega} - \vec{\omega}_0}{t - t_0}$$

$$\Rightarrow \quad \text{if we set } t_0 = 0 \text{ then } \vec{\alpha} = \frac{\vec{\omega} - \vec{\omega}_0}{t - 0}$$

$$\Leftrightarrow \quad \vec{\alpha}t = \vec{\omega} - \vec{\omega}_0$$
$$\Leftrightarrow \quad \vec{\omega} = \vec{\omega}_0 + \vec{\alpha}t, \text{ given } \vec{\alpha} \text{ is constant as a vector.} \qquad \blacksquare$$

We notice how this circular motion's angular velocity equation appears completely analogous to its rectilinear motion's counterpart (see formula (9.6)).

Planar nonuniform circular motion (PNCM)

Returning to our planar circular motion reduces the above general angular velocity equation to 1D vectors (see figure 9.15), which we typeset not-overarrowed as

$$\omega = \omega_0 + \alpha t, \text{ given } \alpha \text{ is constant.} \qquad (9.19)$$

Without further proof, we likewise accept the circular motion's angular location equation as analogous to its rectilinear motion's counterpart (see formula (9.7)) as

$$\theta = \theta_0 + \omega_0 t + \frac{1}{2}\alpha t^2, \text{ given } \alpha \text{ is constant.} \qquad (9.20)$$

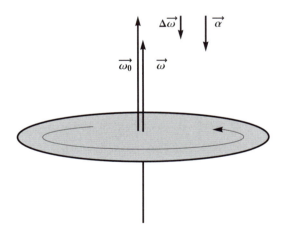

*Figure 9.15*: Planar nonuniform circular motion

In case we prefer a linear acceleration in unit $\frac{m}{s^2}$, based on the linear speed (9.15), we similarly define

$$a_t = r\alpha \text{ in } \frac{m}{s^2}, \qquad (9.21)$$

as the **tangential acceleration** of the nonuniform circular motion. Conclusively, the planar nonuniform circular motion is completely modelled by two mutual orthogonal accelerations: the tangential plus the centripetal acceleration (see figure 9.17).

$$\begin{cases} a_t = r\alpha \\ a_c = r\omega^2 \end{cases} \implies \text{total acceleration vector } \vec{a} = \vec{a}_c + \vec{a}_t \qquad (9.22)$$

We straightforwardly extend this approach to model motions via their combined accelerations in the subsequent section on planar curvilinear motions.

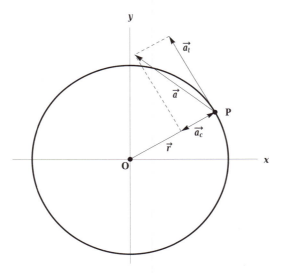

*Figure 9.16*: Nonuniform circular motion as modelled by $\vec{a}_c + \vec{a}_t$

*Example:* Luna stands at the roundabout at $0.5\ m$ from its centre. Taking off at a constant angular acceleration of $0.08\ \frac{rad}{s^2}$, determine her total linear acceleration after 10 seconds.

Applying the above formula (9.22), this involves the vector addition of our tangential and our centripetal acceleration as experienced after 10 seconds. We calculate their magnitudes respectively as

$$\begin{cases} a_t = r\alpha = 0.5 \times 0.08 = 0.04 \ \frac{m}{s^2}, \\ a_c = r\omega^2 = 0.5 \times (0.8)^2 = 0.32 \ \frac{m}{s^2}, \end{cases} \quad \text{via formula (9.19) } \omega = 0 + 0.08 \times 10 = 0.8.$$

Since for circular motion the tangential acceleration is continuously orthogonal to the centripetal acceleration, the Pythagorean Theorem accordingly yields

$$\|\vec{a}\| = \sqrt{a_c^2 + a_t^2} = \sqrt{0.32^2 + 0.04^2} = 0.323 \ \frac{m}{s^2}.$$

Besides its magnitude, we determine the direction of the total acceleration vector $\vec{a}$ by the angle $\tau$ as subtended by its centripetal component $\vec{a}_c$ via the Right Triangle's definition as

$$\tan \tau = \frac{a_t}{a_c} = \frac{0.04}{0.32} = 0.125$$
$$\Rightarrow \tau = \arctan(0.125) = 0.124 \ \text{rad} = 7.13°.$$

## SUMMARY

Since the circular motion with constant angular acceleration's equations of angular velocity and angular location are completely analogous to their RMCA's counterparts, we double our previous advise to know both these equations by heart, because they imply those for uniform circular motion if we simply set $\alpha = 0$:

$$\omega = \omega_0 + \alpha t,$$
$$\theta = \theta_0 + \omega_0 t + \frac{1}{2}\alpha t^2.$$

Finally, we also prove the angular time-independent formula for the circular motion with constant angular acceleration

$$\omega^2 = \omega_0^2 + 2\alpha \cdot \Delta\theta, \quad\quad\quad\quad\quad\quad (9.23)$$

by eliminating the scalar variable $t$ between the angular velocity and the angular location equation.

*Proof:* Therefore we square the angular velocity equation, and substitute.

$$\omega = \omega_0 + \alpha t \quad \Rightarrow \quad \omega^2 = (\omega_0 + \alpha t)^2$$

$$\Leftrightarrow \quad \omega^2 = \omega_0^2 + 2\omega_0 \cdot \alpha t + (\alpha t)^2 \qquad \text{see formula (1.2),}$$

$$\Leftrightarrow \quad \omega^2 = \omega_0^2 + 2\alpha \cdot \omega_0 t + \alpha^2 t^2$$

$$\Leftrightarrow \quad \omega^2 = \omega_0^2 + 2\alpha \cdot \left(\omega_0 t + \frac{1}{2}\alpha t^2\right) \qquad \text{factoring out } 2\alpha$$

$$\theta = \theta_0 + \omega_0 t + \frac{1}{2}\alpha t^2 \quad \Leftrightarrow \quad \theta - \theta_0 = \omega_0 t + \frac{1}{2}\alpha t^2$$

$$\Rightarrow \quad \omega^2 = \omega_0^2 + 2\alpha \cdot (\theta - \theta_0) \qquad \text{angular displacement}$$

$$\Leftrightarrow \quad \omega^2 = \omega_0^2 + 2\alpha \cdot \Delta\theta \qquad \blacksquare$$

For circular motion, we go from angular measures to their linear equivalents by simply multiplying with the constant radius $r$ (see table 9.1).

| measure<br>circular | angular | linear tangential |
|---|---|---|
| location | $\theta$ | $s = r \cdot \theta$ given $\theta$ in radians |
| velocity | $\omega$ | $v = r \cdot \omega$ given $\omega = \frac{2\pi}{T}$ is constant |
| acceleration | $\alpha$ | $a_t = r \cdot \alpha$ given $\alpha = \frac{\Delta\omega}{\Delta t}$ is constant |

*Table 9.1*: Conversion between angular and linear tangential measures

## 9.5 Planar curvilinear motion

The previous part of this chapter dealt only with rectilinear and circular motion. This section briefly outlines the general curvilinear motion in two dimensions as well.

To describe rectilinear motion, we have already introduced location, velocity and acceleration vectors referring to a fixed cartesian coordinate system. To describe circular motion, we similarly introduced the polar coordinate system fixed at the motion centre. Let it be clear that the choice of a specific coordination system can never affect final results, however practically it may simplify the calculations and improve the insights considerably. To describe the curvilinear motion, we define an **intrinsic coordinate system** as a frame of base vectors $\{\vec{e}_1, \vec{e}_2\}$ having its origin $P$ in the moving object, and therefore travelling with it along its path. We picture its corresponding axes as a right-angled bracket moving along with the object. We define **path variables** as measurements within a moving bracket along a trajectory.

## Normal-tangential components

Intrinsic base

Elaborating on the total acceleration of nonuniform circular motion – achieved by combining the centripetal and the tangential contribution (see figure 9.16) – introduces the corresponding base $\{\vec{e}_n, \vec{e}_t\}$ via

$$
\begin{aligned}
\vec{a} &= \vec{a}_c + \vec{a}_t \\
&= a_c \vec{e}_n + a_t \vec{e}_t \\
&= (r\omega^2)\vec{e}_n + (r\alpha)\vec{e}_t.
\end{aligned}
$$

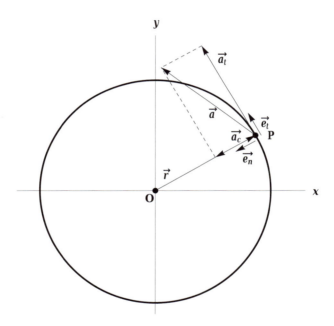

*Figure 9.17*: Nonuniform circular motion as modelled in normal-tangential components

We define its intrinsic normal unit vector $\vec{e}_n$ as directed to the circle centre, whereas its tangential unit vector $\vec{e}_t$ is directed by increasing angular locations $\theta$.

Both unit vectors share the moving point $P$ as initial point (see figure 9.17).

We prove the vector set $\{\vec{e}_n, \vec{e}_t\}$ as an intrinsic 2D base (see page 134).

*Proof:*

▷ The vector set $\{\vec{e}_n, \vec{e}_t\}$ is spanning.

Its linear combination $a_n\vec{e}_n + a_t\vec{e}_t = \vec{a}$ creates any 2D acceleration $\vec{a}$ with its initial point at their intrinsic origin $P$.

▷ The vector set $\{\vec{e}_n, \vec{e}_t\}$ is linearly independent.

Because the tangent is orthogonal to the radius at any location on a circle, we have that $\vec{e}_n \perp \vec{e}_t$ and therefore $\vec{e}_n$ and $\vec{e}_t$ are linearly independent.     ■

We commonly picture their corresponding axes as the $(n,t)$-coordinate system attached to the moving point P.

We define the **normal** and **tangential components** of a motion as we decompose its velocity and acceleration vectors to the respective base vectors $\vec{e}_n$ and $\vec{e}_t$. We extend – without further proof – the use of normal-tangential components to the planar curvilinear motion.

Curvature

Firstly, we redefine the radius $r$ of circular motion as the radius of the '**best fit circle**' to the trajectory at point $P$, which we therefore call the local **radius of curvature** $r_c$ (see figure 9.18). The local **centre of curvature** $C$ is determined as the centre of the instantaneous 'best fit circle' to the path. For a complete outline about curvature, we refer you to the expert literature.

Secondly, we redefine the intrinsic base $\{\vec{e}_n, \vec{e}_t\}$ as unit vectors sharing the initial point $P$, given $\vec{e}_n$ is directed towards the centre of curvature of the path, whereas $\vec{e}_t$ is tangent to the path positive in the direction of the motion.

Curvilinear motion

This general intrinsic normal-tangential base enables us to describe the moving point $P$ – without further proof – via the intrinsic origin's

▷ location: by the arc length along the path anchored to a fixed initial location $\vec{s}_0$, called the **scalar path function** $l(t)$, given $l(0) = 0$ $m$ (see figure 9.18),

▷ velocity: by the vector $\vec{v} = u \cdot \vec{e}_t$ given its linear speed $u = \frac{\Delta l}{\Delta t}$ based on arc length,

▷ acceleration: by the vector

$$\vec{a} = \left(\frac{\Delta u}{\Delta t}\right)\vec{e}_t + \left(\frac{u^2}{r_c}\right)\vec{e}_n$$

given the linear speed $u$ and the local radius of curvature $r_c$ (see figure 9.18).

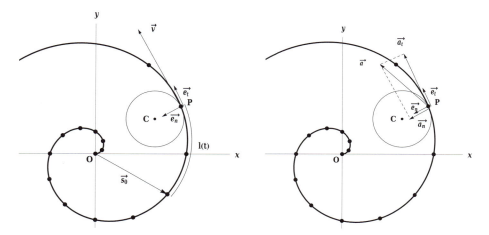

*Figure 9.18*: Location $l(t)$, velocity $\vec{v}$ and acceleration $\vec{a}$ in normal-tangential components

In chapter 6, we defined a parametric curve as the locus of a moving point $P$, so let us refer to its illustrations as exemplary trajectories to be modelled by normal-tangential components (see figures 6.1, 6.2 and 6.4).

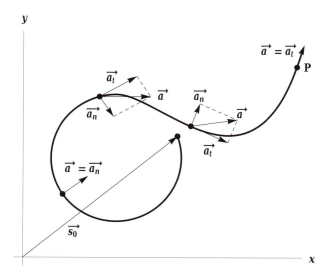

*Figure 9.19*: A curvilinear path shaped by acceleration in normal-tangential components

## Radial-angular components

Similarly to the previously cartesian designed normal-tangential components for a moving point $P$, we can also use fixed polar coordinates to construct such an intrinsic base (see figure 6.5).

### Intrinsic base

Based on polar coordinates, we define the intrinsic radial unit vector $\vec{e}_r$ as directed outwards from the fixed pole $O$, whereas the angular unit vector $\vec{e}_\theta$ is directed counter clockwise from the horizontal polar axis. Both unit vectors share the moving point $P$ as initial point (see figure 9.20).

We prove the vector set $\{\vec{e}_r, \vec{e}_\theta\}$ as an intrinsic 2D base.

*Proof:*

▷ The vector set $\{\vec{e}_r, \vec{e}_\theta\}$ is spanning.

Its linear combination $a_r\vec{e}_r + a_\theta\vec{e}_\theta = \vec{a}$ creates any 2D acceleration $\vec{a}$ with its initial point at their intrinsic origin $P$.

▷ The vector set $\{\vec{e}_r, \vec{e}_\theta\}$ is linearly independent.

As both above defined directions of $\vec{e}_r$ and $\vec{e}_\theta$ cannot be parallel nor antiparallel, $\vec{e}_r$ and $\vec{e}_\theta$ are linearly independent. Since the tangent to a circle is perpendicular to its radius at the tangent piont, we have that $\vec{e}_r \perp \vec{e}_\theta$.  ∎

We commonly picture their corresponding axes as the $(r, \theta)$-coordinate system attached to the moving point P.

We define the **radial** and **angular components** of a motion as we decompose its velocity and acceleration vectors to the respective base vectors $\vec{e}_r$ and $\vec{e}_\theta$.

### Curvilinear motion

This general intrinsic radial-angular base enables us to describe the moving point $P$, without any further proof, via the intrinsic origin's

▷ location: by the vector $\vec{r} = r \cdot \vec{e}_r$ initially starting at $\vec{r}_0$,

▷ velocity: by the vector $\vec{v} = w \cdot \vec{e}_r + (r\omega) \cdot \vec{e}_\theta$ given the radial speed $w = \frac{\Delta r}{\Delta t}$ and the angular speed $\omega = \frac{\Delta \theta}{\Delta t}$,

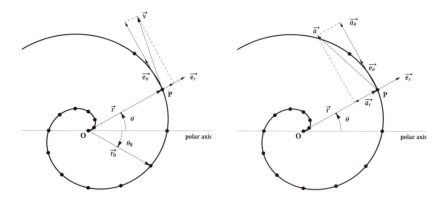

*Figure 9.20*: Location $\vec{r}$, velocity $\vec{v}$ and acceleration $\vec{a}$ in radial-angular components

▷ acceleration: by the vector

$$\vec{a} = \underbrace{\left( \frac{\Delta w}{\Delta t} - r\omega^2 \right) \vec{e}_r}_{\text{radial acceleration } a_r} + \underbrace{(r\alpha + 2\omega \cdot w)\,\vec{e}_\theta}_{\text{polar angular acceleration } a_\theta},$$

given all above and we recall $\alpha = \frac{\Delta \omega}{\Delta t}$ as the angular acceleration (see figure 9.20).

In these radial-angular components, we redefine the coefficients $-r\omega^2$ as the **centripetal** and $+2\omega \cdot w$ as the **coriolis acceleration** respectively. The latter mixed coefficient was named after its discoverer, the French physicist **Caspar de Coriolis** (1792–1843).

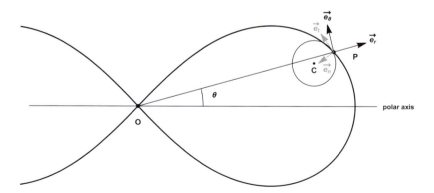

*Figure 9.21*: The radial-angular intrinsic base differs from the normal-tangential

In chapter 6, we defined a polar curve as the locus of a moving point $P$, so let us refer to its illustrations as exemplary trajectories to be modelled by radial-angular components (see figures 6.6, 6.7 and 6.8). For instance, the latter trajectory of figure 6.8 reveals how the radial-angular intrinsic base $\{\vec{e}_r, \vec{e}_\theta\}$ differs essentially from the normal-tangential base $\{\vec{e}_n, \vec{e}_t\}$ (see figure 9.21).

## Planar nonuniform circular motion (PNCM)

We arrive where we started intrinsic coordinate systems, by demonstrating radial-angular components to describe the planar nonuniform circular motion (see figure 9.22). We describe this motion in radial-angular components by rotating the location vector $\vec{r}$ of constant length $r$, over an angle $\theta$. The corresponding radial speed $w = \frac{\Delta r}{\Delta t} = 0$ leads to

$$\begin{cases} \vec{v} = 0 \cdot \vec{e}_r + r\omega \cdot \vec{e}_\theta \\ \vec{a} = (0 - r\omega^2)\vec{e}_r + (r\alpha + 0)\vec{e}_\theta \end{cases} \implies \begin{cases} \vec{v} = r\omega \cdot \vec{e}_\theta \\ \vec{a} = (-r\omega^2)\vec{e}_r + (r\alpha)\vec{e}_\theta \end{cases}$$

$$\implies \begin{cases} \vec{v} = r\omega \cdot \vec{e}_\theta \\ \vec{a} = r\omega^2(-\vec{e}_r) + r\alpha\vec{e}_\theta. \end{cases}$$

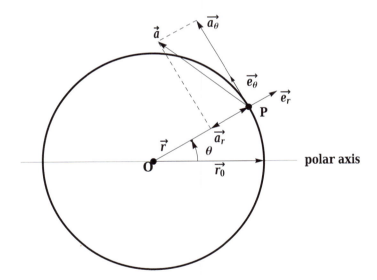

*Figure 9.22*: Planar nonuniform circular motion modelled by radial-angular components

Since for the planar circular motion the *radius of curvature* $r_c$ is at any instant equal to the constant *polar radius* $r$, this case relates the former normal-tangential to the radial-angular components primarily via $-\vec{e}_r = \vec{e}_n$ and consequently $\vec{e}_\theta = \vec{e}_t$. Given $\vec{r}_0 = \vec{s}_0$, we rewrite its radial-angular components into

$$
\begin{cases} \vec{v} = 0 \cdot \vec{e}_r + r\omega \cdot \vec{e}_\theta \\ \vec{a} = (0 - r\omega^2)\vec{e}_r + (r\alpha + 0)\vec{e}_\theta \end{cases} \implies \begin{cases} \vec{v} = r\omega \cdot \vec{e}_t \\ \vec{a} = r\omega^2 \vec{e}_n + r\alpha \vec{e}_t \end{cases}
$$

$$
\implies \begin{cases} \vec{v} = r\omega \cdot \vec{e}_t \text{ a pure tangential velocity} \\ \vec{a} = \vec{a}_c + \vec{a}_t \text{ a composed acceleration,} \end{cases}
$$

which confirms its original formulas (9.15) and (9.22). For a better insight, we compare figure 9.22 to our original figure 9.17.

## 9.6 Independence of Motion

As an alternative to the previous section, curvilinear motions in two (or three) dimensions can be parametrically expressed in cartesian coordinates as well, sometimes relying on the Independence of Motion Principle.

We define the *universal* Independence of Motion Principle as: '*motion-components work independently*' which means the final position, velocity and acceleration reached by any motion through a time interval $[t_0, t]$ remains when we subsequently run its $z$-component after its $y$-component after its $x$-component, each over that same time interval $[t_0, t]$. This componentwise independence is commutative ($z$- after $x$- after $y$-components of motion, $x$- after $y$- after $z$-components of motion, ... all of these yield the same final motion whenever run over that same time interval $[t_0, t]$).

For a better insight in this principle, imagine we slide an object over a frictionless table off the edge, from where it becomes subject to gravity. In the horizontal direction, it continues with constant velocity. In the vertical direction, it speeds up downwards by the constant acceleration due to gravity as described by free fall (see paragraph 9.3). Both perpendicular motions combine to a trajectory that curves downwards, which is but one particular case of what we call **projectile motion**.

### COMBINED RECTILINEAR MOTIONS WITH CONSTANT VELOCITIES

As an Independence of Motion Principle, the following applies for motions with **constant velocity** vectors. The resulting motion with constant velocity $\vec{v}_{\text{ground}}$ of an aeroplane flying through a crosswind is the sum of the two simultaneous constant velocity vectors $\vec{v}_{\text{plane}}$ and $\vec{v}_{\text{wind}}$. By vector addition we find their composite velocity vector $\vec{v}_{\text{ground}}$.
The *other way around*, we may decompose (see paragraph 7.5) the velocity vector $\vec{v}_{\text{ground}}$ equivalently as Rectilinear Motions with Constant Velocities $\vec{v}_x$ plus $\vec{v}_y$.

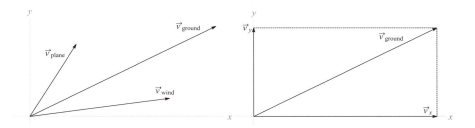

# PROJECTILE MOTION (PM)

We consider an object in a projectile motion, after being launched from an initial position $\vec{s}_0$ by an initial velocity $\vec{v}_0$. We only consider the idealised projectile motion in which air resistance is neglected and only gravity acts upon it via its constant acceleration $\vec{g}$.

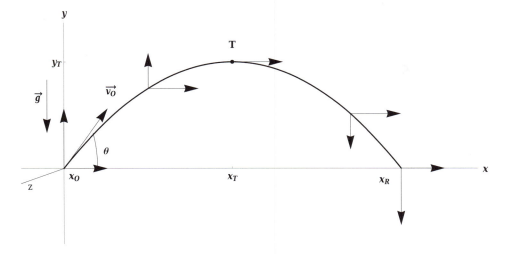

## Location equation

For our convenience, we describe this planar motion in the cartesian $(x, y)$-plane and we place its initial position in the origin, setting $\vec{s}_0 = \vec{o}$. The initial velocity $\vec{v}_0$ subtends some angle $\theta$ with the horizontal $x$-axis. We summarise our approach by its corresponding location equation as

$$\begin{aligned} \vec{s} &= \vec{s}_0 + \vec{v}_0 t + \frac{1}{2}\vec{a}t^2 \\ &= \vec{o} + \vec{v}_0 t + \frac{1}{2}\vec{g}t^2 \end{aligned}$$

We decompose this projectile's motion along each dimension as a horizontal RMCV and a vertical with constant acceleration (due to gravity), with no motion along the $z$-axis.

$$\vec{s} = \begin{pmatrix} \text{rectilinear motion at constant velocity} \\ \text{rectilinear motion at constant acceleration} \\ \text{no motion} \end{pmatrix}$$

Expressing each component's location equation independently, yields

$$\vec{s} = \begin{pmatrix} 0 \\ 0 \\ 0 \end{pmatrix} + \begin{pmatrix} v_0 \cos\theta \\ v_0 \sin\theta \\ 0 \end{pmatrix} t + \frac{1}{2} \begin{pmatrix} 0 \\ -9.81 \\ 0 \end{pmatrix} t^2$$

$$= \begin{pmatrix} (v_0 \cos\theta)t \\ (v_0 \sin\theta)t + \frac{1}{2}(-9.81)t^2 \\ 0 \end{pmatrix}. \tag{9.24}$$

Trajectory

To study this motion, we firstly aim for its *trajectory* drawn in the $(x,y)$-plane via eliminating the variable time $t$ between its location components.

$$x = (v_0 \cos\theta)t \implies t = \frac{x}{v_0 \cos\theta} \quad \text{substituting in}$$

$$y = (v_0 \sin\theta)t + \frac{1}{2}(-9.81)t^2 \quad \text{yields}$$

$$y(x) = (v_0 \sin\theta)\left(\frac{x}{v_0 \cos\theta}\right) + \frac{1}{2}(-9.81)\left(\frac{x}{v_0 \cos\theta}\right)^2$$

$$= (\tan\theta)x + \frac{1}{2}(-9.81)\frac{x^2}{(v_0 \cos\theta)^2}$$

$$= \underbrace{\left(\frac{-9.81}{2(v_0 \cos\theta)^2}\right)}_{a<0} x^2 + \underbrace{(\tan\theta)x}_{b} + \underbrace{0}_{c} \tag{9.25}$$

This teaches us that the projectile motion's trajectory is a parabola opening down, since its leading coefficient $a < 0$ (see figure 4.5). We define its **horizontal range** as the projectile's final horizontal displacement $\Delta x = x_R - x_O$. We define its **apex** $T$ as having the maximum height $y_T$ reached by the projectile, mathematically known as the parabola's vertex.

## Horizontal range

To continue our analysis of this motion, we secondly retrieve its parabolic trajectory's horizontal range and apex (see pages 30 and 72 respectively). Since determining the horizontal range requires the quadratic function's roots, we need to solve $f(x) = 0$ for all $x$. There is however no need to apply the Quadratic Formula, since we easily factor $y(x)$ which immediately yields its roots $x_O$ and $x_R$.

$$0 \;=\; x\left(\left(\frac{-9.81}{2(v_0 \cos\theta)^2}\right)x + \tan\theta\right)$$

$$\Longrightarrow x_O = 0 \;\text{ or }\; \left(\frac{-9.81}{2(v_0 \cos\theta)^2}\right)x_R + \tan\theta = 0 \;\Leftrightarrow$$

$$x_R = -\tan\theta\left(\frac{2(v_0 \cos\theta)^2}{-9.81}\right)$$

$$= \frac{\sin\theta}{\cos\theta}\frac{2v_0^2(\cos\theta)^2}{9.81}$$

$$= \frac{v_0^2(2\sin\theta\cos\theta)}{9.81} = \frac{v_0^2 \sin(2\theta)}{9.81} \qquad (9.26)$$

The trigonometric Sum Identity $\sin(\theta + \theta) = \sin\theta\cos\theta + \cos\theta\sin\theta$ commonly simplifies the second root's expression (see paragraph 3.7). The projectile's horizontal range calculated as $\Delta x = x_R - x_O = \frac{v_0^2 \sin(2\theta)}{9.81} - 0 = \frac{v_0^2 \sin(2\theta)}{9.81}$ can be maximised through maximising the initial speed $v_0$ or assuring $\sin(2\theta) = 1$ and therefore $\theta = \frac{\pi}{4}$. In other words, a projectile's launch with a given initial speed $v_0$ inclined to an angle of 45° assures a maximum horizontal range.

## Apex

Finally, returning the parabola's vertex argument $x_T$, yields the according height $y_T = y(x_T)$ of its projectile's apex $T$ via

$$x_T = \frac{-b}{2a} \;\Longrightarrow\; y(x_T) = \left(\frac{-9.81}{2(v_0 \cos\theta)^2}\right)x_T^2 + (\tan\theta)x_T + 0.$$

However straightforward, the above approach seems too cumbersome to apply.

Alternatively, relying on the symmetry of a parabola and the existence of real roots $x_O$ and $x_R$ we may find this vertex argument $x_T$ simply as their mid point:

$$x_T = \frac{x_O + x_R}{2}$$

$$= \frac{0 + \frac{v_0^2 \sin(2\theta)}{9.81}}{2} = \frac{1}{2}\frac{v_0^2 \sin(2\theta)}{9.81}.$$

Alternatively, relying on the Independence of Motion Principle and the velocity equation $\vec{v} = \vec{v}_0 + \vec{a}t$, we may find the apex height $y_T$ by reasoning along the vertical $y$-dimension only. We most easily find the time $t_T$ when the projectile reaches its maximum height via the $y$-component of its velocity:

$$v_y = v_0 \sin\theta + (-9.81)t$$

reaching zero at maximum height, means we need to solve

$$0 = v_0 \sin\theta + (-9.81)t_T \quad \text{for} \quad t_T = \frac{v_0 \sin\theta}{9.81}.$$

Instead of making use of the parabolic $y(x)$, we simply evaluate this halfway time $t_T$ in the $y$-component of the location equation which returns the according height $y_T = y(t_T)$ as well (see formula (9.24)).

$$
\begin{aligned}
y(t_T) &= (v_0 \sin\theta)t_T + \frac{1}{2}(-9.81)(t_T)^2 \\
y_T &= (v_0 \sin\theta)\frac{v_0 \sin\theta}{9.81} + \frac{1}{2}(-9.81)\left(\frac{v_0 \sin\theta}{9.81}\right)^2 \\
&= \frac{(v_0 \sin\theta)^2}{9.81} - \frac{1}{2}\left(\frac{(v_0 \sin\theta)^2}{9.81}\right) \\
&= \frac{1}{2}\frac{(v_0 \sin\theta)^2}{9.81}
\end{aligned}
$$

This height $y_T$ of the projectile's apex can be maximised through maximising its initial speed $v_0$ or assuring $\sin\theta = 1$ and therefore $\theta = \frac{\pi}{2}$. In other words, a projectile's launch with a given initial speed $v_0$ inclined to an angle of $90°$, assures a maximum height. This perpendicular launch of a projectile returns us to the previously outlined RMCA (see paragraph 9.3).

Euler's trajectory

Applying Euler's method for trajectories (see formula (7.6))

$$\vec{s}_{i+1} = \vec{s}_i + \vec{v}_i \Delta t$$

$$\vec{v}_{i+1} = \vec{v}_i + \vec{a}_i \Delta t$$

to projectile motion, dropping its $z$−component and setting $\vec{a}_0 = \vec{g}$ constant, leads to

$$\vec{s}_1 = \vec{s}_0 + \vec{v}_0 \Delta t = \begin{pmatrix} 0 \\ 0 \end{pmatrix} + \begin{pmatrix} v_0 \cos\theta \\ v_0 \sin\theta \end{pmatrix} \Delta t$$

$$= \begin{pmatrix} (v_0 \cos\theta)\Delta t \\ (v_0 \sin\theta)\Delta t \end{pmatrix}$$

$$\vec{v}_1 = \vec{v}_0 + \vec{g}\Delta t = \begin{pmatrix} v_0 \cos\theta \\ v_0 \sin\theta \end{pmatrix} + \begin{pmatrix} 0 \\ -9.81 \end{pmatrix} \Delta t$$

$$= \begin{pmatrix} v_0 \cos\theta \\ v_0 \sin\theta - 9.81\Delta t \end{pmatrix}$$

and for each successive step size $\Delta t$ recursively to the frame data

$$\vec{s}_2 = \vec{s}_1 + \vec{v}_1 \Delta t = \begin{pmatrix} (v_0 \cos\theta)\Delta t \\ (v_0 \sin\theta)\Delta t \end{pmatrix} + \begin{pmatrix} v_0 \cos\theta \\ v_0 \sin\theta - 9.81\Delta t \end{pmatrix} \Delta t$$

$$= \begin{pmatrix} 2(v_0 \cos\theta)\Delta t \\ 2(v_0 \sin\theta)\Delta t - 9.81\Delta t^2 \end{pmatrix}$$

$$\vec{v}_2 = \vec{v}_1 + \vec{g}\Delta t = \begin{pmatrix} v_0 \cos\theta \\ v_0 \sin\theta - 9.81\Delta t \end{pmatrix} + \begin{pmatrix} 0 \\ -9.81 \end{pmatrix} \Delta t$$

$$= \begin{pmatrix} v_0 \cos\theta \\ v_0 \sin\theta - 2(9.81)\Delta t \end{pmatrix}$$

and

$$\vec{s}_3 = \vec{s}_2 + \vec{v}_2 \Delta t = \begin{pmatrix} 2(v_0 \cos\theta)\Delta t \\ 2(v_0 \sin\theta)\Delta t - 9.81\Delta t^2 \end{pmatrix} + \begin{pmatrix} v_0 \cos\theta \\ v_0 \sin\theta - 2(9.81)\Delta t \end{pmatrix} \Delta t$$

$$= \begin{pmatrix} 3(v_0 \cos\theta)\Delta t \\ 3(v_0 \sin\theta)\Delta t - 3(9.81)\Delta t^2 \end{pmatrix}$$

$$\vec{v}_3 = \vec{v}_2 + \vec{g}\Delta t = \begin{pmatrix} v_0 \cos\theta \\ v_0 \sin\theta - 2(9.81)\Delta t \end{pmatrix} + \begin{pmatrix} 0 \\ -9.81 \end{pmatrix} \Delta t$$

$$= \begin{pmatrix} v_0 \cos\theta \\ v_0 \sin\theta - 3(9.81)\Delta t \end{pmatrix}$$

and so on to a (limited) list of subsequent frames.

## 9.7 Exercises

**Exercise 78** Find its linear location $s$ given the RMCV taking off at $s_0 = 20\ m$ with a constant velocity $v = 15\ \frac{m}{s}$. Outline $s(t)$ completely as a real function by drawing its graph and determining its domain, range and possibly root(s).

**Exercise 79** Find its quadratic location $s$ given the RMCA taking off at height $s_0 = 20\ m$ with an initial upward velocity $v_0 = +15\ \frac{m}{s}$ and therefore subject to free fall above Earth. Outline $s(t)$ completely as a real function by drawing its graph and determining its domain, range and possibly root(s).

**Exercise 80** For a ball launched (vertically) upwards from Earth with an initial *speed* of $8.2\ \frac{m}{s}$ calculate its *velocities* after respectively:

1)  half a second,

2)  just one second.

**Exercise 81** Determine which UCM has been vertically projected onto the y-axis.

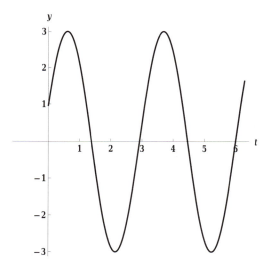

**Exercise 82** A car wheel of radius $30\ cm$ was set to a uniform circular motion with a tangential speed of 72 km per hour.

1)  Determine its angular velocity $\vec{\omega}$, consisting of angular speed $\omega$ and direction.

2)  How many (wheel) turns per second does this angular speed require?

3)  How many radians does this consume in 15 seconds?

4) Calculate its centripetal acceleration's magnitude $a_c$ as well.

**Exercise 83** Track a stunt skier as pictured in three locations $T$ (Top), $U$ (U-dale) and $J$ (Jump) (see figure). Neglecting surface friction and air resistance, determine the direction of his acceleration in respectively $T$, $U$ and $J$.

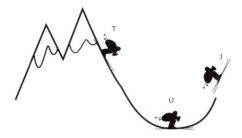

**Exercise 84** We have put baby Aiden at 2.7 $m$ from the centre of a merry-go-round at rest. At $t_0 = 0$ this carousel accelerates by a constant angular acceleration $\alpha = 0.1 \frac{rad}{s^2}$. Calculate at instant $t = 10 \ s$ our baby's

1) linear speed $v$ and tangential acceleration $a_t$,

2) total linear acceleration $\|\vec{a}\|$ based on their vector sum $\vec{a}_t + \vec{a}_c$.

**Exercise 85** A Ferris wheel is turning at a constant rate. Centre the polar coordinate frame at the wheel's axis. Express one cart's position $\vec{s}$, velocity $\vec{v}$ and acceleration $\vec{a}$ in the normal-tangential as well as the radial-angular components.

**Exercise 86** Legend has it that the Italian scientist **Galileo Galilei** (1564–1642) tested out his newly developed ideas about motion by (amongst other experiments) dropping masses from the top of the Leaning Tower of Pisa, from the lower side at a height of 55.86 $m$. Suppose that he simultaneously dropped a first mass just straight down, whereas he threw a second mass horizontally with a velocity of 10 $\frac{m}{s}$. Answer the following questions assuming a constant acceleration due to gravity of 9.81 $\frac{m}{s^2}$ and neglecting air resistance.

1) Which mass struck the ground first?

2) How long (after releasing) did it take to hit the ground?

3) What horizontal distance was travelled by each?

**Exercise 87** As a rider on a 120 $\frac{km}{h}$ speeding motorcycle jumps off a 30°-slope at a height of 6 $m$ above ground level, he remains in the air for 7 seconds. Picture this situation. Calculate the horizontal distance and the max height he and the bike flew before touchdown.

# Chapter 10 · Collision detection

Programming animations or games, we often meet the need to detect collisions between moving objects. For instance, characters are not allowed to walk through walls or objects.

This chapter outlines some efficient techniques to detect collisions on the screen. Therefore we circumscribe moving 2D objects by a single circle. This obviously is a rough approximation for a targeted object, but its results are satisfactory. Whenever there is any need to, we can refine the approximation using a composite of smaller circles. Similarly, we can circumscribe targeted 3D objects by (a number of) spheres.

In this book we distinguish collision detection techniques using circles from those using vectors. However we do not cover collision prediction or collision physics, for which we recommend the specialised literature.

## 10.1 Collision detection using circles and spheres

CIRCLES AND SPHERES

Of course, using intersecting circles to detect collisions implies having their equations. For this reason, it is important to realise that a circle pictured on our screen is the locus of all points at equal distance or **radius** to a **centre**. In other words, we only need those two elements, the circle's radius $r$ and centre $M(m_1, m_2)$, to find its equation.

> ▷ in words: The **circle** $C(M,r)$ given its centre $M = (m_1, m_2)$ and radius $r$, is the locus of all points $P$ at equal distance $r$ to $M$.

> ▷ in symbols: Given $P(x,y)$ we define the circle by $d(P,M) = r$, which leads to its equation
> $$\sqrt{(x-m_1)^2 + (y-m_2)^2} = r,$$
> or after squaring both sides:
> $$(x-m_1)^2 + (y-m_2)^2 = r^2. \tag{10.1}$$

All points $(x,y)$ satisfying this equation are lying on the circle $C(M,r)$ and the other way around, all points of the circle will satisfy this equation. We hereby emphasise that a circle is not a function since most arguments $x$ produce two return values $y$.

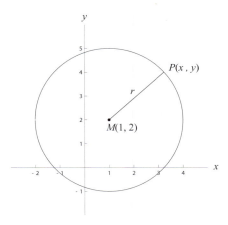

*Figure 10.1*: Circle $C((1,2),3)$

*Example*:  In figure 10.1 we draw the circle $C(M,3)$ with radius 3 around centre $M(1,2)$. Its equation is $(x-1)^2 + (y-2)^2 = 9$. Expanding brackets and grouping terms leads to its **implicit** shape $x^2 + y^2 - 2x - 4y - 4 = 0$.

Generalising this implicit result, we can write any circle equation as

$$x^2 + y^2 + ax + by + c = 0, \tag{10.2}$$

given $a,b,c \in \mathbb{R}$. Though not every equation of this shape necessarily defines a circle.

*Example*:   We study the implicit equation $x^2 + y^2 - 4x - 6y + 14 = 0$. Twice, we try to rewrite a Perfect Square (see formula (1.2)), step by step:

$$x^2 + y^2 - 4x - 6y + 14 = 0$$
$$\Leftrightarrow (x^2 - 4x) + (y^2 - 6y) = -14$$
$$\Leftrightarrow (x^2 - 2 \cdot 2x + \mathbf{2^2}) - \mathbf{2^2} + (y^2 - 2 \cdot 3y + \mathbf{3^2}) - \mathbf{3^2} = -14$$
$$\Leftrightarrow (x-2)^2 + (y-3)^2 - 2^2 - 3^2 = -14$$
$$\Leftrightarrow (x-2)^2 + (y-3)^2 = -1.$$

Theoretically the right hand side is the square of a radius, which has to be non-negative. Considering the left hand side, as it is a sum of two squares, it should be non-negative too. In conclusion, no single point $(x,y)$ can ever satisfy this equation: it corresponds to the empty set instead of to a circle.

Therefore we recall it is good practice to always check whether an implicit equation $x^2 + y^2 + ax + by + c = 0$ corresponds to a circle, before we start from it.

In 3D space we consequently use a sphere to detect collision between interacting screen objects, whenever such an approximation suits the context. The **sphere** $B(M,r)$ with centre $M = (m_1, m_2, m_3)$ and radius $r$ is the locus of all points $P = (x, y, z)$ at equal distance $d(P,M) = r$ to the centre of the sphere. This definition leads to the final equation of a three-dimensional sphere

$$(x - m_1)^2 + (y - m_2)^2 + (z - m_3)^2 = r^2, \qquad (10.3)$$

which is a straightforward extension of the equation of a plane circle.

*Example*: We consider the sphere with radius $r = 8$ around centre $M = (-5, 2, -3)$. Its equation appears to be $(x+5)^2 + (y-2)^2 + (z+3)^2 = 64$. We can rewrite it to its implicit shape $x^2 + y^2 + z^2 + 10x - 4y + 6z - 26 = 0$.

### Intersecting line and circle

Imagine we need to prevent a character from walking through a wall. In order to capture this situation in source code, we need to test for the collision between a circumscribing circle (the character) and a line (the wall). We can test a given line $t$ and circle $C(M,r)$ for collision, by comparing the distance between the centre of the circle and the line to the circle's radius. The line and the circle can have zero, one or two intersection points
(see figure 10.2). In the case of one intersection point, the line is tangent to the circle and therefore colliding. We also have a collision in the case of two intersection points.

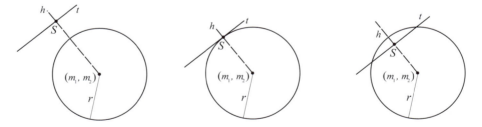

*Figure 10.2*: Intersecting a line $t$ and a circle $C(M,r)$

Let us design the above outlined test for collision between a straight line and a circle. We therefore determine the altitude $h$ through the circle centre $M$ perpendicular to the line $t$ given by its recipe $y = mx + c$.

The slopes of perpendicular lines relate as $m \cdot u = -1$ and evaluating $M(m_1, m_2) \in h$ yields an intercept $v$, completing the recipe for the altitude $h$ as $y = ux + v$. This enables us to find the altitude's foot $S$ on the given line $t$ as the intersection point of the straight lines $h$ and $t$ by solving their linear system

$$\begin{cases} y = mx + c \\ y = ux + v \end{cases}$$

for $x$ and $y$, which yields the cartesian coordinates of $S(s_1, s_2)$. Finally, comparing the distance $d(M, S)$ to the radius $r$ reveals or excludes a collision between the given line $t$ and the circle $C(M, r)$. Applying the distance formula (3.3) we need to compare $\sqrt{(m_1 - s_1)^2 + (m_2 - s_2)^2}$ to the radius $r$. As it is good practice to avoid square roots in computer code because of their computational overhead and precision issues, we square both sides of our collision test for implementation purposes.

We finalise the test for collision between a line and a circle as:

$$\begin{array}{lll} \text{not colliding} & (m_1 - s_1)^2 + (m_2 - s_2)^2 > r^2 & \\ \text{tangent line} & (m_1 - s_1)^2 + (m_2 - s_2)^2 = r^2 & (10.4) \\ \text{intersecting} & (m_1 - s_1)^2 + (m_2 - s_2)^2 < r^2 & \end{array}$$

*Example:* We want to test a wall on the line $y = 2x - 3$ and an object circumscribed by the circle $C\left((-4, 4), \sqrt{5}\right)$ for collision.

We determine the altitude $y = ux + v$ through the circle centre perpendicular to the given line. The slopes of both perpendicular lines relate as $2 \cdot u = -1$ which yields $u = -\frac{1}{2}$ for the altitude's slope. Evaluating $M(-4, 4) \in h$ means $4 = -\frac{1}{2}(-4) + v$, which yields the altitude's intercept as $v = 2$.

This enables us to find the altitude's foot $S$ solving the linear system

$$\begin{cases} y & = & 2x - 3 \\ y & = & -\frac{1}{2}x + 2 \end{cases} \Longleftrightarrow \begin{cases} 2x - y & = 3 \\ x + 2y & = 4 \end{cases} \Longleftrightarrow \begin{cases} x & = 2 \\ y & = 1 \end{cases}$$

for $x$ and $y$, which yields the intersection point $S(2, 1)$.

This allows us to evaluate the collision test based on $d(M, S)$:

$$\begin{array}{rcll} (m_1 - s_1)^2 + (m_2 - s_2)^2 & = & & \\ (-4 - s_1)^2 + (4 - s_2)^2 & = & & \\ (-4 - 2)^2 + (4 - 1)^2 & = & & \\ (-6)^2 + 3^2 & > & (\sqrt{5})^2 & \Longrightarrow \quad 45 > 5 \quad \text{leads to no collision.} \end{array}$$

INTERSECTING CIRCLES AND SPHERES

The collision test for circles is surprisingly easier than the previous collision test (for line and circle). Apart from the obvious
simplification a circumscribed circle or a sphere offers, their test
for collision is also beyond comparison the most efficient. Of
course, circumscribing targeted objects by a circle or sphere can
only approximate interactions on the screen, but they offer suffi-
cient 'quick and dirty' collision tests.

To test for collision between two objects, we need to check for an
intersection of their circumscribing circles. Two different circles
can have zero, one or two intersection points (see figure 10.3). In
the case of exactly one intersection point, both circles are tangent
and therefore colliding. In the case of two intersection points,
both circles are without any doubt colliding.

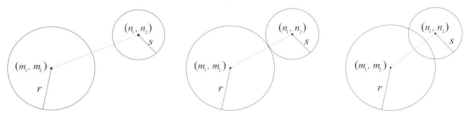

*Figure 10.3*: Intersecting two circles $C(M,r)$ and $C(N,s)$

Given the frame rate types (see page 216) and the collision prediction techniques (byeond
the scope of this book), we will not discuss the collision issues outlined in figure 10.4.
We choose to focus on the regular collisions as sketched in figure 10.3.

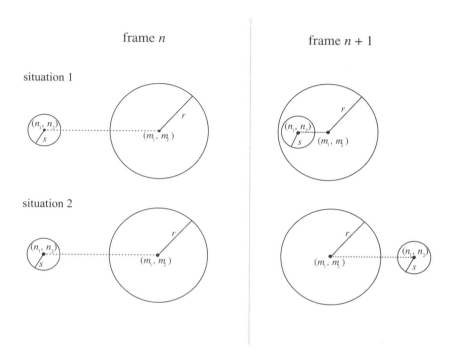

*Figure 10.4*: Collision detection issues with circles

Two circles do not intersect if the distance between their centres exceeds the sum of their radii. Two circles are tangent if the distance between their centres equals the sum of their radii. Two circles have two intersection points if the distance between their centres is smaller than the sum of their radii. Applying distance formula (3.3), we need to compare $\sqrt{(m_1 - n_1)^2 + (m_2 - n_2)^2}$ to the sum of the radii $r + s$. Square roots cause computational overhead and precision issues when processed. To overcome those drawbacks, we square both sides of our collision test before implementation. In other words, we finalise the test for collision between two circles as:

$$
\begin{aligned}
&\text{not colliding} && (m_1 - n_1)^2 + (m_2 - n_2)^2 > (r+s)^2 \\
&\text{tangent circles} && (m_1 - n_1)^2 + (m_2 - n_2)^2 = (r+s)^2 \qquad (10.5) \\
&\text{intersecting} && (m_1 - n_1)^2 + (m_2 - n_2)^2 < (r+s)^2
\end{aligned}
$$

In conclusion, we only need to implement the latter intersection test: whether the distance between the centres is smaller than or equal to the sum of the radii. Each time when this is ths case, a collision occurs.

*Example*: We want to test the circles $C((1,2),3)$ and $C\left((-1,1),\sqrt{2}\right)$ for collision. For a collision to occur, the distance between their centres ought to be smaller than or equal to the sum of their radii. For that reason, we need to calculate $(m_1 - n_1)^2 + (m_2 - n_2)^2 = (1+1)^2 + (2-1)^2 = 5$ and $(r+s)^2 = (3+\sqrt{2})^2 \approx 19,5$. Both circles are colliding because $(m_1 - n_1)^2 + (m_2 - n_2)^2 < (r+s)^2$.

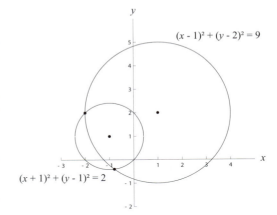

In case we would like to inspect their exact intersection points, we need the circle's equations

$$(x-1)^2 + (y-2)^2 = 9$$

and

$$(x+1)^2 + (y-1)^2 = 2.$$

Each intersection point $(x,y)$ ought to satisfy both equations

$$\begin{cases} x^2 + y^2 - 2x - 4y - 4 = 0, \\ x^2 + y^2 + 2x - 2y = 0. \end{cases}$$

To solve this nonlinear system for $x$ and $y$, we linearly combine the first equation with factor 1 and the second with factor $-1$ (in other words: we subtract the second equation from the first), which yields the linear equation $-4x - 2y - 4 = 0$, simplified to $y = -2x - 2$. Substituting $y = -2x - 2$ in one of both circles, for instance in $C((1,2),3)$, yields

$$(x-1)^2 + (-2x - 2 - 2)^2 = 9 \iff (x-1)^2 + (-2x - 4)^2 = 9$$
$$\iff 5x^2 + 14x + 8 = 0.$$

This is a quadratic equation in $x$, for which the positive discriminant leads to the solutions $x_1 = -2$ and $x_2 = \frac{-4}{5}$. Because each solution $(x,y)$ also needs to satisfy the former linear substitute $y = -2x - 2$, we calculate $y_1 = -2(-2) - 2 = 2$ and $y_2 = -2\left(\frac{-4}{5}\right) - 2 = \frac{-2}{5}$ respectively. We conclude that these circles intersect at the two points $(-2,2)$ and $\left(\frac{-4}{5}, \frac{-2}{5}\right)$. We realise that a negative discriminant would correspond to no intersection and a zero discriminant to exactly one solution (being the tangent point).

We can straightforwardly extend the collision test for circles with one dimension, leading to the similar collision test for spheres in 3D space. Again ignoring frame rate issues and collision prediction, we test two spheres: one centred on $M(m_1, m_2, m_3)$ having radius $r$ and one with centre $N(n_1, n_2, n_3)$ and radius $s$ for collision via

$$\begin{aligned}
\text{not colliding} \quad & (m_1 - n_1)^2 + (m_2 - n_2)^2 + (m_3 - n_3)^2 > (r+s)^2 \\
\text{tangent spheres} \quad & (m_1 - n_1)^2 + (m_2 - n_2)^2 + (m_3 - n_3)^2 = (r+s)^2 \\
\text{intersecting} \quad & (m_1 - n_1)^2 + (m_2 - n_2)^2 + (m_3 - n_3)^2 < (r+s)^2
\end{aligned}$$

## 10.2 Collision detection using vectors

We recall 3D objects are completely built up by polygons. We define a **polygon** as a convex cutout from a plane by segments $[P_0 P_1], [P_1 P_2], [P_2 P_3], \ldots, [P_{n-1} P_n]$ and $[P_n P_0]$, which connect a sequence of points $P_0, P_1, P_2, \ldots, P_n$, given $n \geqslant 3$.

*Figure 10.5*: Some convex polygons

Most render engines make use of triangular polygons to build up 3D objects. We will further outline how to test whether a sphere collides with a polygon, by the use of vectors. In this book we typeset a triangular polygon as $ABC$ and the plane where it resides as $v_A$.

### LOCATION OF A POINT WITH RESPECT TO OTHER POINTS

We define **collinear** points as points lying on the same straight line.

Given three collinear points $P, Q, R$, we find the point $Q$ lying in between the two remaining points $P$ and $R$ if

$$(\vec{v} \cdot \vec{p} - \vec{v} \cdot \vec{q})(\vec{v} \cdot \vec{r} - \vec{v} \cdot \vec{q}) \leqslant 0.$$

This test requires the dot products of the location vectors $\vec{p} = \overrightarrow{OP}$, $\vec{q} = \overrightarrow{OQ}$ and $\vec{r} = \overrightarrow{OR}$ with a freely chosen constant vector $\vec{v} \neq \vec{o}$. This test holds irrespective the position of the straight line $PR$ with respect to the origin $O$, and irrespective of the direction of the constant vector $\vec{v} \neq \vec{o}$.

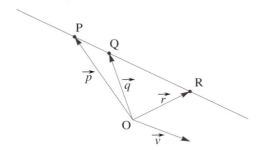

*Figure 10.6*: Mutual location of collinear points

*Proof*:  We assume the three points $P, Q$ and $R$ to be collinear. Consequently, the vectors $\overrightarrow{QP}$ and $\overrightarrow{QR}$ are parallel or antiparallel. In other words, the internal angle between the vectors $\overrightarrow{QP}$ and $\overrightarrow{QR}$ equals $0°$ or $180°$.

We rewrite the test $(\vec{v} \cdot \vec{p} - \vec{v} \cdot \vec{q})(\vec{v} \cdot \vec{r} - \vec{v} \cdot \vec{q}) \leqslant 0$ as

$$
\begin{aligned}
(\vec{v} \cdot \vec{p} - \vec{v} \cdot \vec{q})(\vec{v} \cdot \vec{r} - \vec{v} \cdot \vec{q}) &= (\vec{v} \cdot \overrightarrow{OP} - \vec{v} \cdot \overrightarrow{OQ})(\vec{v} \cdot \overrightarrow{OR} - \vec{v} \cdot \overrightarrow{OQ}) \\
&= \vec{v} \cdot (\overrightarrow{OP} - \overrightarrow{OQ}) \cdot \vec{v} \cdot (\overrightarrow{OR} - \overrightarrow{OQ}) \\
&= \vec{v}^2 (\overrightarrow{OP} - \overrightarrow{OQ}) \cdot (\overrightarrow{OR} - \overrightarrow{OQ}) \\
&= \vec{v}^2 (\overrightarrow{QP} \cdot \overrightarrow{QR}).
\end{aligned}
$$

In other words, $\vec{v}^2 (\overrightarrow{QP} \cdot \overrightarrow{QR}) \leqslant 0$ should hold. Since all squares are non-negative, $\overrightarrow{QP} \cdot \overrightarrow{QR}$ should be negative. Applying the dot product (7.9) yields $||\overrightarrow{QP}|| \, ||\overrightarrow{QR}|| \cos\theta \leqslant 0$. Consequently, $\cos\theta$ is negative. This can only be achieved by an internal angle between the vectors $\overrightarrow{QP}$ and $\overrightarrow{QR}$ equal to $180°$, which guarantees that the point $Q$ lies between the points $P$ and $R$. ∎

### Altitude to a straight line

We consider the straight line $r$ running through two points $A$ and $B$. For this line, we can suggest for instance $\overrightarrow{AB}$ as direction vector. The distance from a point $P$ to this straight line $r$ is the distance from the point $P$ to the foot of its altitude perpendicular on $r$.

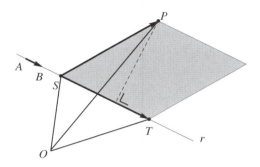

*Figure 10.7*: Distance from a point to a straight line

We therefore choose two points $S$ and $T$ on the straight line $r$ in such a way that the sense of their vector $\overrightarrow{ST}$ equals the sense of the direction vector $\overrightarrow{AB}$ and with length $||\overrightarrow{ST}|| = 1$. In other words, the free vector $\overrightarrow{ST}$ is a unit vector sharing the same sense with the direction vector $\overrightarrow{AB}$. Consequently, we can express this unit vector as $\overrightarrow{ST} = \hat{e}_{AB} = \frac{\overrightarrow{AB}}{||\overrightarrow{AB}||}$.

Figure 10.7 shows how the inclining vectors $\overrightarrow{ST}$ and $\overrightarrow{SP}$ span the grey parallelogram. The height $d$ of this parallelogram is exactly the distance we are aiming for. The area of a parallelogram equals its base times its height. We set its base as the length of the unit vector $\overrightarrow{ST}$. In other words,

$$\text{area}_{parallelogram} = ||\overrightarrow{ST}|| \cdot d = 1 \cdot d = d.$$

Moreover, we recall that the area of a parallelogram spanned by two vectors equals the length of their cross product (see page 140). In other words, our parallelogram area equals $||\overrightarrow{ST} \times \overrightarrow{SP}||$. Since we have already found that this area also equals the distance $d$, this leads us to

$$d = ||\overrightarrow{ST} \times \overrightarrow{SP}|| = \left|\left|\frac{\overrightarrow{AB}}{||\overrightarrow{AB}||} \times \overrightarrow{SP}\right|\right|.$$

We finally calculate the distance $d$ from the point $P$ to the line $r$, given $\overrightarrow{AB}$ as a direction vector for the line $r$ and an arbitrary point $S$ on the line $r$ by

$$d(P,r) = \frac{||\overrightarrow{AB} \times \overrightarrow{SP}||}{||\overrightarrow{AB}||}. \tag{10.6}$$

We mention that the arbitrary point $S$ can also be one of the given points $A$ or $B$.

*Example*:   We calculate the distance from the point $P(5, 3, -2)$ to the straight line $r$, which runs through the points $A(1, 0, 1)$ and $B(3, 5, 3)$. The $x$-, $y$- and $z$-axis are put in centimetres.

We need an arbitrary point on the straight line $r$, for instance the point $A$. We also need a direction vector of the straight line $r$, for instance $\overrightarrow{AB} = \begin{pmatrix} 2 \\ 5 \\ 2 \end{pmatrix}$. We then evaluate the above distance formula as

$$d(P, r) = \frac{||\overrightarrow{AB} \times \overrightarrow{AP}||}{||\overrightarrow{AB}||}.$$

In this formula we set $\overrightarrow{AP} = \begin{pmatrix} 5 \\ 3 \\ -2 \end{pmatrix} - \begin{pmatrix} 1 \\ 0 \\ 1 \end{pmatrix} = \begin{pmatrix} 4 \\ 3 \\ -3 \end{pmatrix}$ and as cross product we calculate

$$\overrightarrow{AB} \times \overrightarrow{AP} = \begin{pmatrix} 2 \\ 5 \\ 2 \end{pmatrix} \times \begin{pmatrix} 4 \\ 3 \\ -3 \end{pmatrix} = \begin{pmatrix} -21 \\ 14 \\ -14 \end{pmatrix}.$$ Consequently, this distance formula for point-to-line evaluates as

$$d(P, r) = \frac{\sqrt{(-21)^2 + 14^2 + (-14)^2}}{\sqrt{2^2 + 5^2 + 2^2}} \approx 5.02 \text{ cm.}$$

We mention obtaining the same result when choosing for the arbitrary point $S = B$ and the direction vector $\overrightarrow{AB}$. And of course, we could also have chosen $\overrightarrow{BA}$ for the direction vector.

## ALTITUDE TO A PLANE

Assume that we know the normal vector $\hat{n}$ on the plane $v_A$. This plane $v_A$ holds a given point $A$ in 3D space. We assume that $\hat{n}$ was normalised into a unit normal on its plane ($||\hat{n}|| = 1$). We adjust the sense of this unit normal $\hat{n}$ according to our 3D software. We typeset the distance from a point $P(p_1, p_2, p_3)$ to a plane $v_A$, given $A(a_1, a_2, a_3)$ simply as $d(P, v_A)$.

We set the vector $\vec{v} = \overrightarrow{AP}$ running from the reference point $A$ of the plane $v_A$ to the head $P$, from which we want to measure the distance of the point $P$ to the plane $v_A$. The components of this free vector $\vec{v} = \overrightarrow{AP}$ are $\vec{v} = \begin{pmatrix} p_1 - a_1 \\ p_2 - a_2 \\ p_3 - a_3 \end{pmatrix}$. We now dot product this free vector $\vec{v}$ with the unit normal $\hat{n}$ in order to calculate the distance from the point $P$ to

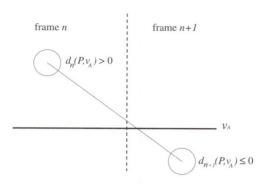

Figure 10.9: Collision detection failure

## LOCATION OF A POINT WITH RESPECT TO A POLYGON

For more accurate collision detection, we need to be sure whether objects really make contact or not. We will develop a test using vectors, for an object to hit a certain polygon $ABC$. We assume we have already measured the distance from a point $P$ to the polygon's plane $v_A$ (see page 214).

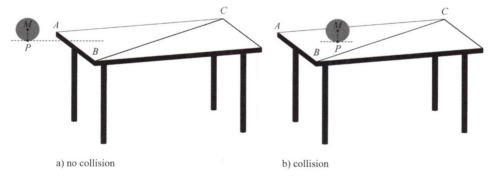

a) no collision                                        b) collision

Figure 10.10: Location of a point $P$ within a polygon $ABC$

When we for instance make use of a sphere, we assume we already know that the distance from its centre $M$ to the polygon's plane $v_A$ equals its radius. In figure 10.10 we witness that there is no collision in situation $a$, whereas situation $b$ shows collision. Despite the point $P$ being located in the polygon's plane $v_A$, it is not located inside the polygon $ABC$ in situation $a$. This confirms the need for determining where a point is located with respect to a polygon.

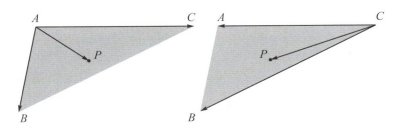

*Figure 10.11*: The point *P* is located inside of the polygon *ABC*

We outline such a location test which involves determining the sense of normal vectors on the polygon. In the case of three times equality of senses, then the point *P* is located inside the polygon *ABC*. In all other cases, the point *P* is located outside the polygon *ABC*.

Applying the right-hand rule or corkscrew rule, for the point *P* and each polygon edge, we test whether the sense of two normal vectors equals.

1) Is the sense of $\overrightarrow{AB} \times \overrightarrow{AC}$ equal to the sense of $\overrightarrow{AP} \times \mathbf{\overrightarrow{AC}}$?

2) Is the sense of $\overrightarrow{AC} \times \overrightarrow{AB}$ equal to the sense of $\overrightarrow{AP} \times \mathbf{\overrightarrow{AB}}$?

3) Is the sense of $\overrightarrow{CA} \times \overrightarrow{CB}$ equal to the sense of $\overrightarrow{CP} \times \mathbf{\overrightarrow{CB}}$?

Since we encounter an equal sense for the normal vectors three times, we conclude the point *P* is located inside the polygon *ABC*.

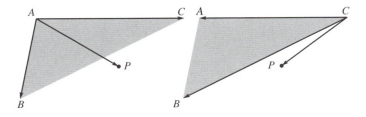

*Figure 10.12*: The point *P* is located outside of the polygon *ABC*

To make it even more clear, we illustrate in figure 10.12 the situation of a point *P* lying in the plane spanned by *A, B* and *C*, but being located outside of the polygon *ABC*. Initially the test seems to hold: $\overrightarrow{AB} \times \overrightarrow{AC}$ has a sense equal to the sense of $\overrightarrow{AP} \times \mathbf{\overrightarrow{AC}}$. But applying the right-hand rule or corkscrew rule, we find the normal vector $\overrightarrow{CA} \times \overrightarrow{CB}$ having a sense opposite to the sense of the normal vector $\overrightarrow{CP} \times \mathbf{\overrightarrow{CB}}$. For this reason, we conclude that the point *P* is located outside of the polygon *ABC*.

As a comment to the above test, we warn that it fails for points collinear with the vertices $A$ and $C$, for instance. Because in this case the cross product leads to $\overrightarrow{AP} \times \overrightarrow{AC} = \vec{o}$, and the null vector $\vec{o}$ is without direction.

Combining the above 'location-in-polygon' test with the 'point-line' distance formula, we obtain a kind of collision detection model like the one used in the 3D game 'DooM'. Even though 'DooM' (1993) may sound somewhat outdated, combining even the most elementary collision detection techniques already assures cool runtime effects. Astonishingly, usually the most basic collision detections offer the best runtime results. Given these facts, we kindly recommend that the professional game programmer starts to write or to collect a personal collision detection library, containing at least the methods outlined in this chapter.

## 10.3 Exercises

**Exercise 88**  Verify whether the equations below represent circles or spheres. In the case of a circle or a sphere, determine its centre and radius.

1)  $x^2 + y^2 + z^2 - 2x - 6y - 10z + 19 = 0$.

2)  $x^2 + y^2 + z^2 - 6x - 12y + 6z + 50 = 0$.

3)  $x^2 - 12x + y^2 - 4y + 40 = 0$.

**Exercise 89**  Determine the centre and radius of the circle containing the points $(0,0), (2,0)$ and $(0,4)$,

1)  by applying algebra.

2)  by a ruler-and-compass construction.

**Exercise 90**  We decided to make use of a circumscribed circle for collision detection in a car game. The car's centre is $(20,50)$ and the car's most extreme edge point has cartesian coordinates $(60,80)$. Find the circumscribed circle for this car by its cartesian equation.

**Exercise 91**  We decided to make use of circumscribed circles for collision detection of players in a 2D baseball game. In the actual frame one player is circumscribed by the circle $(x - 50)^2 + (y - 20)^2 = 900$ and the other player by $(x + 10)^2 + (y - 10)^2 = 400$. Do they collide in this frame?

**Exercise 92**  A car which is circumscribed by a circle with centre $(x, -x)$ and radius $\frac{1}{\sqrt{2}}$ moves during seven successive frames on the straight line $y = -x$, for each new frame incrementing $x$ with $\frac{1}{2}$. Does this car collide with the parked lorry circumscribed by the circle with radius $\sqrt{2}$ around centre $(2, -2)$, when it takes off in frame 0 from the initial point $(-3, 3)$? In the case of collision, calculate their intersection points.

**Exercise 93**  Calculate the distance from point $P(1, 0, 1)$ to the straight line $a$, which is running through the origin $O(0, 0, 0)$ in the direction of the vector $\vec{a} = (2, -1, 3)$.

**Exercise 94**  Test for collision between the squash ball $S((-1, 2, -2), 5)$ and the plane wall through the points $A(-2, 0, 1)$, $B(1, 5, 2)$ and $C(2, 7, 0)$.

**Exercise 95**  Prove that the point $Q(1, 0, 1)$ is lying on the line $PR$ in between its points $P(-1, 0, -1)$ and $R(3, 0, 3)$.

**Exercise 96** A landing strip $POQ$ at the stern of an aircraft carrier in a flight simulator is marked by the origin $O(0,0,0)$ and two remaining vertices $P(-5,0,0)$ and $Q(-3,0,-5)$. We now test two touchdown moves from a jet $V$.

1) Calculate the height above the landing platform of this jet in point $V(10,11,12)$.

2) Is the landing attempt successful at location $V(-1,0,\frac{-5}{2})$?

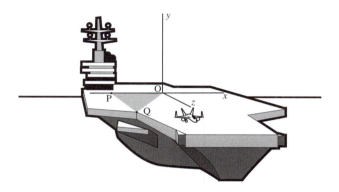

**Exercise 97** Use vectors to determine whether the centre $S(\frac{17}{3},5,\sqrt{15})$ of a snooker ball ended up in the triangle $ABC$ filled with numbered balls. The plane of the pool table is described by $y = 5$ and the vertices of the billiard triangle are having the cartesian coordinates $A(3,5,0)$, $B(0,5,4)$ and $C(6,5,4)$ respectively.

**Exercise 98** Detect between which frame numbers the projectile $P$ modelled by a variable point hits the wall modelled by the plane $v_A$ through the point $A(1,1,1)$ with unit normal $\hat{n} = (-\frac{4}{5},0,\frac{3}{5})$. During four successive frames, the projectile $P$ receives the following actual coordinate values:

1) $P(-2,1,0)$             3) $P(2,7,2)$

2) $P(0,4,1)$              4) $P(4,10,3)$

**Exercise 99** Detect between which frame numbers the tank shell modelled by the unit sphere $E((t_1,t_2,t_3),1)$ hits the enemy tank circumscribed by the sphere $B((v_1,v_2,v_3),14)$. During four successive frames, both centres $T$ and $V$ receive the following actual coordinate values:

1) $T(78,61,40)$ and $V(11,8,10)$      3) $T(-2,1,0)$ and $V(9,6,8)$

2) $T(38,31,20)$ and $V(10,7,9)$      4) $T(-42,-29,-20)$ and $V(9,6,8)$

# Chapter 11 · Matrices

We often make use of 'rectangles' of numbers, for instance when we declare nested *arrays* during (game) programming (see page 123). Such rectangular configurations of numbers or symbols are called matrices. Matrices are extremely useful as they are capable of just holding data, moving screen pixels, describing Bézier curves and much more. For this reason we study them here in great detail, in order to apply matrices optimally in our final chapters. For our convenience, we even turn to the matrix exponential function, bridging us to the chapters about transformation.

## 11.1 The concept of a matrix

The concept of a **matrix** in all its straightforward simplicity was introduced by the British mathematician **James Joseph Sylvester** (1814–1897). A matrix is a rectangle of numbers

$$A = \begin{pmatrix} a_{11} & a_{12} & \cdots & a_{1n} \\ a_{21} & a_{22} & \cdots & a_{2n} \\ \vdots & \vdots & \vdots & \vdots \\ a_{m1} & a_{m2} & \cdots & a_{mn} \end{pmatrix},$$

of which $a_{ij}$ are called its **matrix elements**. We typeset a matrix using an upper case Latin character and its indexed elements by the same lower case character.

Horizontally we count $m$ matrix **rows** and vertically, $n$ matrix **columns**. This way we find element $a_{ij}$ residing in the $i^{th}$ row and $j^{th}$ column.

$$j^{th}\text{column}$$

$$A = \begin{pmatrix} & \vdots & \\ \cdots & a_{ij} & \cdots \\ & \vdots & \end{pmatrix} i^{th}\text{row}$$

In other words, the first index $i$ denotes the row of the element $a_{ij}$ and the second index $j$ denotes its column. We typeset a matrix of $m$ rows and $n$ columns as $A_{m \times n}$ or for instance $A \in \mathbb{R}^{m \times n}$ in case its elements are real numbers. We call a matrix of this size an '$m$ by $n$' matrix.

If the number of rows is 1, in other words for $m = 1$, we have a **row matrix** or **row vector**.

$$B = \begin{pmatrix} 1 & 0 & 1 & 1 & 1 & 0 & 1 & 0 & 0 \end{pmatrix} \text{ is a } 1 \times 9 \text{ row vector.}$$

If the number of columns is 1, in other words for $n = 1$, we have a **column matrix** or **column vector**.

$$C = \begin{pmatrix} x \\ y \\ z \end{pmatrix} \text{ is a } 3 \times 1 \text{ column vector.}$$

We define a **zero matrix** as a matrix for which all elements equal zero. We typeset such a zero matrix or null matrix using the upper case character $O$ for varying matrix sizes $O_{m \times n}$.

If the number of rows equals the number of columns, in other words for $m = n$, we have a **square-sized** matrix.

$$D = \begin{pmatrix} 41 & 32 \\ 10 & 16 \end{pmatrix} \text{ is a square } 2 \times 2 \text{ matrix.}$$

We define the sequence of elements $e_{11}, e_{22}, \ldots, e_{nn}$ of a square matrix as its **main diagonal**. The main diagonal runs from the upper left to the lower right element. The antidiagonal runs from the lower left to the upper right element.

$$E = \begin{pmatrix} e_{11} & & & \\ & e_{22} & & \\ & & \ddots & \\ & & & e_{nn} \end{pmatrix}$$

We define an **identity matrix** as a square matrix for which all elements are zero, except for all main diagonal elements which are one. We typeset an identity matrix using its capital first character $I$ for varying square sizes $I_n$.

$$I_2 = \begin{pmatrix} 1 & 0 \\ 0 & 1 \end{pmatrix} \text{ is a } 2 \times 2 \text{ identity matrix.}$$

Two matrices $A$ and $B$ are **equal** when they are of the same size and all of their corresponding elements are equal.

## 11.2 Determinant of a square matrix

Each square matrix can produce a return value called the **determinant** of the matrix, which we typeset as $\det A$ or $|A|$. We calculate this scalar value based upon the elements of the square matrix. Its determinant determines whether the square matrix has an inverse or not.

The determinant of a $1 \times 1$ matrix, in other words one scalar, is simply equal to this scalar.

The determinant of a $2 \times 2$ matrix is

$$\begin{vmatrix} a_{11} & a_{12} \\ a_{21} & a_{22} \end{vmatrix} = a_{11}a_{22} - a_{21}a_{12}. \tag{11.1}$$

We can easily memorise this calculation for a determinant of a $2 \times 2$ matrix as its 'main diagonal minus antidiagonal'.

When we delete the row and column of a square matrix element $a_{ij}$ and calculate the determinant of its underlying submatrix, we define this return value as the **minor** of the element $a_{ij}$. We define the **cofactor** of the element $a_{ij}$ as the multiplication of its minor with $(-1)^{i+j}$, given $i$ the row and $j$ the column index of the element. Based upon these definitions, we develop the determinant of an $n \times n$ matrix as the sum of the products of the elements of a single arbitrary row (or column) with their respective cofactors.

The determinant of a $3 \times 3$ matrix is, developed to its first row,

$$\begin{vmatrix} a_{11} & a_{12} & a_{13} \\ a_{21} & a_{22} & a_{23} \\ a_{31} & a_{32} & a_{33} \end{vmatrix}$$

$$= (-1)^{1+1}a_{11} \begin{vmatrix} a_{22} & a_{23} \\ a_{32} & a_{33} \end{vmatrix} + (-1)^{1+2}a_{12} \begin{vmatrix} a_{21} & a_{23} \\ a_{31} & a_{33} \end{vmatrix}$$

$$+ (-1)^{1+3}a_{13} \begin{vmatrix} a_{21} & a_{22} \\ a_{31} & a_{32} \end{vmatrix}$$

$$= a_{11} \begin{vmatrix} a_{22} & a_{23} \\ a_{32} & a_{33} \end{vmatrix} - a_{12} \begin{vmatrix} a_{21} & a_{23} \\ a_{31} & a_{33} \end{vmatrix} + a_{13} \begin{vmatrix} a_{21} & a_{22} \\ a_{31} & a_{32} \end{vmatrix} \tag{11.2}$$

$$= a_{11}(a_{22}a_{33} - a_{32}a_{23}) - a_{12}(a_{21}a_{33} - a_{31}a_{23}) + a_{13}(a_{21}a_{32} - a_{31}a_{22}).$$

*Example 1*:

$$\begin{vmatrix} 3 & 2 & 1 \\ 7 & 4 & 2 \\ -2 & 0 & 5 \end{vmatrix} = 3 \begin{vmatrix} 4 & 2 \\ 0 & 5 \end{vmatrix} - 2 \begin{vmatrix} 7 & 2 \\ -2 & 5 \end{vmatrix} + 1 \begin{vmatrix} 7 & 4 \\ -2 & 0 \end{vmatrix}$$

$$= 3(4 \cdot 5 - 0 \cdot 2) - 2(7 \cdot 5 - (-2) \cdot 2) + 1(7 \cdot 0 - (-2) \cdot 4)$$

$$= -10$$

*Example 2:* We can even apply the above formula to easily memorise the component formula of the cross product of two vectors (see formula (7.12)). We rewrite this formula as the determinant of a $3 \times 3$ matrix, given $\vec{e}_1, \vec{e}_2$ and $\vec{e}_3$ the unit vectors spanning the $x$-, $y$- and $z$-axis respectively.

$$\begin{pmatrix} a_1 \\ a_2 \\ a_3 \end{pmatrix} \times \begin{pmatrix} b_1 \\ b_2 \\ b_3 \end{pmatrix} = \begin{vmatrix} \vec{e}_1 & \vec{e}_2 & \vec{e}_3 \\ a_1 & a_2 & a_3 \\ b_1 & b_2 & b_3 \end{vmatrix}$$

$$= (a_2 b_3 - b_2 a_3)\vec{e}_1 + (-a_1 b_3 + b_1 a_3)\vec{e}_2 \\ + (a_1 b_2 - b_1 a_2)\vec{e}_3$$

$$= \begin{pmatrix} a_2 b_3 - b_2 a_3 \\ -a_1 b_3 + b_1 a_3 \\ a_1 b_2 - b_1 a_2 \end{pmatrix} \tag{11.3}$$

We realise we may only use the above determinant to memorise the cross product formula (7.12). We are aware that the components of the first vector $\vec{a}$ strictly belong to the second determinant row and those of the second vector $\vec{b}$ to the last determinant row, since the cross product is not commutative.

## 11.3 Addition and scalar multiplication of matrices

### ADDITION OF MATRICES

Adding matrices is as straightforward as it is important. We can only add or subtract two matrices $A$ and $B$ of the same size. As a result, their **matrix sum** $A + B$ or **difference** $A - B$ inherits this size.

▷  in words: We add two matrices by adding their corresponding matrix elements.

▷  in symbols: For $A, B \in \mathbb{R}^{m \times n}$ we calculate their sum $A + B \in \mathbb{R}^{m \times n}$ as

$$\begin{pmatrix} a_{11} & a_{12} & \cdots & a_{1n} \\ a_{21} & a_{22} & \cdots & a_{2n} \\ \vdots & \vdots & \vdots & \vdots \\ a_{m1} & a_{m2} & \cdots & a_{mn} \end{pmatrix} + \begin{pmatrix} b_{11} & b_{12} & \cdots & b_{1n} \\ b_{21} & b_{22} & \cdots & b_{2n} \\ \vdots & \vdots & \vdots & \vdots \\ b_{m1} & b_{m2} & \cdots & b_{mn} \end{pmatrix}$$

$$= \begin{pmatrix} a_{11}+b_{11} & a_{12}+b_{12} & \cdots & a_{1n}+b_{1n} \\ a_{21}+b_{21} & a_{22}+b_{22} & \cdots & a_{2n}+b_{2n} \\ \vdots & \vdots & \vdots & \vdots \\ a_{m1}+b_{m1} & a_{m2}+b_{m2} & \cdots & a_{mn}+b_{mn} \end{pmatrix}.$$

We define the difference $A - B$ of two matrices $A$ and $B$ as the sum of $A$ with the opposite matrix of $B$. We subtract two matrices by subtracting their corresponding matrix elements.

*Example*: $A = \begin{pmatrix} -1 & 5 & \sqrt{2} \\ 4 & -7 & \sqrt{3} \end{pmatrix}$ $\qquad$ $B = \begin{pmatrix} 3 & 2 & -1 \\ 0 & -1 & -2 \end{pmatrix}$

Since $A \in \mathbb{R}^{2\times3}$ and $B \in \mathbb{R}^{2\times3}$ are of the same size, we can calculate $A+B$ and $A-B$.

$$A+B = \begin{pmatrix} -1 & 5 & \sqrt{2} \\ 4 & -7 & \sqrt{3} \end{pmatrix} + \begin{pmatrix} 3 & 2 & -1 \\ 0 & -1 & -2 \end{pmatrix}$$

$$= \begin{pmatrix} -1+3 & 5+2 & \sqrt{2}+(-1) \\ 4+0 & -7+(-1) & \sqrt{3}+(-2) \end{pmatrix}$$

$$= \begin{pmatrix} 2 & 7 & \sqrt{2}-1 \\ 4 & -8 & \sqrt{3}-2 \end{pmatrix}$$

$$A-B = \begin{pmatrix} -1 & 5 & \sqrt{2} \\ 4 & -7 & \sqrt{3} \end{pmatrix} - \begin{pmatrix} 3 & 2 & -1 \\ 0 & -1 & -2 \end{pmatrix}$$

$$= \begin{pmatrix} -1-3 & 5-2 & \sqrt{2}-(-1) \\ 4-0 & -7-(-1) & \sqrt{3}-(-2) \end{pmatrix}$$

$$= \begin{pmatrix} -4 & 3 & \sqrt{2}+1 \\ 4 & -6 & \sqrt{3}+2 \end{pmatrix}$$

*Additive group*: Given $A$, $B$ and $C \in \mathbb{R}^{m\times n}$ matrices of the same size, we summarise all properties of $+$ in $\mathbb{R}^{m\times n}$:

| | |
|---|---|
| closure | $A+B \in \mathbb{R}^{m\times n}$, |
| associative property | $A+(B+C) = (A+B)+C$, |
| zero matrix | $A+O_{m\times n} = A = O_{m\times n}+A$, |
| opposite matrix | $A+(-A) = O_{m\times n} = -A+A$, |
| commutative property | $A+B = B+A$. |

We conclude $(\mathbb{R}^{m \times n}, +)$ is an additive **group** with the zero matrix as the neutral element and featuring commutativity.

## SCALAR MULTIPLICATION OF A MATRIX

We define the **scalar multiplication** of a matrix as a real number times the matrix.

▷ in words: We scalar multiply a matrix with a real number $\lambda$ by multiplying all matrix elements with $\lambda$.

▷ in symbols: We multiply the matrix $A \in \mathbb{R}^{m \times n}$ with a number $\lambda \in \mathbb{R}$ as

$$\lambda \begin{pmatrix} a_{11} & a_{12} & \cdots & a_{1n} \\ a_{21} & a_{22} & \cdots & a_{2n} \\ \vdots & \vdots & \vdots & \vdots \\ a_{m1} & a_{m2} & \cdots & a_{mn} \end{pmatrix} = \begin{pmatrix} \lambda a_{11} & \lambda a_{12} & \cdots & \lambda a_{1n} \\ \lambda a_{21} & \lambda a_{22} & \cdots & \lambda a_{2n} \\ \vdots & \vdots & \vdots & \vdots \\ \lambda a_{m1} & \lambda a_{m2} & \cdots & \lambda a_{mn} \end{pmatrix}.$$

*Example 1*: If $A = \begin{pmatrix} -1 & 5 & \sqrt{2} \\ 4 & -7 & \sqrt{3} \end{pmatrix}$, we can for instance calculate its scalar multiple

$$-2A = -2 \begin{pmatrix} -1 & 5 & \sqrt{2} \\ 4 & -7 & \sqrt{3} \end{pmatrix} = \begin{pmatrix} -2 \cdot (-1) & -2 \cdot 5 & -2 \cdot \sqrt{2} \\ -2 \cdot 4 & -2 \cdot (-7) & -2 \cdot \sqrt{3} \end{pmatrix}$$

$$= \begin{pmatrix} 2 & -10 & -2\sqrt{2} \\ -8 & 14 & -2\sqrt{3} \end{pmatrix}.$$

*Example 2*: For $A = \begin{pmatrix} -1 & 5 & \sqrt{2} \\ 4 & -7 & \sqrt{3} \end{pmatrix}$, we define its **opposite matrix** $-A$ as the scalar multiple

$$(-1)A = (-1) \begin{pmatrix} -1 & 5 & \sqrt{2} \\ 4 & -7 & \sqrt{3} \end{pmatrix} = \begin{pmatrix} (-1) \cdot (-1) & (-1) \cdot 5 & (-1) \cdot \sqrt{2} \\ (-1) \cdot 4 & (-1) \cdot (-7) & (-1) \cdot \sqrt{3} \end{pmatrix}$$

$$= \begin{pmatrix} 1 & -5 & -\sqrt{2} \\ -4 & 7 & -\sqrt{3} \end{pmatrix}.$$

*Properties*: Given $\lambda, \mu \in \mathbb{R}$ and $A, B \in \mathbb{R}^{m \times n}$, we summarise their arithmetic:

$$\begin{array}{rcl} \text{mixed distributive law} \quad \lambda(A + B) & = & \lambda A + \lambda B, \\ (\lambda + \mu)A & = & \lambda A + \mu A, \\ \text{mixed associative law} \quad (\lambda \cdot \mu)A & = & \lambda(\mu A). \end{array}$$

## 11.4 Transpose of a matrix

To transpose a matrix is to tilt its rectangular shape, which we define in detail below.

▷ in words: We define the **transpose** of an $m \times n$ matrix $A$ as the $n \times m$ matrix obtained by rewriting all rows of $A$ into columns, and consequently the columns of $A$ into rows. We typeset this tilted version of the matrix $A$ as $A^T$, realising this is just a notation and not a matrix power. In other words, the first row of $A$ corresponds to the first column of $A^T$, the second row of $A$ corresponds to the second column of $A^T$, ....

▷ in symbols: We tilt a matrix $A \in \mathbb{R}^{m \times n}$ into its transposed matrix $A^T$ as

$$\begin{pmatrix} a_{11} & a_{12} & \cdots & a_{1n} \\ a_{21} & a_{22} & \cdots & a_{2n} \\ \vdots & \vdots & \vdots & \vdots \\ a_{m1} & a_{m2} & \cdots & a_{mn} \end{pmatrix}^T = \begin{pmatrix} a_{11} & a_{21} & \cdots & a_{m1} \\ a_{12} & a_{22} & \cdots & a_{m2} \\ \vdots & \vdots & \vdots & \vdots \\ a_{1n} & a_{2n} & \cdots & a_{mn} \end{pmatrix} \in \mathbb{R}^{n \times m}.$$

*Example*: Given a matrix $A = \begin{pmatrix} -1 & 5 & \sqrt{2} \\ 4 & -7 & \sqrt{3} \end{pmatrix} \in \mathbb{R}^{2 \times 3}$, we transpose it as

$$A^T = \begin{pmatrix} -1 & 4 \\ 5 & -7 \\ \sqrt{2} & \sqrt{3} \end{pmatrix} \in \mathbb{R}^{3 \times 2}.$$

*Properties*: Given $A, B \in \mathbb{R}^{m \times n}$, we mention

$$\begin{aligned} (A^T)^T &= A, \\ (A + B)^T &= A^T + B^T. \end{aligned}$$

We define a **symmetric matrix** as a square matrix $A$ for which $A^T = A$.

## 11.5 Dot product of matrices

INTRODUCTION

A computer store sells three different models of laptops: model 1, model 2 and model 3. A matrix $L$ is holding a double order, meant for two different destinations, of a major customer.

$$L = \begin{pmatrix} 60 & 10 & 0 \\ 100 & 50 & 150 \end{pmatrix} \begin{matrix} \leftarrow 1^{st} \text{ order} \\ \leftarrow 2^{nd} \text{ order} \end{matrix}$$

$$\begin{matrix} \uparrow & \uparrow & \uparrow \\ \text{model 1} & \text{model 2} & \text{model 3} \end{matrix}$$

A column matrix $S$ is meanwhile holding all model sale prices in euro.

$$S = \begin{pmatrix} 900 \\ 1200 \\ 1500 \end{pmatrix} \begin{matrix} \leftarrow \text{model 1} \\ \leftarrow \text{model 2} \\ \leftarrow \text{model 3} \end{matrix}$$

For his first order, the customer pays the full amount of

$$60 \cdot 900 \quad + \quad 10 \cdot 1200 \quad + \quad 0 \cdot 1500 \quad = 66000$$

and for his second order

$$100 \cdot 900 \quad + \quad 50 \cdot 1200 \quad + \quad 150 \cdot 1500 \quad = 375000.$$

Putting both subtotals in a column matrix

$$P = \begin{pmatrix} 66000 \\ 375000 \end{pmatrix},$$

we can interpret $P$ as a product of the matrices $L$ and $S$. We realise how the elements of $P$ were calculated: for the first order

$$\begin{pmatrix} 60 & 10 & 0 \\ 100 & 50 & 150 \end{pmatrix} \cdot \begin{pmatrix} 900 \\ 1200 \\ 1500 \end{pmatrix} = \begin{pmatrix} 60 \cdot 900 + 10 \cdot 1200 + 0 \cdot 1500 \\ \cdots \end{pmatrix}$$

$$= \begin{pmatrix} 66000 \\ \cdots \end{pmatrix}$$

and for the second order

$$\begin{pmatrix} 60 & 10 & 0 \\ 100 & 50 & 150 \end{pmatrix} \cdot \begin{pmatrix} 900 \\ 1200 \\ 1500 \end{pmatrix} = \begin{pmatrix} \cdots \\ 100 \cdot 900 + 50 \cdot 1200 + 150 \cdot 1500 \end{pmatrix}$$

$$= \begin{pmatrix} \cdots \\ 375000 \end{pmatrix}.$$

We obtain the element on the first row of $P$ by multiplying the elements of the first row of $L$ with the corresponding elements of $S$ and then adding all these products. We realise how multiplying a **2** $\times$ **3** matrix with a **3** $\times$ **1** matrix results into a **2** $\times$ **1** product matrix.

## CONDITION

We multiply two matrices by repeatedly multiplying a row of the first matrix with a column of the second matrix.

$$\begin{pmatrix} 1 & 2 \\ 3 & 4 \\ 5 & 6 \end{pmatrix} \cdot \begin{pmatrix} 1 & 2 & 3 \\ 4 & 5 & 6 \end{pmatrix} = \begin{pmatrix} 9 & 12 & 15 \\ 19 & 26 & 33 \\ 29 & 40 & 51 \end{pmatrix}$$

For instance, the element 33 in the product matrix residing on the second row and third column is the sum of the products of the elements of the second row of the first matrix with the corresponding elements of the third column of the second matrix: $3 \cdot 3 + 4 \cdot 6 = 33$.

Consequently, in order to multiply two matrices, the number of columns of the first matrix needs to equal the number of rows of the second matrix. The multiplication of a $3 \times 2$ matrix with a compatible $2 \times 5$ matrix results in a $3 \times 5$ product matrix. We can highlight this condition to easily memorise it by matrix sizes: $3 \times$ 2 and 2 $\times 5$ yields $3 \times 5$.

## DEFINITION

We can multiply two matrices $A$ and $B$ *in this order* if the number of columns of $A$ equals the number of rows of $B$. The number of rows of their product matrix $A \cdot B$ corresponds to the number of rows of $A$; the number of columns of $A \cdot B$ corresponds to the number of columns of $B$.

We typeset the **dot** or **matrix product** of $A \in \mathbb{R}^{m \times n}$ and $B \in \mathbb{R}^{n \times p}$ *in this order* as $A \cdot B = C$. Consequently their **product matrix** $C$ will belong to $\mathbb{R}^{m \times p}$. We hereby define this dot product for matrices

▷ in words: Product matrix $A \cdot B$ holds at its $i^{th}$ *row and its* $j^{th}$ *column* the result of multiplying all elements of the $i^{th}$ *row of matrix A* with the corresponding elements of the $j^{th}$ *column of matrix B* and then adding these products.

▷ in symbols:

$$\begin{pmatrix} a_{i1} & a_{i2} & a_{i3} & \cdots & a_{in} \end{pmatrix} \cdot \begin{pmatrix} b_{1j} \\ b_{2j} \\ b_{3j} \\ \vdots \\ b_{nj} \end{pmatrix} = \begin{pmatrix} & \vdots & \\ \cdots & c_{ij} & \cdots \\ & \vdots & \end{pmatrix}$$

expressing

$$c_{ij} = a_{i1}b_{1j} + a_{i2}b_{2j} + a_{i3}b_{3j} + \cdots + a_{in}b_{nj} = \sum_{k=1}^{n} a_{ik}b_{kj}$$

for each $1 \leqslant i \leqslant m$ and $1 \leqslant j \leqslant p$.

*Example*: We declare two matrices

$$A = \begin{pmatrix} -1 & 5 & \sqrt{2} \\ 4 & -7 & \sqrt{3} \end{pmatrix} \in \mathbb{R}^{2\times 3} \text{ and } B = \begin{pmatrix} 2 & 8 & 12 \\ -3 & -1 & 5 \\ 0 & 4 & 10 \end{pmatrix} \in \mathbb{R}^{3\times 3}.$$

We are able to calculate their dot product $A \cdot B$ because the matrix sizes $2 \times$ 3 and 3 $\times 3$ are compatible and the size of the product matrix will be $2 \times 3$.

$$\begin{pmatrix} -1 & 5 & \sqrt{2} \\ 4 & -7 & \sqrt{3} \end{pmatrix} \cdot \begin{pmatrix} 2 & 8 & 12 \\ -3 & -1 & 5 \\ 0 & 4 & 10 \end{pmatrix}$$

$$= \begin{pmatrix} -1\cdot 2+5\cdot(-3)+\sqrt{2}\cdot 0 & -1\cdot 8+5\cdot(-1)+\sqrt{2}\cdot 4 & -1\cdot 12+5\cdot 5+\sqrt{2}\cdot 10 \\ 4\cdot 2-7\cdot(-3)+\sqrt{3}\cdot 0 & 4\cdot 8-7\cdot(-1)+\sqrt{3}\cdot 4 & 4\cdot 12-7\cdot 5+\sqrt{3}\cdot 10 \end{pmatrix}$$

$$= \begin{pmatrix} -17 & -13+4\sqrt{2} & 13+10\sqrt{2} \\ 29 & 39+4\sqrt{3} & 13+10\sqrt{3} \end{pmatrix}$$

It is impossible to calculate the dot product $B \cdot A$ as the number of columns of $B$ does not equal the number of rows of $A$. Their matrix sizes $3 \times$ 3 and 2 $\times 3$ are incompatible for a dot product.

## PROPERTIES

1) A product matrix $A \cdot B$ is in general not equal to the product matrix $B \cdot A$. We have already experienced that one of them can even not exist. In other words, for the dot product of matrices we have **no** *commutativity*. As a consequence, respecting the order of the matrices is of major importance to multiply them.

2) Assuming the matrix sizes of $A, B$ and $C$ allow for all their dot products, we have:

   associative property     $(A \cdot B) \cdot C = A \cdot (B \cdot C)$,

   right distributive law     $(A + B) \cdot C = A \cdot C + B \cdot C$,

   left distributive law     $C \cdot (A + B) = C \cdot A + C \cdot B$.

3) Zero divisors

   Given $A = \begin{pmatrix} 1 & -2 \\ -1 & 2 \end{pmatrix}$ and $B = \begin{pmatrix} 2 & -10 \\ 1 & -5 \end{pmatrix}$, their dot product $A \cdot B$ is

$$A \cdot B = \begin{pmatrix} 1 & -2 \\ -1 & 2 \end{pmatrix} \cdot \begin{pmatrix} 2 & -10 \\ 1 & -5 \end{pmatrix} = \begin{pmatrix} 0 & 0 \\ 0 & 0 \end{pmatrix}.$$

We have to conclude that the dot product of two matrices can result in the zero matrix, without any of the initial matrices being the zero matrix. We define a **zero divisor** as a matrix $A \neq O$ for which $A \cdot B = O$ or $B \cdot A = O$, for some matrix $B$ different from the zero matrix.

While for scalar numbers $a, b \in \mathbb{R}$ we can deduct from their product $a \cdot b = 0$ either $a = 0$ (and)or $b = 0$, it does not apply at all to matrices. A dot product $A \cdot B = O_{m \times n}$ offers no conclusion about $A$ (and)or $B$ being zero matrices.

4) We can only apply a **matrix power** to square matrices.

$$\begin{pmatrix} 1 & -2 & 3 \\ -1 & 8 & 6 \\ 0 & -5 & 2 \end{pmatrix}^2 = \begin{pmatrix} 1 & -2 & 3 \\ -1 & 8 & 6 \\ 0 & -5 & 2 \end{pmatrix} \cdot \begin{pmatrix} 1 & -2 & 3 \\ -1 & 8 & 6 \\ 0 & -5 & 2 \end{pmatrix}$$
$$= \begin{pmatrix} 3 & -33 & -3 \\ -9 & 36 & 57 \\ 5 & -50 & -26 \end{pmatrix}$$

5) An identity matrix $I_n$ is the matrix equivalent of the number $1 \in \mathbb{R}$, which means $I_n \cdot A = A \cdot I_n = A$, for any $n \times n$ matrix $A$.

$$\begin{pmatrix} 1 & 0 & 0 \\ 0 & 1 & 0 \\ 0 & 0 & 1 \end{pmatrix} \cdot \begin{pmatrix} 1 & -2 & 3 \\ -1 & 8 & 6 \\ 0 & -5 & 2 \end{pmatrix} = \begin{pmatrix} 1 & -2 & 3 \\ -1 & 8 & 6 \\ 0 & -5 & 2 \end{pmatrix} \cdot \begin{pmatrix} 1 & 0 & 0 \\ 0 & 1 & 0 \\ 0 & 0 & 1 \end{pmatrix}$$
$$= \begin{pmatrix} 1 & -2 & 3 \\ -1 & 8 & 6 \\ 0 & -5 & 2 \end{pmatrix}$$

6) The transpose of a matrix product reverses its order:

$$(S \cdot B)^T = B^T \cdot S^T.$$

We can easily memorise this effect as the 'Socks-and-Boots' rule. In the morning we first put on our Socks, then we put on our Boots. We 'Transpose' this in the evening by first Taking off our Boots, then Taking off our Socks.

7) We have the mixed associative property for scalar multiplication and the matrix product. For a scalar $\lambda \in \mathbb{R}$ and assuming the matrix sizes of $A$ and $B$ allow for their dot product, we express this property as

$$\lambda (A \cdot B) = \lambda A \cdot B = A \cdot (\lambda B).$$

## 11.6 Inverse of a matrix

### INTRODUCTION

For instance, we consider the inverse of 0.6 to be 1.666... because their product

$$0.6 \times 1.666... = 1.$$

We say $b$ is the inverse number of $a$ if and only if $a \cdot b = 1$, and we typeset $b = \frac{1}{a} = a^{-1}$.

Nextly, we consider two $3 \times 3$ matrices

$$A = \begin{pmatrix} 1 & 5 & 2 \\ 1 & 1 & 7 \\ 0 & -3 & 4 \end{pmatrix} \qquad \text{and} \qquad B = \begin{pmatrix} -25 & 26 & -33 \\ 4 & -4 & 5 \\ 3 & -3 & 4 \end{pmatrix}.$$

At a first glance we do not spot a special relation between the matrix $A$ and the matrix $B$, until we look at their dot product $A \cdot B$

$$A \cdot B = \begin{pmatrix} 1 & 0 & 0 \\ 0 & 1 & 0 \\ 0 & 0 & 1 \end{pmatrix} = I_3.$$

Similarly to numbers, we typeset $B = A^{-1}$, calling it the inverse matrix of the matrix $A$.

### DEFINITION

Consider an $n \times n$ matrix $A$. We define the unique $n \times n$ matrix $B$ for which

$$A \cdot B = B \cdot A = I_n$$

as the **inverse matrix** of the matrix $A$, typeset as $A^{-1}$ if this matrix $B$ exists. In this respect, we call the matrix $A$ **invertible**. In other words, if the matrix $A \in \mathbb{R}^{n \times n}$ is invertible, then the unique matrix $A^{-1}$ exists for which

$$A \cdot A^{-1} = A^{-1} \cdot A = I_n.$$

CONDITIONS

▷ The inverse matrix can only exist for square matrices.

▷ Many square matrices do not have an inverse matrix. (In real numbers only zero does not have an inverse number. Instead it leads to the concept of *infinity* which is not a number.)

► In case the determinant of an $n \times n$ matrix $A$ equals zero, there is no inverse matrix of $A$. In this case we call the matrix $A$ **singular**.

► In case the determinant of an $n \times n$ matrix $A$ is not zero, there is a unique inverse matrix of $A$. In this case we call the matrix $A$ **invertible**.

▷ For an invertible square matrix $A$, its inverse matrix is unique. In other words, there is exactly one matrix $A^{-1}$ for which $A \cdot A^{-1} = A^{-1} \cdot A = I$.

▷ For real numbers, we equivalently typeset $a^{-1} = \frac{1}{a}$. The former notation is used but the latter is invalid for matrices. We should not write divisions of matrices; putting multiplications with inverse matrices is much better.

▷ $\left(A^{-1}\right)^{-1} = A$, given $A^{-1}$ exists.

ROW REDUCTION

Three elementary row operations can be used on matrices (see also page 41). We perform these row operations for matrix inversion, as the process of finding the inverse of a square matrix.

1) Swap the positions of two rows.

This row operation simply reorders the rows of a given matrix. Take for instance the matrix

$$\begin{pmatrix} 0 & 1 & 0 \\ 1 & 0 & 0 \\ 0 & 0 & 1 \end{pmatrix}.$$

Swap its first two rows in order to acquire the identity matrix

$$\begin{pmatrix} 0 & 1 & 0 \\ 1 & 0 & 0 \\ 0 & 0 & 1 \end{pmatrix} \overset{R_1 \leftrightarrow R_2}{\sim} \begin{pmatrix} 1 & 0 & 0 \\ 0 & 1 & 0 \\ 0 & 0 & 1 \end{pmatrix}.$$

2) Multiply a row by a nonzero scalar $\lambda \neq 0$.

We sometimes need or like to multiply each element of a certain matrix row by a nonzero number.

For instance, we multiply the first row of this matrix

$$\begin{pmatrix} \frac{3}{2} & 0 & 0 \\ 0 & 1 & 0 \\ 0 & 0 & 1 \end{pmatrix}$$

by $\frac{2}{3}$ in order to acquire the identity matrix.

$$\begin{pmatrix} \frac{3}{2} & 0 & 0 \\ 0 & 1 & 0 \\ 0 & 0 & 1 \end{pmatrix} \underset{\sim}{R_1 \to \frac{2}{3} R_1} \begin{pmatrix} 1 & 0 & 0 \\ 0 & 1 & 0 \\ 0 & 0 & 1 \end{pmatrix}$$

3) Adding two rows.

This row operation adds a certain row to another row, which we overwrite by the result. Let us for instance consider this matrix

$$\begin{pmatrix} 1 & -1 & 0 \\ 0 & 1 & 0 \\ 0 & 0 & 1 \end{pmatrix}.$$

We add the second row to the first row, overwriting the first row by their sum, in order to acquire the identity matrix once more.

$$\begin{pmatrix} 1 & -1 & 0 \\ 0 & 1 & 0 \\ 0 & 0 & 1 \end{pmatrix} \underset{\sim}{R_1 \to R_1 + R_2} \begin{pmatrix} 1 & 0 & 0 \\ 0 & 1 & 0 \\ 0 & 0 & 1 \end{pmatrix}$$

Matrix inversion involves applying a strategic sequence of those elementary row operations.

## MATRIX INVERSION

There are different ways to calculate the inverse matrix of a given square matrix. In case of an invertible square matrix, we can for instance choose to calculate its inverse matrix by gaussian elimination.

1) We augment the given square matrix $A \in \mathbb{R}^{n \times n}$ to the right by the identity matrix $I_n$, which results in an $n \times 2n$ **block matrix** $(A|I_n)$ .

2) We then perform elementary row operations to reduce the block matrix $(A|I_n)$ until the initial matrix $A$ in its left block turns into the identity matrix.

3) When this identity matrix can be achieved in the left block, the matrix $A$ is invertible and its inverse matrix appears in the right block of the augmented matrix. We have reduced the initial augmented matrix $(A|I_n)$ to the final block matrix $(I_n|A^{-1})$.

4)  Alternatively, a row full of zeros appearing in the left block means the matrix $A$ is not invertible but singular.

*Example*:  We calculate the inverse of this matrix

$$A = \begin{pmatrix} 1 & 5 & 2 \\ 1 & 1 & 7 \\ 0 & -3 & 4 \end{pmatrix}.$$

Its inverse matrix $A^{-1}$ exists, because of

$$\det A = 1(1 \cdot 4 - (-3) \cdot 7) - 5(1 \cdot 4 - 0 \cdot 7) + 2(1 \cdot (-3) - 0 \cdot 1) = -1.$$

We now perform row operations on the augmented matrix $(A|I_3)$ until its left block displays the identity matrix $I_3$. As there is a multitude of routes to achieve this, we like to demonstrate two of them. We kick off with outlining a first approach along three stages.

Step 1: we aim for zeros in the column of the first element on the first row.

$$\begin{pmatrix} 1 & 5 & 2 & | & 1 & 0 & 0 \\ 1 & 1 & 7 & | & 0 & 1 & 0 \\ 0 & -3 & 4 & | & 0 & 0 & 1 \end{pmatrix}$$

$$\sim \begin{pmatrix} 1 & 5 & 2 & | & 1 & 0 & 0 \\ 0 & -4 & 5 & | & -1 & 1 & 0 \\ 0 & -3 & 4 & | & 0 & 0 & 1 \end{pmatrix} \qquad (R_2 \rightarrow R_2 - R_1)$$

Step 2: we tune the element on the second row, second column, to 1 and turn the remaining elements of that column to zeros.

$$\sim \begin{pmatrix} 1 & 5 & 2 & | & 1 & 0 & 0 \\ 0 & 1 & \frac{-5}{4} & | & \frac{1}{4} & \frac{-1}{4} & 0 \\ 0 & -3 & 4 & | & 0 & 0 & 1 \end{pmatrix} \qquad (R_2 \rightarrow \frac{-1}{4} R_2)$$

$$\sim \begin{pmatrix} 1 & 0 & \frac{33}{4} & | & \frac{-1}{4} & \frac{5}{4} & 0 \\ 0 & 1 & \frac{-5}{4} & | & \frac{1}{4} & \frac{-1}{4} & 0 \\ 0 & -3 & 4 & | & 0 & 0 & 1 \end{pmatrix} \qquad (R_1 \rightarrow R_1 - 5R_2)$$

$$\sim \begin{pmatrix} 1 & 0 & \frac{33}{4} & | & \frac{-1}{4} & \frac{5}{4} & 0 \\ 0 & 1 & \frac{-5}{4} & | & \frac{1}{4} & \frac{-1}{4} & 0 \\ 0 & 0 & \frac{1}{4} & | & \frac{3}{4} & \frac{-3}{4} & 1 \end{pmatrix} \qquad (R_3 \rightarrow R_3 + 3R_2)$$

Step 3: we tune the element on the third row, third column, to 1 and turn the remaining elements of that column to zeros.

$$\sim \left( \begin{array}{ccc|ccc} 1 & 0 & \frac{33}{4} & \frac{-1}{4} & \frac{5}{4} & 0 \\ 0 & 1 & \frac{-5}{4} & \frac{1}{4} & \frac{-1}{4} & 0 \\ 0 & 0 & 1 & 3 & -3 & 4 \end{array} \right) \qquad (R_3 \to 4R_3)$$

$$\sim \left( \begin{array}{ccc|ccc} 1 & 0 & 0 & -25 & 26 & -33 \\ 0 & 1 & \frac{-5}{4} & \frac{1}{4} & \frac{-1}{4} & 0 \\ 0 & 0 & 1 & 3 & -3 & 4 \end{array} \right) \qquad (R_1 \to R_1 - \frac{33}{4} R_3)$$

$$\sim \left( \begin{array}{ccc|ccc} 1 & 0 & 0 & -25 & 26 & -33 \\ 0 & 1 & 0 & 4 & -4 & 5 \\ 0 & 0 & 1 & 3 & -3 & 4 \end{array} \right) \qquad (R_2 \to R_2 + \frac{5}{4} R_3)$$

Finally, we may conclude

$$\left( \begin{array}{ccc} 1 & 5 & 2 \\ 1 & 1 & 7 \\ 0 & -3 & 4 \end{array} \right)^{-1} = \left( \begin{array}{ccc} -25 & 26 & -33 \\ 4 & -4 & 5 \\ 3 & -3 & 4 \end{array} \right)$$

as it is confirmed by the dot product

$$\left( \begin{array}{ccc} 1 & 5 & 2 \\ 1 & 1 & 7 \\ 0 & -3 & 4 \end{array} \right) \cdot \left( \begin{array}{ccc} -25 & 26 & -33 \\ 4 & -4 & 5 \\ 3 & -3 & 4 \end{array} \right) = \left( \begin{array}{ccc} 1 & 0 & 0 \\ 0 & 1 & 0 \\ 0 & 0 & 1 \end{array} \right).$$

An alternative route, avoiding fractions, leads of course to the same inverse matrix.

$$\left( \begin{array}{ccc|ccc} 1 & 5 & 2 & 1 & 0 & 0 \\ 1 & 1 & 7 & 0 & 1 & 0 \\ 0 & -3 & 4 & 0 & 0 & 1 \end{array} \right)$$

step 1

$$\sim \begin{pmatrix} 1 & 5 & 2 & 1 & 0 & 0 \\ 0 & -4 & 5 & -1 & 1 & 0 \\ 0 & -3 & 4 & 0 & 0 & 1 \end{pmatrix} \qquad (R_2 \to R_2 - R_1)$$

$$\sim \begin{pmatrix} 1 & 5 & 2 & 1 & 0 & 0 \\ 0 & -1 & 1 & -1 & 1 & -1 \\ 0 & -3 & 4 & 0 & 0 & 1 \end{pmatrix} \qquad (R_2 \to R_2 - R_3)$$

step 2

$$\sim \begin{pmatrix} 1 & 5 & 2 & 1 & 0 & 0 \\ 0 & 1 & -1 & 1 & -1 & 1 \\ 0 & -3 & 4 & 0 & 0 & 1 \end{pmatrix} \qquad (R_2 \to -R_2)$$

$$\sim \begin{pmatrix} 1 & 5 & 2 & 1 & 0 & 0 \\ 0 & 1 & -1 & 1 & -1 & 1 \\ 0 & 0 & 1 & 3 & -3 & 4 \end{pmatrix} \qquad (R_3 \to R_3 + 3R_2)$$

$$\sim \begin{pmatrix} 1 & 0 & 7 & -4 & 5 & -5 \\ 0 & 1 & -1 & 1 & -1 & 1 \\ 0 & 0 & 1 & 3 & -3 & 4 \end{pmatrix} \qquad (R_1 \to R_1 - 5R_2)$$

$$\sim \begin{pmatrix} 1 & 0 & 0 & -25 & 26 & -33 \\ 0 & 1 & -1 & 1 & -1 & 1 \\ 0 & 0 & 1 & 3 & -3 & 4 \end{pmatrix} \qquad (R_1 \to R_1 - 7R_3)$$

step 3

$$\sim \begin{pmatrix} 1 & 0 & 0 & -25 & 26 & -33 \\ 0 & 1 & 0 & 4 & -4 & 5 \\ 0 & 0 & 1 & 3 & -3 & 4 \end{pmatrix} \qquad (R_2 \to R_2 + R_3)$$

INVERSE OF A PRODUCT

The inverse of a matrix product reverses its order:

$$(S \cdot B)^{-1} = B^{-1} \cdot S^{-1}.$$

We recall this effect as the 'Socks-and-Boots' rule, which we have already encountered at the transpose of a matrix product.

*Example*: Given the matrices $A = \begin{pmatrix} 3 & 2 \\ 4 & 3 \end{pmatrix}$ and $B = \begin{pmatrix} 1 & -2 \\ 1 & 0 \end{pmatrix}$.

Since $\det A = 1$ and $\det B = 2$ we are able to calculate their inverse matrices:

$$A \cdot B = \begin{pmatrix} 5 & -6 \\ 7 & -8 \end{pmatrix}$$

$$A^{-1} = \begin{pmatrix} 3 & -2 \\ -4 & 3 \end{pmatrix}$$

$$B^{-1} = \begin{pmatrix} 0 & 1 \\ -\frac{1}{2} & \frac{1}{2} \end{pmatrix}$$

$$(A \cdot B)^{-1} = \begin{pmatrix} -4 & 3 \\ -\frac{7}{2} & \frac{5}{2} \end{pmatrix}$$

$$A^{-1} \cdot B^{-1} = \begin{pmatrix} 1 & 2 \\ -\frac{3}{2} & -\frac{5}{2} \end{pmatrix}$$

$$B^{-1} \cdot A^{-1} = \begin{pmatrix} -4 & 3 \\ -\frac{7}{2} & \frac{5}{2} \end{pmatrix}$$

## SOLVING SYSTEMS OF LINEAR EQUATIONS

We are able to express linear systems as matrix products. Expanding the dot product and applying the equality of matrices yields the two equations of the given $2 \times 2$ system.

$$\begin{cases} -2x + y & = & -2 \\ x + y & = & 10 \end{cases} \quad \Longleftrightarrow \quad \begin{pmatrix} -2 & 1 \\ 1 & 1 \end{pmatrix} \cdot \begin{pmatrix} x \\ y \end{pmatrix} = \begin{pmatrix} -2 \\ 10 \end{pmatrix}$$

After trimming a given $m \times n$ system into its default shape, we perceive three matrices to express it as a matrix product.

▷ We define the **coefficient matrix** as the $m \times n$ matrix $A$ taking the coefficients $a_{ij}$ as its elements.

▷ The $n \times 1$ column matrix $X$ contains the unknown quantities $x_i$ at the left hand side.

▷ The $m \times 1$ column matrix $T$ contains the constant terms $t_i$ at the right hand side.

This allows us to express a linear system

$$\begin{cases} a_{11}x_1 + a_{12}x_2 + \cdots + a_{1n}x_n & = & t_1 \\ a_{21}x_1 + a_{22}x_2 + \cdots + a_{2n}x_n & = & t_2 \\ & \vdots & \\ a_{m1}x_1 + a_{m2}x_2 + \cdots + a_{mn}x_n & = & t_m \end{cases}$$

as the matrix product

$$
\Longleftrightarrow \quad
\begin{pmatrix}
a_{11} & a_{12} & \cdots & a_{1n} \\
a_{21} & a_{22} & \cdots & a_{2n} \\
\vdots & \vdots &        & \vdots \\
a_{m1} & a_{m2} & \cdots & a_{mn}
\end{pmatrix}
\cdot
\begin{pmatrix}
x_1 \\ x_2 \\ \vdots \\ x_n
\end{pmatrix}
=
\begin{pmatrix}
t_1 \\ t_2 \\ \vdots \\ t_m
\end{pmatrix},
$$

or even more efficiently as

$$
A \cdot X = T.
$$

We call the above expression the **matrix equation** of the given $m \times n$ system. In case its coefficient matrix $A$ is an invertible square matrix, we can choose to solve the $n \times n$ system using its inverse matrix $A^{-1}$.

$$
\begin{aligned}
A \cdot X = T &\Leftrightarrow A^{-1} \cdot (A \cdot X) = A^{-1} \cdot T \\
&\Leftrightarrow I_n \cdot X = A^{-1} \cdot T \\
&\Leftrightarrow X = A^{-1} \cdot T
\end{aligned}
$$

*Example*:  Let us solve the linear system

$$
\begin{cases}
-2x + y &= -2 \\
x + y &= 10
\end{cases}.
$$

First of all, we express the system in its matrix form.

$$
\begin{pmatrix} -2 & 1 \\ 1 & 1 \end{pmatrix} \cdot \begin{pmatrix} x \\ y \end{pmatrix} = \begin{pmatrix} -2 \\ 10 \end{pmatrix}
$$

The determinant of this coefficient matrix $A$ equalling $-3$ assures that the inverse coefficient matrix exists.

$$
A^{-1} = \begin{pmatrix} -2 & 1 \\ 1 & 1 \end{pmatrix}^{-1} = \begin{pmatrix} -\frac{1}{3} & \frac{1}{3} \\ \frac{1}{3} & \frac{2}{3} \end{pmatrix}
$$

Consequently, we find the solution to this system as

$$
\begin{pmatrix} x \\ y \end{pmatrix} = \begin{pmatrix} -\frac{1}{3} & \frac{1}{3} \\ \frac{1}{3} & \frac{2}{3} \end{pmatrix} \cdot \begin{pmatrix} -2 \\ 10 \end{pmatrix} = \begin{pmatrix} 4 \\ 6 \end{pmatrix}.
$$

## 11.7 The Fibonacci operator

We can generate the Fibonacci sequence (see paragraph 5.3) using matrices. Defining the **Fibonacci operator** as the $2 \times 2$ matrix $F = \begin{pmatrix} 0 & 1 \\ 1 & 1 \end{pmatrix}$ and its initial Fibonacci vector as the column matrix $\vec{f}_0 = \begin{pmatrix} 0 \\ 1 \end{pmatrix}$, we subsequently calculate these dot products:

$$\vec{f}_1 = F \cdot \vec{f}_0 = \begin{pmatrix} 0 & 1 \\ 1 & 1 \end{pmatrix} \cdot \begin{pmatrix} 0 \\ 1 \end{pmatrix} = \begin{pmatrix} 1 \\ 1 \end{pmatrix},$$

$$\vec{f}_2 = F \cdot \vec{f}_1 = \begin{pmatrix} 0 & 1 \\ 1 & 1 \end{pmatrix} \cdot \begin{pmatrix} 1 \\ 1 \end{pmatrix} = \begin{pmatrix} 1 \\ 2 \end{pmatrix},$$

$$\vec{f}_3 = F \cdot \vec{f}_2 = \begin{pmatrix} 0 & 1 \\ 1 & 1 \end{pmatrix} \cdot \begin{pmatrix} 1 \\ 2 \end{pmatrix} = \begin{pmatrix} 2 \\ 3 \end{pmatrix},$$

$$\vec{f}_4 = F \cdot \vec{f}_3 = \begin{pmatrix} 0 & 1 \\ 1 & 1 \end{pmatrix} \cdot \begin{pmatrix} 2 \\ 3 \end{pmatrix} = \begin{pmatrix} 3 \\ 5 \end{pmatrix}.$$

We rewrite this fourth Fibonacci vector in its matrix form as

$$\vec{f}_4 = F \cdot \vec{f}_3 = F \cdot F \cdot \vec{f}_2 = F \cdot F \cdot F \cdot \vec{f}_1 = F \cdot F \cdot F \cdot F \cdot \vec{f}_0$$

$$= F^4 \cdot \vec{f}_0 = \begin{pmatrix} 0 & 1 \\ 1 & 1 \end{pmatrix}^4 \cdot \begin{pmatrix} 0 \\ 1 \end{pmatrix} = \begin{pmatrix} 3 \\ 5 \end{pmatrix}.$$

Next, we apply this matrix form to all natural matrix powers

$$\vec{f}_k = F^k \cdot \vec{f}_0,$$

given their exponents $k \in \mathbb{N}$.

There is an amazing relationship between the matrix powers of the Fibonacci operator $F$ and the golden number $\Phi$ (see paragraph **??**), which we outline without proof:

$$F^k = \begin{pmatrix} 1 & 1 \\ \Phi & \Phi' \end{pmatrix} \cdot \begin{pmatrix} \Phi^k & 0 \\ 0 & \Phi'^k \end{pmatrix} \cdot \begin{pmatrix} 1 & 1 \\ \Phi & \Phi' \end{pmatrix}^{-1}, \qquad (11.4)$$

given $\Phi = \frac{1+\sqrt{5}}{2}$ (see definition (5.1)) and $\Phi' = \frac{1-\sqrt{5}}{2}$ (see definition (5.3)) as both roots of the quadratic equation describe the golden number.

Matrix algebra simplifies the matrix power $F^k$ into a single $2 \times 2$ product matrix.

$$F^k = \begin{pmatrix} 1 & 1 \\ \Phi & \Phi' \end{pmatrix} \cdot \begin{pmatrix} \Phi^k & 0 \\ 0 & \Phi'^k \end{pmatrix} \cdot \begin{pmatrix} \frac{-1+\sqrt{5}}{2\sqrt{5}} & \frac{1}{\sqrt{5}} \\ \frac{1+\sqrt{5}}{2\sqrt{5}} & -\frac{1}{\sqrt{5}} \end{pmatrix}$$

$$= \begin{pmatrix} \Phi^k & \Phi'^k \\ \Phi^{k+1} & \Phi'^{k+1} \end{pmatrix} \cdot \begin{pmatrix} \frac{-1+\sqrt{5}}{2\sqrt{5}} & \frac{1}{\sqrt{5}} \\ \frac{1+\sqrt{5}}{2\sqrt{5}} & -\frac{1}{\sqrt{5}} \end{pmatrix}$$

$$= \begin{pmatrix} \Phi^k & \Phi'^k \\ \Phi^{k+1} & \Phi'^{k+1} \end{pmatrix} \cdot \frac{1}{\sqrt{5}} \cdot \begin{pmatrix} -\Phi' & 1 \\ \Phi & -1 \end{pmatrix}$$

$$= \frac{1}{\sqrt{5}} \cdot \begin{pmatrix} -\Phi^k \cdot \Phi' + \Phi \cdot \Phi'^k & \Phi^k - \Phi'^k \\ -\Phi^{k+1} \cdot \Phi' + \Phi \cdot \Phi'^{k+1} & \Phi^{k+1} - \Phi'^{k+1} \end{pmatrix}$$

Applying this single matrix expression of $F^k$ for calculating the $k$-th Fibonacci vector $\vec{f_k}$, yields

$$\vec{f_k} = F^k \cdot \vec{f_0}$$

$$= \frac{1}{\sqrt{5}} \cdot \begin{pmatrix} -\Phi^k \cdot \Phi' + \Phi \cdot \Phi'^k & \Phi^k - \Phi'^k \\ -\Phi^{k+1} \cdot \Phi' + \Phi \cdot \Phi'^{k+1} & \Phi^{k+1} - \Phi'^{k+1} \end{pmatrix} \cdot \begin{pmatrix} 0 \\ 1 \end{pmatrix}$$

$$= \frac{1}{\sqrt{5}} \cdot \begin{pmatrix} \Phi^k - \Phi'^k \\ \Phi^{k+1} - \Phi'^{k+1} \end{pmatrix}.$$

This reformulation of the link between Fibonacci numbers and the golden number by means of matrices reveals **Binet's formula**. The first component of the Fibonacci vector $\vec{f_k}$ is the $k$-th Fibonacci number, for which this holds

$$f_k = \frac{(\Phi)^k - (\Phi')^k}{\sqrt{5}}. \tag{11.5}$$

In this formula, $\Phi$ is the golden number and $\Phi' = \frac{1-\sqrt{5}}{2}$, given natural exponents $k \in \mathbb{N}$.

This formula had already been discovered in 1730 by the French mathematician **Abraham de Moivre** (1667–1754) but was only proven in 1843 by his compatriot **Jacques Binet** (1786–1856).

When we for instance determine the first Fibonacci number $f_1$ applying Binet's formula, we evaluate it as $f_1 = \frac{1}{\sqrt{5}} \cdot \left( \left( \frac{1+\sqrt{5}}{2} \right)^1 - \left( \frac{1-\sqrt{5}}{2} \right)^1 \right)$ which simplifies to $f_1 = 1$.

## 11.8 The matrix exponential

*Multiplicative group*: given the invertible matrices $A$, $B$ and $C \in \mathbb{R}^{n \times n}$, we summarise all properties of the matrix product $\cdot$ in invertible $\mathbb{R}^{n \times n}$ as:

| | |
|---|---|
| closure | $A \cdot B \in$ invertible $\mathbb{R}^{n \times n}$, |
| associative property | $A \cdot (B \cdot C) = (A \cdot B) \cdot C$, |
| identity matrix | $A \cdot I_n = A = I_n \cdot A$, |
| inverse matrix | $A \cdot A^{-1} = I_n = A^{-1} \cdot A$, |
| noncommutative ! | $A \cdot B \neq B \cdot A$ in general. |

We conclude the multiplicative group of (invertible $\mathbb{R}^{n \times n}$, $\cdot$) with the identity matrix $I_n$ as the neutral element.

We define a **matrix lie group** as a group built by **continuous symmetries** meaning that is has elements arbitrarily close to the neutral element. For instance, the standard 2D rotations around the origin $O$ by revolving angles $\theta$ build a matrix lie group under the matrix product $(R_{O,\theta}, \cdot)$ because $R_{O,\theta} \approx I_2$ when $\theta \approx 0$.

We define the **lie algebra** (here without its operation) as the matrix set generating the matrix lie group by exponentiating its elements. A matrix lie group as generated by its lie algebra were both pioneered by the Norwegian mathematician **Marius Sophus Lie** (1842–1899). For the lie algebra's operation (called the lie bracket) we kindly refer you to the specialised literature [21].

We define the **matrix exponential** on square matrices in $X \in \mathbb{R}^{n \times n}$ based on the maclaurin expansion of the natural exponential function $\exp(x) = e^x$ in $\mathbb{R}$ (see formula (4.5)) as

$$\exp(X) = I_n + X + \frac{1}{2} X \cdot X + \frac{X^3}{3!} + \frac{X^4}{4!} + \frac{X^5}{5!} + \ldots \quad \text{valid for all } X \in \mathbb{R}^{n \times n}. \quad (11.6)$$

The exponential of a square matrix $\exp(X) = e^X$ is based on the matrix product in $\mathbb{R}^{n \times n}$ and well defined given the above maclaurin series always converges. The matrix exponential is in use to solve systems of linear differential equations, for which we refer you to the specialised literature. Apart from its calculus applications, the matrix exponential is also powerful in mapping matrix lie algebras to their corresponding matrix lie groups.

*Example*: The matrix lie algebra spanned by these three basis elements

$$L_x = \begin{pmatrix} 0 & 0 & 0 \\ 0 & 0 & -1 \\ 0 & 1 & 0 \end{pmatrix} \text{ and } L_y = \begin{pmatrix} 0 & 0 & 1 \\ 0 & 0 & 0 \\ -1 & 0 & 0 \end{pmatrix} \text{ and } L_z = \begin{pmatrix} 0 & -1 & 0 \\ 1 & 0 & 0 \\ 0 & 0 & 0 \end{pmatrix}$$

generates the lie group of the 3D rotators $(R_\theta, \cdot)$. For instance, exponentiating the first generator $L_x$ can be done by exponentiating only its $2 \times 2$ square nonzero outtake, the matrix $L_2 = \begin{pmatrix} 0 & -1 \\ 1 & 0 \end{pmatrix}$. We split the list of the subsequent matrix powers of $L_2$ into

$$\begin{aligned} &\text{the even powers} \quad L_2^2 = (-1)I_2, \quad L_2^4 = I_2, \quad L_2^6 = (-1)I_2, \quad \dots \\ &\text{the odd powers} \quad L_2^3 = (-1)L_2, \quad L_2^5 = L_2, \quad L_2^7 = (-1)L_2, \quad \dots \end{aligned}$$

to evaluate the matrix exponential of a scalar multiple $\theta L_2$ by its maclaurin expansion:

$$\begin{aligned}
\exp(\theta L_2) &= I_2 + (\theta L_2) + \frac{1}{2!}(\theta L_2)^2 + \frac{1}{3!}(\theta L_2)^3 + \frac{1}{4!}(\theta L_2)^4 + \frac{1}{5!}(\theta L_2)^5 + \dots \text{ in } \mathbb{R}^{2\times 2} \\
&= I_2 + \theta L_2 + \frac{1}{2!}\theta^2 L_2^2 + \frac{1}{3!}\theta^3 L_2^3 + \frac{1}{4!}\theta^4 L_2^4 + \frac{1}{5!}\theta^5 L_2^5 + \dots \\
&= I_2 + \theta L_2 + \frac{1}{2!}\theta^2(-1)I_2 + \frac{1}{3!}\theta^3(-1)L_2 + \frac{1}{4!}\theta^4 I_2 + \frac{1}{5!}\theta^5 L_2 + \dots \\
&= I_2 - \frac{1}{2!}\theta^2 I_2 + \frac{1}{4!}\theta^4 I_2 - \dots \quad + \theta L_2 - \frac{1}{3!}\theta^3 L_2 + \frac{1}{5!}\theta^5 L_2 - \dots \\
&= \left(1 - \frac{1}{2!}\theta^2 + \frac{1}{4!}\theta^4 - \dots\right)I_2 + \left(\theta - \frac{1}{3!}\theta^3 + \frac{1}{5!}\theta^5 - \dots\right)L_2 \\
&= (\cos\theta)I_2 + (\sin\theta)L_2 \quad\quad\quad\quad\quad \text{see formulas (4.6) and (4.7)} \\
&= \cos\theta\begin{pmatrix} 1 & 0 \\ 0 & 1 \end{pmatrix} + \sin\theta\begin{pmatrix} 0 & -1 \\ 1 & 0 \end{pmatrix} \quad\quad \text{inserting the square matrices} \\
&= \begin{pmatrix} \cos\theta & 0 \\ 0 & \cos\theta \end{pmatrix} + \begin{pmatrix} 0 & -\sin\theta \\ \sin\theta & 0 \end{pmatrix} = \begin{pmatrix} \cos\theta & -\sin\theta \\ \sin\theta & \cos\theta \end{pmatrix} = R_{O,\theta}.
\end{aligned}$$

This downsized outcome $R_{O,\theta}$ is actually an element of the 2D rotations around the origin $O$ over the angle $\theta$ which build a lie group. Restoring it back to the $3 \times 3$ matrices, it shows how the lie algebra element $L_x$ generates the lie group element $R_{x,\theta}$ which is the standard 3D rotator around the $x-$axis over the angle $\theta$.

$$L_x = \begin{pmatrix} 0 & 0 & 0 \\ 0 & 0 & -1 \\ 0 & 1 & 0 \end{pmatrix} \xrightarrow{\text{matrix exp}} R_{x,\theta} = \begin{pmatrix} 1 & 0 & 0 \\ 0 & \cos\theta & -\sin\theta \\ 0 & \sin\theta & \cos\theta \end{pmatrix}$$

There is one lie algebra to each lie group, but there can be many lie groups sharing the same lie algebra. Therefore reversing this **exponential map** is – in general – not possible.

## 11.9 Exercises

**Exercise 100** Given $A = \begin{pmatrix} 5 & 10 & 0 & 1 \\ 10 & 200 & 10 & 20 \\ 20 & 10 & 2 & 1 \end{pmatrix}$ and $B = \begin{pmatrix} 4 & 10 & 3 & 0 \\ 10 & 80 & 40 & 50 \\ 10 & 30 & 2 & 0 \end{pmatrix}$,

calculate the following matrices:

1) $A + B$

2) $2B - 3A$

3) $A^T$

4) $B^T$

5) $(B - A)^T$

6) $B^T - A^T$

**Exercise 101** Calculate, whenever possible, given the matrices

$$A = \begin{pmatrix} 2 & 1 & 3 \\ -1 & 0 & 1 \end{pmatrix}, B = \begin{pmatrix} 1 & -1 \\ 0 & 1 \\ 2 & 3 \end{pmatrix}, C = \begin{pmatrix} 1 & 0 & -2 & 1 \\ 0 & 3 & 1 & 4 \end{pmatrix},$$

the following dot products:

1) $A \cdot B$

2) $B \cdot A$

3) $B \cdot C$

4) $C \cdot B$

5) $(A \cdot B) \cdot C$

6) $A \cdot (B \cdot C)$

7) $(A \cdot B)^T$

8) $B^T \cdot A^T$

**Exercise 102** Calculate, whenever possible, given the matrices

$$A = \begin{pmatrix} 8 & 2 \\ 3 & 2 \end{pmatrix}, B = \begin{pmatrix} 3 & 4 \\ 2 & 3 \end{pmatrix}, C = \begin{pmatrix} 1 & 2 & 3 \\ 2 & 5 & 3 \\ 1 & 0 & 8 \end{pmatrix}, D = \begin{pmatrix} 1 & 2 & 3 \\ 2 & 5 & 3 \\ 1 & 0 & 9 \end{pmatrix},$$

the following matrix expressions:

1) $A^2$

2) $A^{-1}$

3) $B^{-1}$

4) $C^{-1}$

5) $D^{-1}$

6) $A^{-1}B^{-1}$

7) $B^{-1}A^{-1}$

8) $(A \cdot B)^{-1}$

**Exercise 103**  Solve the following linear system using the inverse matrix method.

$$\begin{cases} -x & + & y & - & 4z & = 5 \\ 2x & + & 2y & & & = 4 \\ 3x & + & 3y & + & 2z & = 2 \end{cases}$$

**Exercise 104**  Prove that for all natural exponents $n \in \mathbb{N}$ the dot product $F^n \cdot \vec{f_0}$ generates all Fibonacci vectors $\vec{f_n}$. Hint: apply 'mathematical induction' in $\mathbb{N}$.

1) Verify the conjecture for a small value of $n$, for instance $n = 1$.

2) Accept the conjecture for an arbitrary value of $n$, given $n \in \mathbb{N}$.

3) Prove the 'inheritance' of it to the next natural index $n + 1$, based only upon both previous induction steps 1 and 2.

**Exercise 105**  Verify formula (11.4), which links the Fibonacci operator $F$ to the golden number $\Phi$, for exponent values $k = 1$ and $k = 2$.

**Exercise 106**  Apply Binet's formula (11.5) to generate the first four Fibonacci numbers $f_0, f_1, f_2$ and $f_3$.

**Exercise 107**  We define **idempotent** matrices $P$ as square matrices reproducing themselves by applying the dot product, i.e. $P^2 = P$. We can find more amazing examples of idempotent matrixes than just the identity matrices: $I_n \cdot I_n = I_n$. Verify for instance that all of the following real matrices are idempotent matrices:

$$K = \begin{pmatrix} 1 & 0 \\ 0 & 0 \end{pmatrix}, L = \begin{pmatrix} 1-a & \frac{(1-a)a}{b} \\ b & a \end{pmatrix}, M = \begin{pmatrix} 1 & a & 0 \\ 0 & 0 & 0 \\ 0 & 0 & 1 \end{pmatrix},$$

given $a \in \mathbb{R}, b \in \mathbb{R} \setminus \{0\}$.

**Exercise 108**  We define **nilpotent** matrices $N$ as square zero divisors, producing the zero matrix $O$ by matrix exponentiating themselves, i.e. $N^k = O$. Verify that all of the following real matrices are nilpotent matrices, and seek their corresponding exponent $k$:

$$N = \begin{pmatrix} 0 & 1 \\ 0 & 0 \end{pmatrix}, P = \begin{pmatrix} 12 & -18 \\ 8 & -12 \end{pmatrix}, R = \begin{pmatrix} 1 & 2 & 5 \\ 2 & 4 & 10 \\ -1 & -2 & -5 \end{pmatrix}.$$

# Chapter 12 · Bezier curves

Programming multimedia applications often involves plotting lines or curves from point to point. Amongst the variety of methods to achieve this, the most popular and practical algorithm is based on Bezier or B-spline segments. Those segments are named after the French engineer **Pierre Bézier** (1910–1999) who discovered a mathematical approach to design smooth curves for the company Renault, which ran on their first industrial computers. Simultaneously, the French mathematician **Paul de Casteljau** (1930) developed a similar plot algorithm for the competing company Citroën, to digitally plot smooth curves.

Ever since, apart from plotting smooth lines, Bezier segments are in use for 2D and 3D shapes, to model letter fonts and to design timelines for animation. All professional design, graphical and animation software makes use of Bezier segments or B-splines.

## 12.1 Vector equation of segments

Renault engineer Pierre Bézier developed the following parametrical segments for the company's computer aided design during the sixties.

LINEAR BEZIER SEGMENT

We recall the vector equation of a line as $\vec{x} = \vec{b} + \lambda(\vec{a} - \vec{b})$, given parameter $\lambda$ or often $t \in \mathbb{R}$ (see figure 8.1). We also recall that in euclidean space every two different points span one straight line.

If we limit the values for the parameter $t$ to the **unity interval** $[0,1] \subset \mathbb{R}$, we limit the head point $X$ to the segment $[BA]$. Let us explore this by rearranging the vector equation to

$$\vec{x} = (1-t)\vec{b} + t\vec{a},$$

which for the parameter sequence $t = 0, \frac{1}{2}, 1$, heads its location vector $\vec{x}$ respectively to the points $B, \dfrac{B+A}{2}, A$. We define such a line segment as a **linear** or **2-point Bezier segment**. We express the initial and the terminal point of the line segment $[P_0 P_1]$ by their respective location vectors $\vec{p}_0$ and $\vec{p}_1$ and typeset this Bezier segment, a function of $t$, as $\vec{b}_{01}(t)$. Its indices 0 and 1 refer to its initial and terminal point $P_0$ and $P_1$. We summarise the above into the parametric equation of the linear Bezier segment from $P_0$ to $P_1$:

$$\vec{b}_{01}(t) = (1-t)\vec{p}_0 + t\vec{p}_1 \quad \text{given} \quad t \in [0,1] \subset \mathbb{R}. \tag{12.1}$$

The parametric equation of $\vec{b}_{01}(t)$ confirms it indeed is linear in the parameter $t$ and it runs through its initial point $B_{01}(0) = P_0$ and through its terminal point $B_{01}(1) = P_1$.

QUADRATIC BEZIER SEGMENT

Plotting a Bezier segment running from $P_0$ to $P_2$ and curved by one control point $P_1$ is somehow more involving. For a better insight, we sample the continuous parameter $t \in [0,1] \subset \mathbb{R}$ by a long sequence of discrete parameter values $t_0 = 0, t_1, t_2, t_3, \ldots, t_{n-1}, t_n = 1$. Starting at the parameter value $t_0 = 0$, we plot

▷ for every subsequent parameter value $t_1$ the point $B_{01}(t_1)$ as well as the point $B_{12}(t_1)$,

▷ and on their fresh Bezier segment $[B_{01}B_{12}]$ again the point $B_{012}(t_1)$ corresponding to the parameter value $t_1$

and we iterate both steps for the incrementing parameter values $t_2, t_3, \ldots, t_{n-1}$ up to $t_n = 1$.

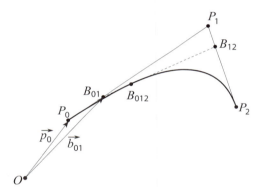

*Figure 12.1*: Quadratic or 3-point Bezier segment

We define the obtained (and theoretically continuous) curve $\vec{b}(t)$ as the **quadratic** or **3-point Bezier segment** $\vec{b}_{012}(t)$ defined by its **control points** $P_0, P_1$ and $P_2$. These control points $P_0, P_1$ and $P_2$ are indexed by the Bezier's indices $0, 1$ and $2$.

Based on the above plot algorithm, we elaborate its parametric equation as:

$$
\begin{aligned}
\vec{b}_{012}(t) &= (1-t) \cdot \vec{b}_{01}(t) + t \cdot \vec{b}_{12}(t) \\
&= (1-t)\left((1-t)\vec{p}_0 + t\vec{p}_1\right) + t\left((1-t)\vec{p}_1 + t\vec{p}_2\right) \\
&= (1-t)^2 \vec{p}_0 + 2(1-t)t\vec{p}_1 + t^2 \vec{p}_2.
\end{aligned}
\tag{12.2}
$$

The parametric equation of $\vec{b}_{012}(t)$ confirms it is indeed quadratic in the parameter $t$ and it runs through its initial point $B_{012}(0) = P_0$ and through its terminal point $B_{012}(1) = P_2$, whilst curved by the attracting control point $P_1$.

We recall that three noncollinear points define exactly one plane (see figure 12.1). In general, the control points $P_0, P_1$ and $P_2$ are noncollinear, which implies that their corresponding location vectors $\vec{p}_0, \vec{p}_1$ and $\vec{p}_2$ are at least plane vectors:

$$\vec{b}_{012}(t) = (1-t)^2 \begin{pmatrix} x_0 \\ y_0 \end{pmatrix} + 2(1-t)t \begin{pmatrix} x_1 \\ y_1 \end{pmatrix} + t^2 \begin{pmatrix} x_2 \\ y_2 \end{pmatrix}.$$

The plane linear and quadratic Bezier segments are default built-in in the 2D software Adobe Flash®.

### CUBIC BEZIER SEGMENT

Let us plot a Bezier segment running from $P_0$ to $P_3$ and curved by two control points $P_1$ and $P_2$. Again starting at the parameter value $t_0 = 0$, we plot

▷ for every subsequent parameter value $t_1$ the points $B_{01}(t_1)$, $B_{12}(t_1)$ and $B_{23}(t_1)$,

▷ and on their fresh Bezier segments $[B_{01}B_{12}]$ and $[B_{12}B_{23}]$ the point $B(t_1)$ corresponding to the parameter value $t_1$, which we typeset as $B_{012}$ and $B_{123}$ respectively,

▷ and finally on the line segment $[B_{012}B_{123}]$ the point $B_{0123}(t_1)$ corresponding to the parameter value $t_1$,

and we iterate all three steps for the incrementing parameter values $t_2, t_3, \ldots, t_{n-1}$ up to $t_n = 1$.

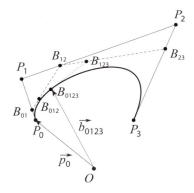

*Figure 12.2*: Cubic or 4-point Bezier segment

We define the obtained path $\vec{b}_{0123}(t)$ as the **cubic** or **4-point Bezier segment** defined by its control points $P_0, P_1, P_2$ and $P_3$. The control points are indexed by the Bezier indices $0, 1, 2$ and $3$.

Based on the above plot algorithm, we elaborate its parametric equation as:

$$\begin{aligned} \vec{b}_{0123}(t) &= (1-t) \cdot \vec{b}_{012}(t) + t \cdot \vec{b}_{123}(t) \qquad\qquad \text{and via formula (12.2)} \\ &= (1-t)\left((1-t)^2\vec{p}_0 + 2(1-t)t\vec{p}_1 + t^2\vec{p}_2\right) + t\left((1-t)^2\vec{p}_1 + 2(1-t)t\vec{p}_2 + t^2\vec{p}_3\right) \\ &= (1-t)^3\vec{p}_0 + 3(1-t)^2t\vec{p}_1 + 3(1-t)t^2\vec{p}_2 + t^3\vec{p}_3. \qquad\qquad (12.3) \end{aligned}$$

The parametric equation of $\vec{b}_{0123}(t)$ confirms it is indeed a cubic polynomial in $t$, running from its initial point $B_{0123}(0) = P_0$ to its terminal point $B_{0123}(1) = P_3$, whilst curved by the attracting control points $P_1$ and $P_2$.

We define four points $P, Q, U$ and $V$ as **coplanar** if they lie in a common plane. In general, the four control points $P_0, P_1, P_2$ and $P_3$ are noncollinear and noncoplanar, which implies that their corresponding location vectors $\vec{p}_0, \vec{p}_1, \vec{p}_2$ and $\vec{p}_3$ are at least 3D vectors:

$$\vec{b}_{0123}(t) = (1-t)^3 \begin{pmatrix} x_0 \\ y_0 \\ z_0 \end{pmatrix} + 3(1-t)^2 t \begin{pmatrix} x_1 \\ y_1 \\ z_1 \end{pmatrix} + 3(1-t)t^2 \begin{pmatrix} x_2 \\ y_2 \\ z_2 \end{pmatrix} + t^3 \begin{pmatrix} x_3 \\ y_3 \\ z_3 \end{pmatrix}.$$

The 3D Bezier segments are naturally built-in in 3D software such as CAD applications and 3ds Max®. Such segments offer – even in case of coplanarity – a much wider variety of curves, up to looped profiles. We note the order of the four control points indeed matters (see figure 12.3). The cubic plot routine features reusability for quadratic and linear Bezier segments, by selecting their control points backwards compatibly.

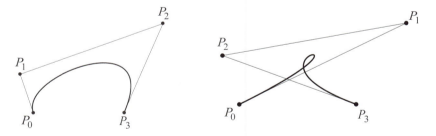

*Figure 12.3*: Cubic plane Bezier segment, regular versus looped

*Example:* The parametric equation of the cubic Bezier segment in 3D defined by the location vectors $\vec{p}_0 = \begin{pmatrix} 1 \\ 0 \\ 0 \end{pmatrix}, \vec{p}_1 = \begin{pmatrix} 0 \\ 1 \\ 0 \end{pmatrix}, \vec{p}_2 = \begin{pmatrix} 0 \\ 0 \\ 1 \end{pmatrix}$ and $\vec{p}_3 = \begin{pmatrix} 1 \\ 1 \\ 1 \end{pmatrix}$ simplifies to

$$\vec{b}_{0123}(t) = (1-t)^3 \begin{pmatrix} 1 \\ 0 \\ 0 \end{pmatrix} + 3(1-t)^2 t \begin{pmatrix} 0 \\ 1 \\ 0 \end{pmatrix} + 3(1-t)t^2 \begin{pmatrix} 0 \\ 0 \\ 1 \end{pmatrix} + t^3 \begin{pmatrix} 1 \\ 1 \\ 1 \end{pmatrix}$$

$$= \begin{cases} x(t) = (1-t)^3 + t^3 \\ y(t) = 3(1-t)^2 t + t^3 \\ z(t) = 3(1-t)t^2 + t^3 \end{cases}$$

$$= \begin{cases} x(t) = 1 - 3t + 3t^2 \\ y(t) = 3t - 6t^2 + 4t^3 \\ z(t) = 3t^2 - 2t^3 \end{cases} \tag{12.4}$$

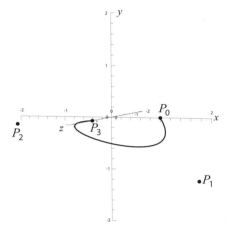

*Figure 12.4*: Cubic Bezier segment in 3D

BEZIER SEGMENTS OF HIGHER DEGREE

We can extend the above routine to generally plot $(n+1)$-point Bezier segments running from $P_0$ to $P_n$ and curved by intermediate control points $P_1$, $P_2$, $P_3$, ..., $P_{n-1}$ by applying Pierre Bézier's method. Skipping its details, we note that the resulting polynomial will keep satisfying Pascal's Triangle (see page 28):

$$\vec{b}_{01234}(t) = \mathbf{1}(1-t)^4 \vec{p}_0 + \mathbf{4}(1-t)^3 t \vec{p}_1 + \mathbf{6}(1-t)^2 t^2 \vec{p}_2 + \mathbf{4}(1-t)t^3 \vec{p}_3 + \mathbf{1}t^4 \vec{p}_4.$$

## 12.2 De Casteljau algorithm

Simultaneously to Pierre Bézier at Renault, Paul de Casteljau developed his own construction algorithm for Bezier segments at Citroën in 1959. We outline **de Casteljau construction** for a cubic Bezier segment defined by the four control points $P_0$, $P_1$, $P_2$ and $P_3$.

*Construction*:   We plot the Bezier segment from its initial point $P_0$ to its terminal point $P_3$, curved by its control points $P_1$ and $P_2$. Both the points $P_0$ and $P_3$ lie by design on the Bezier segment.

1)   We construct six midpoints (see page 51) like this:

$A$ as the midpoint of the line segment $[P_0P_1]$,

$B$ as the midpoint of the line segment $[P_2P_3]$,

$C$ as the midpoint of the line segment $[P_1P_2]$,

$A'$ as the midpoint of the line segment $[AC]$,

$B'$ as the midpoint of the line segment $[BC]$,

$C'$ as the midpoint of the line segment $[A'B']$.

2)   This first step corresponds to Pierre Bézier's plot method for parameter value $t = \frac{1}{2}$. Its final midpoint $C'$ equals the point $B_{0123}\left(\frac{1}{2}\right)$ and is it the only midpoint lying on the Bezier segment. Meanwhile the original set of control points $P_0, P_1, P_2, P_3$ splits into two new sets $P_0, A, A', C'$ and $C', B', B, P_3$. For each of them we repeat the first step:

   ▷   plot the Bezier segment from the initial point $P_0$ to the terminal point $C'$, curved by the control points $A$ and $A'$,

   ▷   plot the Bezier segment from the initial point $C'$ to the terminal point $P_3$, curved by the control points $B'$ and $B$.

   The above outlined de Casteljau construction is clearly recursive (see page 132).

3)   Iterate the above steps until the distance between the initial and the terminal control points becomes smaller than a fixed minimum $\varepsilon > 0$, and then connect them both by a line segment.

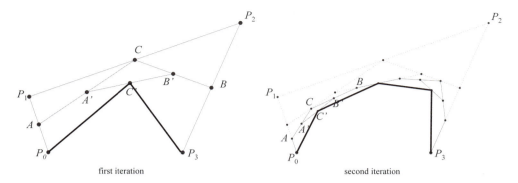

*Figure 12.5*: First two iterations of $\vec{b}_{0123}(t)$ via de Casteljau construction

Via de Casteljau construction we acquire the Bezier segment as a sequence of hundreds of connected points, starting at $P_0$ and ending at $P_3$. It can be proved that this recursively generated sequence of hundreds of connected midpoints equals the Bezier segment $\vec{b}_{0123}(t)$ as defined by the corresponding location vectors $\vec{p}_0, \vec{p}_1, \vec{p}_2$ and $\vec{p}_3$ (see formula (12.3)). We visualise the first two iterations of $\vec{b}_{0123}(t)$ by de Casteljau construction in figure 12.5. For the complete proof of its equivalence with figure 12.2, we refer you to the expert literature. We note that the de Casteljau construction is **numerically stable**.

## 12.3 Bezier curves

CONCATENATION

When we smoothly concatenate two or more Bezier segments, for instance a cubic segment defined by $P_0, P_1, P_2$ and $P_3$, and a quadratic segment defined by $Q_0, Q_1$ and $Q_2$, by coinciding the terminal point $P_3$ and the initial point $Q_0$ as the contact point, we obtain a **Bezier curve**.

We define the concatenation having **positional** or **zeroth order continuity** in case the contact point $P_3 = Q_0$ is not lying on the line segment $[P_2 Q_1]$. In this situation both segments meet continuously, however each segment has a different tangent line at the contact point $P_3 = Q_0$. Alternatively, we may call such a positional continuous concatenation **non smooth**. For a correct understanding of all continuity concepts, we emphasise that $P_2 P_3$ is the tangent line to the first segment at $P_3$ and $Q_0 Q_1$ is the tangent line to the second segment at $Q_0$. Knowing this immediately offers a deeper insight into the geometrical meaning of the intermediate control points such as $P_2$ and $Q_1$. Each Bezier segment is at its initial point $P_0$ tangent to the line $P_0 P_1$ and at its terminal point $P_n$ tangent to $P_{n-1} P_n$.

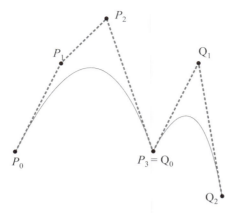

*Figure 12.6*: Positional or zeroth order continuous join of two Bezier segments

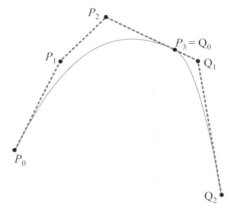

*Figure 12.7*: Tangential or first order continuous join of two Bezier segments

We define the concatenation having **tangential** or **first order continuity** in the case that junction $P_3 = Q_0$ is lying on the line segment $[P_2Q_1]$. In this situation both segments meet continuously and each segment shares the same tangent line $P_2P_3 = Q_0Q_1$ at the junction $P_3 = Q_0$. Alternatively, we may call such a tangential continuous concatenation **smooth**.

For an outline of higher order continuities such as **curvature** or **second order continuity**, we refer you to the expert literature on this topic.

LINEAR TRANSFORMATIONS

Linearly transforming (reflecting, scaling, rotating and shearing) an object made out of Bezier segments only requires transforming all of the control points of its Bezier segments.

*Proof:*

We prove this property for instance for a quadratic Bezier segment $\vec{b}_{012}(t)$, based on the linearity of the transformation operator $L$.

$$
\begin{aligned}
\vec{b}_{012}(t)' &= L(\vec{b}_{012}(t)) \\
&= L\big((1-t)^2 \vec{p}_0 + 2(1-t)t\vec{p}_1 + t^2 \vec{p}_2\big) \\
&= (1-t)^2 L(\vec{p}_0) + 2(1-t)tL(\vec{p}_1) + t^2 L(\vec{p}_2) \\
&= (1-t)^2 \vec{p}_0' + 2(1-t)t\vec{p}_1' + t^2 \vec{p}_2' \qquad \blacksquare
\end{aligned}
$$

Amongst all linear transformations $L$, especially the scaling $S$, realises a powerful graphical advantage through this property (see page 277). Due to their internal use of location vectors, applications and file types which visualise via Bezier segments are called respectively vector graphics software (e.g. InkScape, Adobe Illustrator® and CorelDRAW®) and vector image formats (for instance .svg or scalable vector graphics). Scalable vector formats do not suffer **aliasing** by zooming, given the redrawing of their Bezier segments does not suffer any quality losses (see figure 12.8).

*Figure 12.8*: Scaling with and without loss of quality

ILLUSTRATIONS

*Example 1:* Continuous function graphs consist of an uncountable infinite amount of points $(x, f(x))$, impossible to display on a discrete screen. Therefore we need to sample a function $f$ by a finite sequence of its points $(x_0, f(x_0)), (x_1, f(x_1)), (x_2, f(x_2)), \ldots, (x_n, f(x_n))$, which we display assembled by linear Bezier segments. In figure 12.9 we approximate the graph of $\sin x$ in this way, initially by a sequence of only ten sample points $(x_0, \sin x_0), (x_1, \sin x_1), (x_2, \sin x_2), \ldots, (x_9, \sin x_9)$, which we subsequently double up to a Bezier curve of forty segments.

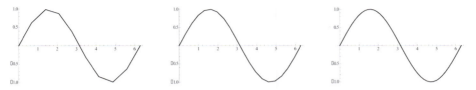

*Figure 12.9*: Approximating a continuous function graph by Bezier curves

*Example 2:* We illustrate **vector** or **outline fonts** such as TrueType® and OpenType® by designing a capital letter 'A' as a concatenation of ten linear and four quadratic Bezier segments. We recall that a linear transformation $L$ of this vector font only requires the linear transformation $L$ of all of its control points. This powerful property allows for reflecting, rotating or scaling such fonts without any loss of graphical quality. For instance, italicising this vector font only requires the shearing of all of its control points along the $x$-axis.

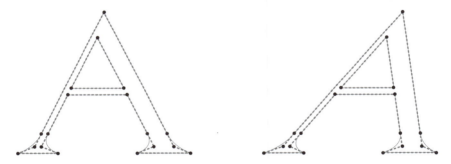

*Figure 12.10*: An outline or vector font is a Bezier curve

*Example 3:* We illustrate Bezier based shapes by designing a 'circle' approximated by a smoothly closed concatenation of four cubic Bezier segments (see figure 12.11). Of course, a mathematical circle is far from a Bezier curve, but a good approximation might convince and may offer a better plot performance. Moreover, Bezier based default shapes guarantee efficient transformation without loss of graphical quality.

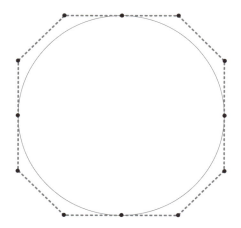

*Figure 12.11*: Approximating a circle by a closed Bezier curve

## 12.4 Matrix representation

LINEAR BEZIER SEGMENT

We present the matrix representation of the linear Bezier segment defined by two control points $P_0$ and $P_1$ from its parametric equation.

$$
\begin{aligned}
\vec{b}_{01}(t) &= (1-t)\vec{p_0} + t\vec{p_1} \quad \text{with} \quad t \in [0,1] \subset \mathbb{R} \\
&= \left(\ (1-t) \quad t\ \right) \cdot \begin{pmatrix} \vec{p_0} \\ \vec{p_1} \end{pmatrix} \\
&= \left(\ 1-t \quad t\ \right) \cdot \begin{pmatrix} \vec{p_0} \\ \vec{p_1} \end{pmatrix} \\
&= \left(\ 1 \quad t\ \right) \cdot \begin{pmatrix} 1 & 0 \\ -1 & 1 \end{pmatrix} \cdot \begin{pmatrix} \vec{p_0} \\ \vec{p_1} \end{pmatrix}
\end{aligned}
$$

We express the linear Bezier segment as the dot product of three matrices $\vec{b}_{01}(t) = T \cdot B \cdot P$, given

$$
T = \left(\ 1 \quad t\ \right)
$$

containing Bezier's **parameter base** and

$$
B = \begin{pmatrix} 1 & 0 \\ -1 & 1 \end{pmatrix}
$$

which we define as its **characteristic coefficient matrix**.

Expanding this matrix product via $\vec{b}_{01}(t) = T \cdot (B \cdot P)$, we encounter

$$
\begin{aligned}
\vec{b}_{01}(t) &= \begin{pmatrix} 1 & t \end{pmatrix} \cdot \left( \begin{pmatrix} 1 & 0 \\ -1 & 1 \end{pmatrix} \cdot \begin{pmatrix} \vec{p}_0 \\ \vec{p}_1 \end{pmatrix} \right) \\
&= \begin{pmatrix} 1 & t \end{pmatrix} \cdot \begin{pmatrix} \vec{p}_0 \\ -\vec{p}_0 + \vec{p}_1 \end{pmatrix} \\
&= \vec{p}_0 + t(\vec{p}_1 - \vec{p}_0) \\
&= \vec{p}_0 + t\overrightarrow{P_0 P_1}
\end{aligned}
$$

as the parametric equation of the line segment $[P_0 P_1]$ (see page 250).

## QUADRATIC BEZIER SEGMENT

We present the matrix representation of the quadratic Bezier segment defined by three control points $P_0, P_1$ and $P_2$ from its parametric equation.

$$
\begin{aligned}
\vec{b}_{012}(t) &= (1-t)^2 \vec{p}_0 + 2(1-t)t \vec{p}_1 + t^2 \vec{p}_2 \quad \text{with} \quad t \in [0,1] \subset \mathbb{R} \\
&= \begin{pmatrix} (1-t)^2 & 2(1-t)t & t^2 \end{pmatrix} \cdot \begin{pmatrix} \vec{p}_0 \\ \vec{p}_1 \\ \vec{p}_2 \end{pmatrix} \\
&= \begin{pmatrix} 1 - 2t + t^2 & 2t - 2t^2 & t^2 \end{pmatrix} \cdot \begin{pmatrix} \vec{p}_0 \\ \vec{p}_1 \\ \vec{p}_2 \end{pmatrix} \\
&= \begin{pmatrix} 1 & t & t^2 \end{pmatrix} \cdot \begin{pmatrix} 1 & 0 & 0 \\ -2 & 2 & 0 \\ 1 & -2 & 1 \end{pmatrix} \cdot \begin{pmatrix} \vec{p}_0 \\ \vec{p}_1 \\ \vec{p}_2 \end{pmatrix}
\end{aligned}
$$

We express the quadratic Bezier segment as the dot product of three matrices $\vec{b}_{012}(t) = T \cdot B \cdot P$, given its parameter base

$$
T = \begin{pmatrix} 1 & t & t^2 \end{pmatrix} \quad \text{and} \quad B = \begin{pmatrix} 1 & 0 & 0 \\ -2 & 2 & 0 \\ 1 & -2 & 1 \end{pmatrix}
$$

known as its characteristic coefficient matrix.

Expanding the matrix product in a reverse order as $\vec{b}_{012}(t) = T \cdot (B \cdot P)$, leads to

$$\vec{b}_{012}(t) = \left(\begin{array}{ccc} 1 & t & t^2 \end{array}\right) \cdot \left( \left(\begin{array}{ccc} 1 & 0 & 0 \\ -2 & 2 & 0 \\ 1 & -2 & 1 \end{array}\right) \cdot \left(\begin{array}{c} \vec{p}_0 \\ \vec{p}_1 \\ \vec{p}_2 \end{array}\right) \right)$$

$$= \left(\begin{array}{ccc} 1 & t & t^2 \end{array}\right) \cdot \left(\begin{array}{c} \vec{p}_0 \\ -2\vec{p}_0 + 2\vec{p}_1 \\ \vec{p}_0 - 2\vec{p}_1 + \vec{p}_2 \end{array}\right)$$

$$= \vec{p}_0 + 2(-\vec{p}_0 + \vec{p}_1)t + (\vec{p}_0 - 2\vec{p}_1 + \vec{p}_2)t^2.$$

We define such a reverse expansion of the matrix product as the **performant equivalent** of the quadratic Bezier segment,

$$\vec{b}_{012}(t) = \vec{p}_0 + 2(-\vec{p}_0 + \vec{p}_1)t + (\vec{p}_0 - 2\vec{p}_1 + \vec{p}_2)t^2.$$

## CUBIC BEZIER SEGMENT

To present the matrix representation of the cubic Bezier segment defined by four control points $P_0, P_1, P_2$ and $P_3$, we rewrite its parametric equation

$$\vec{b}_{0123}(t) = (1-t)^3 \vec{p}_0 + 3(1-t)^2 t \vec{p}_1 + 3(1-t)t^2 \vec{p}_2 + t^3 \vec{p}_3$$

to the matrix product

$$\vec{b}_{0123}(t) = \left(\begin{array}{cccc} (1-t)^3 & 3t(1-t)^2 & 3t^2(1-t) & t^3 \end{array}\right) \cdot \left(\begin{array}{c} \vec{p}_0 \\ \vec{p}_1 \\ \vec{p}_2 \\ \vec{p}_3 \end{array}\right).$$

We expand the elements of the row matrix to

$$\vec{b}_{0123}(t) = \left(\begin{array}{cccc} 1 - 3t + 3t^2 - t^3 & 3t - 6t^2 + 3t^3 & 3t^2 - 3t^3 & t^3 \end{array}\right) \cdot \left(\begin{array}{c} \vec{p}_0 \\ \vec{p}_1 \\ \vec{p}_2 \\ \vec{p}_3 \end{array}\right).$$

Consequently, we again rewrite this row matrix into a dot product.

$$\vec{b}_{0123}(t) = \left( \left(\begin{array}{cccc} 1 & t & t^2 & t^3 \end{array}\right) \cdot \left(\begin{array}{cccc} 1 & 0 & 0 & 0 \\ -3 & 3 & 0 & 0 \\ 3 & -6 & 3 & 0 \\ -1 & 3 & -3 & 1 \end{array}\right) \right) \cdot \left(\begin{array}{c} \vec{p}_0 \\ \vec{p}_1 \\ \vec{p}_2 \\ \vec{p}_3 \end{array}\right). \tag{12.5}$$

The first matrix factor is the parameter base $T$. The second matrix factor is the characteristic coefficient matrix of a cubic Bezier segment. Subsequently, we apply the associative property of the dot product $(T \cdot B) \cdot P = T \cdot (B \cdot P)$, to further rewrite this matrix representation.

$$\vec{b}_{0123}(t) = \begin{pmatrix} 1 & t & t^2 & t^3 \end{pmatrix} \cdot \left( \begin{pmatrix} 1 & 0 & 0 & 0 \\ -3 & 3 & 0 & 0 \\ 3 & -6 & 3 & 0 \\ -1 & 3 & -3 & 1 \end{pmatrix} \cdot \begin{pmatrix} \vec{p}_0 \\ \vec{p}_1 \\ \vec{p}_2 \\ \vec{p}_3 \end{pmatrix} \right)$$

$$= \begin{pmatrix} 1 & t & t^2 & t^3 \end{pmatrix} \cdot \begin{pmatrix} \vec{p}_0 \\ -3(\vec{p}_0 - \vec{p}_1) \\ 3(\vec{p}_0 - 2\vec{p}_1 + \vec{p}_2) \\ -\vec{p}_0 + 3\vec{p}_1 - 3\vec{p}_2 + \vec{p}_3 \end{pmatrix}$$

This reordered expansion of the matrix representation eventually yields the performant equivalent of the cubic Bezier segment,

$$\vec{b}_{0123}(t) = \vec{p}_0 - 3(\vec{p}_0 - \vec{p}_1)t + 3(\vec{p}_0 - 2\vec{p}_1 + \vec{p}_2)t^2 + (-\vec{p}_0 + 3\vec{p}_1 - 3\vec{p}_2 + \vec{p}_3)t^3.$$

At a first glance, the performant equivalent equation does not simplify the initial parametric equation, but it may offer some plot performance compared to the previous methods (see pages 253 and 255).

For each parameter value $t_0 = 0, t_1, t_2, t_3, \ldots, t_{n-1}, t_n = 1$ during the plot session, a smaller number of arithmetic operations is required. This is due to the stored constant coefficients $-3(\vec{p}_0 - \vec{p}_1)$ and $3(\vec{p}_0 - 2\vec{p}_1 + \vec{p}_2)$ and $(-\vec{p}_0 + 3\vec{p}_1 - 3\vec{p}_2 + \vec{p}_3)$, which only need to be multiplied with powers of $t$. We recall that the parametric equation (12.3) on the contrary requires a complete recalculation for each new parameter value $t$. Even the pen and paper practising of Bezier segments will benefit from the performant equivalent equation, because of the easy evaluation of its control point location vectors $\vec{p}_0, \vec{p}_1, \vec{p}_2$ and $\vec{p}_3$.

*Example:* We evaluate the performant equivalent equation of the cubic Bezier segment in 3D for the location vectors $\vec{p}_0 = \begin{pmatrix} 1 \\ 0 \\ 0 \end{pmatrix}, \vec{p}_1 = \begin{pmatrix} 0 \\ 1 \\ 0 \end{pmatrix}, \vec{p}_2 = \begin{pmatrix} 0 \\ 0 \\ 1 \end{pmatrix}$ and $\vec{p}_3 = \begin{pmatrix} 1 \\ 1 \\ 1 \end{pmatrix}$.

$$\vec{b}_{0123}(t) = \begin{pmatrix} 1 \\ 0 \\ 0 \end{pmatrix} - 3\left( \begin{pmatrix} 1 \\ 0 \\ 0 \end{pmatrix} - \begin{pmatrix} 0 \\ 1 \\ 0 \end{pmatrix} \right) t$$

$$+ 3\left( \begin{pmatrix} 1 \\ 0 \\ 0 \end{pmatrix} - 2\begin{pmatrix} 0 \\ 1 \\ 0 \end{pmatrix} + \begin{pmatrix} 0 \\ 0 \\ 1 \end{pmatrix} \right) t^2$$

$$+ \left( -\begin{pmatrix} 1 \\ 0 \\ 0 \end{pmatrix} + 3\begin{pmatrix} 0 \\ 1 \\ 0 \end{pmatrix} - 3\begin{pmatrix} 0 \\ 0 \\ 1 \end{pmatrix} + \begin{pmatrix} 1 \\ 1 \\ 1 \end{pmatrix} \right) t^3$$

$$= \begin{pmatrix} 1 \\ 0 \\ 0 \end{pmatrix} - 3\begin{pmatrix} 1 \\ -1 \\ 0 \end{pmatrix} t + 3\begin{pmatrix} 1 \\ -2 \\ 1 \end{pmatrix} t^2 + \begin{pmatrix} 0 \\ 4 \\ -2 \end{pmatrix} t^3$$

$$= \begin{cases} x(t) = 1 - 3t + 3t^2 \\ y(t) = 3t - 6t^2 + 4t^3 \\ z(t) = 3t^2 - 2t^3 \end{cases}$$

We note that the evaluated result of the performant equivalent equation indeed corresponds to the parametric equation (12.4) of the same Bezier segment $\vec{b}_{0123}(t)$.

## 12.5 B-splines

Bezier segments make a subset of the so-called **B-splines**, referring to Bezier and a spline, meaning flexible strip (of wood or rubber). In other words, there are many more techniques to assemble a finite sequence of points into a smooth curve.

### CUBIC B-SPLINES

Cubic B-splines generally need to contain not even one of their four control points $P_0, P_1, P_2$ or $P_3$. Constructing a B-spline $\vec{s}_{0123}(t)$ is analogous to the former construction of a 4-point Bezier segment $\vec{b}_{0123}(t)$ and for which we refer you to further expert literature. Defining B-splines exactly is technically involved, and we once more refer you to the specific literature. In this book, we illustrate B-splines by constructing only one cubic B-spline through its matrix representation.

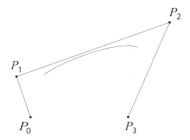

*Figure 12.12*: Cubic B-spline defined by four control points

## Matrix representation

Skipping its exact definition, we are able to express a cubic B-spline $\vec{s}_{0123}(t)$ defined by the control points $P_0, P_1, P_2$ and $P_3$, analogously to the former matrix product for ordinary Bezier segments (12.5),

$$\vec{s}_{0123}(t) = \begin{pmatrix} 1 & t & t^2 & t^3 \end{pmatrix} \cdot \frac{1}{6} \begin{pmatrix} 1 & 4 & 1 & 0 \\ -3 & 0 & 3 & 0 \\ 3 & -6 & 3 & 0 \\ -1 & 3 & -3 & 1 \end{pmatrix} \cdot \begin{pmatrix} \vec{p}_0 \\ \vec{p}_1 \\ \vec{p}_2 \\ \vec{p}_3 \end{pmatrix}. \qquad (12.6)$$

Completely similar to the Bezier segment, the parameter $t$ runs through the real unity interval $t \in [0,1] \subset \mathbb{R}$, while the characteristic coefficient matrix

$$B = \frac{1}{6} \begin{pmatrix} 1 & 4 & 1 & 0 \\ -3 & 0 & 3 & 0 \\ 3 & -6 & 3 & 0 \\ -1 & 3 & -3 & 1 \end{pmatrix}, \qquad (12.7)$$

determines a specified B-spline $\vec{s}_{0123}(t)$. Similarly, the row matrix

$$T = \begin{pmatrix} 1 & t & t^2 & t^3 \end{pmatrix}$$

contains the parameter base for a 4-point B-spline.

By expanding the above dot product of matrices as $T \cdot (B \cdot P)$, we obtain the performant equivalent equation of the B-spline $\vec{s}_{0123}(t)$ defined by $\vec{p}_0, \vec{p}_1, \vec{p}_2$ and $\vec{p}_3$.

$$\vec{s}_{0123}(t) = \begin{pmatrix} 1 & t & t^2 & t^3 \end{pmatrix} \cdot \left( \frac{1}{6} \begin{pmatrix} 1 & 4 & 1 & 0 \\ -3 & 0 & 3 & 0 \\ 3 & -6 & 3 & 0 \\ -1 & 3 & -3 & 1 \end{pmatrix} \cdot \begin{pmatrix} \vec{p}_0 \\ \vec{p}_1 \\ \vec{p}_2 \\ \vec{p}_3 \end{pmatrix} \right)$$

$$= \frac{1}{6} \begin{pmatrix} 1 & t & t^2 & t^3 \end{pmatrix} \cdot \begin{pmatrix} \vec{p}_0 + 4\vec{p}_1 + \vec{p}_2 \\ -3\vec{p}_0 + 3\vec{p}_2 \\ 3\vec{p}_0 - 6\vec{p}_1 + 3\vec{p}_2 \\ -\vec{p}_0 + 3\vec{p}_1 - 3\vec{p}_2 + \vec{p}_3 \end{pmatrix}$$

We finalise the matrix expansion which results in the performant equivalent equation,

$$\begin{aligned} \vec{s}_{0123}(t) &= \frac{1}{6}(\vec{p}_0 + 4\vec{p}_1 + \vec{p}_2) + \frac{1}{2}(-\vec{p}_0 + \vec{p}_2)t \\ &+ \frac{1}{2}(\vec{p}_0 - 2\vec{p}_1 + \vec{p}_2)t^2 + \frac{1}{6}(-\vec{p}_0 + 3\vec{p}_1 - 3\vec{p}_2 + \vec{p}_3)t^3. \end{aligned} \quad (12.8)$$

The performant equivalent expression reveals this B-spline $\vec{s}_{0123}(t)$ does not contain either its initial control point $P_0 \neq S_{0123}(0)$ nor its terminal control point $P_3 \neq S_{0123}(1)$. Its stored constant coefficients $\frac{1}{6}(\vec{p}_0 + 4\vec{p}_1 + \vec{p}_2)$, $\frac{1}{2}(-\vec{p}_0 + \vec{p}_2)$, $\frac{1}{2}(\vec{p}_0 - 2\vec{p}_1 + \vec{p}_2)$ and $\frac{1}{6}(-\vec{p}_0 + 3\vec{p}_1 - 3\vec{p}_2 + \vec{p}_3)$ are calculated independently from the running parameter $t$.

*Example:* We evaluate the performant equivalent equation of the cubic B-spline in 3D for the location vectors $\vec{p}_0 = \begin{pmatrix} 1 \\ 0 \\ 0 \end{pmatrix}$, $\vec{p}_1 = \begin{pmatrix} 0 \\ 1 \\ 0 \end{pmatrix}$, $\vec{p}_2 = \begin{pmatrix} 0 \\ 0 \\ 1 \end{pmatrix}$ and $\vec{p}_3 = \begin{pmatrix} 1 \\ 1 \\ 1 \end{pmatrix}$.

$$\begin{aligned} \vec{s}_{0123}(t) &= \frac{1}{6}\left( \begin{pmatrix} 1 \\ 0 \\ 0 \end{pmatrix} + 4\begin{pmatrix} 0 \\ 1 \\ 0 \end{pmatrix} + \begin{pmatrix} 0 \\ 0 \\ 1 \end{pmatrix} \right) + \frac{1}{2}\left( -\begin{pmatrix} 1 \\ 0 \\ 0 \end{pmatrix} + \begin{pmatrix} 0 \\ 0 \\ 1 \end{pmatrix} \right) t \\ &+ \frac{1}{2}\left( \begin{pmatrix} 1 \\ 0 \\ 0 \end{pmatrix} - 2\begin{pmatrix} 0 \\ 1 \\ 0 \end{pmatrix} + \begin{pmatrix} 0 \\ 0 \\ 1 \end{pmatrix} \right) t^2 \\ &+ \frac{1}{6}\left( -\begin{pmatrix} 1 \\ 0 \\ 0 \end{pmatrix} + 3\begin{pmatrix} 0 \\ 1 \\ 0 \end{pmatrix} - 3\begin{pmatrix} 0 \\ 0 \\ 1 \end{pmatrix} + \begin{pmatrix} 1 \\ 1 \\ 1 \end{pmatrix} \right) t^3 \end{aligned}$$

$$\vec{s}_{0123}(t) \;=\; \frac{1}{6}\begin{pmatrix} 1 \\ 4 \\ 1 \end{pmatrix} + \frac{1}{2}\begin{pmatrix} -1 \\ 0 \\ 1 \end{pmatrix} t + \frac{1}{2}\begin{pmatrix} 1 \\ -2 \\ 1 \end{pmatrix} t^2 + \frac{1}{6}\begin{pmatrix} 0 \\ 4 \\ -2 \end{pmatrix} t^3$$

$$= \quad \begin{cases} x(t) = \frac{1}{6} - \frac{1}{2}t + \frac{1}{2}t^2 \\[4pt] y(t) = \frac{2}{3} - t^2 + \frac{2}{3}t^3 \\[4pt] z(t) = \frac{1}{6} + \frac{1}{2}t + \frac{1}{2}t^2 - \frac{1}{3}t^3 \end{cases}$$

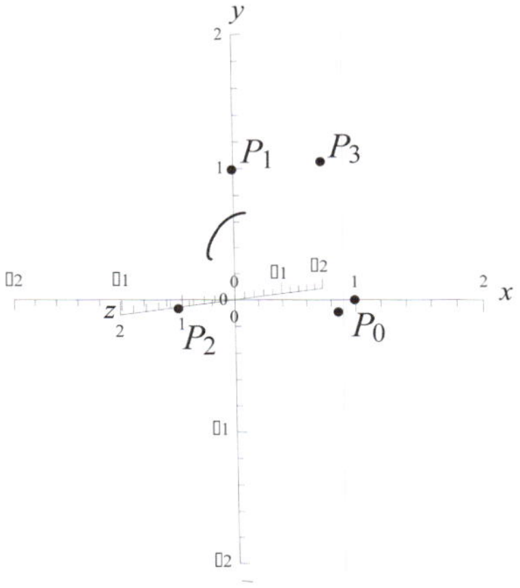

*Figure 12.13*: Cubic B-spline in 3D space

We note that our exemplary B-spline $\vec{s}_{0123}(t)$ differs from the ordinary Bezier segment (12.4) defined by the same control points as headed by their location vectors

$$\vec{p}_0 = \begin{pmatrix} 1 \\ 0 \\ 0 \end{pmatrix}, \vec{p}_1 = \begin{pmatrix} 0 \\ 1 \\ 0 \end{pmatrix}, \vec{p}_2 = \begin{pmatrix} 0 \\ 0 \\ 1 \end{pmatrix} \text{ and } \vec{p}_3 = \begin{pmatrix} 1 \\ 1 \\ 1 \end{pmatrix}.$$

DE BOOR'S ALGORITHM

The German-American mathematician **Carl de Boor** (1937) generalised the popular de Casteljau construction to his **de Boor construction** as the numerically stable alternative to draw or to split B-splines. We skip a complete outline of the de Boor construction, for which we refer to the expert literature. B-splines and B-spline surfaces still reign in our contemporary computer aided design (see figures 12.14 and 12.15), ending this chapter where it historically started.

*Figure 12.14*: CAD using linear, quadratic and cubic B-spline surfaces

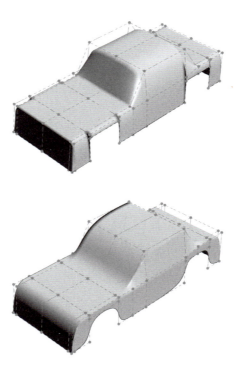

*Figure 12.15*: CAD applying linear and cubic B-splines orthogonally

## 12.6 Exercises

**Exercise 109**   Verify that we can express a linear Bezier segment $\vec{b}_{01}(t)$ as a quadratic Bezier segment $\vec{b}_{021}(t)$ by locating its *intermediate* location vector $\vec{p}_2 = \dfrac{\vec{p}_0 + \vec{p}_1}{2}$ at the midpoint of $[P_0 P_1]$.

**Exercise 110**   Calculate the parametric components $x(t), y(t)$ and $z(t)$ of the cubic Bezier segment described by the parametric equation

$$\vec{b}_{0123}(t) = (1-t)^3 \vec{p}_0 + 3(1-t)^2 t \vec{p}_1 + 3(1-t)t^2 \vec{p}_2 + t^3 \vec{p}_3$$

and the location vectors $\vec{p}_0 = \begin{pmatrix} 0 \\ 3 \\ 0 \end{pmatrix}, \vec{p}_1 = \begin{pmatrix} 1 \\ -1 \\ 1 \end{pmatrix}, \vec{p}_2 = \begin{pmatrix} -2 \\ 4 \\ 0 \end{pmatrix}$ and $\vec{p}_3 = \begin{pmatrix} 3 \\ 0 \\ 2 \end{pmatrix}$.

Use a computer to plot this Bezier segment in 3D space.

**Exercise 111**   For the parametric equations of Bezier segments, the sum of their co-efficients in $t$ equals 1. This is easily verified in case of linear Bezier segments via $(1-t) + t = 1$. Verify this property in the case of quadratic and cubic Bezier segments.

**Exercise 112**   Draw the set of control points $P_0(-2, -1)$, $P_1(0, 3)$, $P_2(4, 6)$ and $P_3(5, 0)$ in three different $(x, y)$−frames. Dedicate an $(x, y)$−frame to draw each of the following:

1)   the linear Bezier segment $\vec{b}_{03}(t)$ from $P_0$ to $P_3$,

2)   the third iteration from the de Casteljau construction of the quadratic Bezier segment from $P_0$ to $P_3$ curved by the control point $P_2$,

3)   the second iteration from the de Casteljau construction of the cubic Bezier segment $\vec{b}_{0123}(t)$ defined by the control points $P_0$, $P_1$, $P_2$ and $P_3$.

**Exercise 113**   Calculate the 2D components $x(t)$ and $y(t)$ based on the performant equivalent equations of:

1)   the linear Bezier segment $\vec{b}_{03}(t)$ from $P_0$ to $P_3$,

2)   the quadratic Bezier segment from $P_0$ to $P_3$ curved by the control point $P_2$,

3)   the cubic Bezier segment $\vec{b}_{0123}(t)$ defined by $P_0$, $P_1$, $P_2$ and $P_3$,

and the location vectors $\vec{p}_0 = \begin{pmatrix} -2 \\ -1 \end{pmatrix}, \vec{p}_1 = \begin{pmatrix} 0 \\ 3 \end{pmatrix}, \vec{p}_2 = \begin{pmatrix} 4 \\ 6 \end{pmatrix}$ and $\vec{p}_3 = \begin{pmatrix} 5 \\ 0 \end{pmatrix}$.

Use a computer to plot each of these Bezier segments in 2D space.

**Exercise 114** Calculate the components $x(t)$ and $y(t)$ via the performant equivalents of:

1) $\vec{b}_{0123}(t)$ given $\vec{p}_0 = \begin{pmatrix} -2 \\ -1 \end{pmatrix}$, $\vec{p}_1 = \begin{pmatrix} -3 \\ 3 \end{pmatrix}$, $\vec{p}_2 = \begin{pmatrix} 4 \\ 6 \end{pmatrix}$ and $\vec{p}_3 = \begin{pmatrix} 0 \\ 0 \end{pmatrix}$.

2) $\vec{b}_{0123}(t)$ given $\vec{p}_0 = \begin{pmatrix} -2 \\ -1 \end{pmatrix}$, $\vec{p}_1 = \begin{pmatrix} 4 \\ 6 \end{pmatrix}$, $\vec{p}_2 = \begin{pmatrix} -3 \\ 3 \end{pmatrix}$ and $\vec{p}_3 = \begin{pmatrix} 0 \\ 0 \end{pmatrix}$.

Use a computer to plot each of these Bezier segments in an $(x,y)$-frame. Anything remarkable to notice?

**Exercise 115** Calculate the components $x(t)$ and $y(t)$ of both the cubic Bezier segment $\vec{b}_{0123}(t)$ and of the B-spline $\vec{s}_{0123}(t)$, based on their respective matrix representations (12.5) and (12.6), and given the location vectors

$$\vec{p}_0 = \begin{pmatrix} -2 \\ -1 \end{pmatrix}, \vec{p}_1 = \begin{pmatrix} -3 \\ 3 \end{pmatrix}, \vec{p}_2 = \begin{pmatrix} 4 \\ 6 \end{pmatrix} \text{ and } \vec{p}_3 = \begin{pmatrix} 0 \\ 0 \end{pmatrix}.$$

Use a computer to plot both segments $\vec{b}_{0123}(t)$ and $\vec{s}_{0123}(t)$ in a different colour in the same $(x,y)$-frame.

**Exercise 116** Verify whether this B-spline, in its performant equivalent equation,

$$\vec{s}_{0123}(t) = \frac{1}{6}(\vec{p}_0 + 4\vec{p}_1 + \vec{p}_2) + \frac{1}{2}(-\vec{p}_0 + \vec{p}_2)t$$
$$+ \frac{1}{2}(\vec{p}_0 - 2\vec{p}_1 + \vec{p}_2)t^2 + \frac{1}{6}(-\vec{p}_0 + 3\vec{p}_1 - 3\vec{p}_2 + \vec{p}_3)t^3,$$

contains its initial and its terminal control point.

**Exercise 117** Calculate the components $x(t)$ and $y(t)$ of the cubic B-spline $\vec{s}_{0123}(t)$ described by its performant equivalent equation (12.8), given the location vectors
$$\vec{p}_0 = \begin{pmatrix} 0 \\ 3 \end{pmatrix}, \vec{p}_1 = \begin{pmatrix} 1 \\ 1 \end{pmatrix}, \vec{p}_2 = \begin{pmatrix} -2 \\ 0 \end{pmatrix} \text{ and } \vec{p}_3 = \begin{pmatrix} 1 \\ 0 \end{pmatrix}.$$

**Exercise 118** Given $\vec{p}_0 = \begin{pmatrix} -1 \\ 0 \end{pmatrix}, \vec{p}_1 = \begin{pmatrix} 1 \\ 2 \end{pmatrix}, \vec{p}_2 = \begin{pmatrix} -2 \\ 1 \end{pmatrix}$ and $\vec{p}_3 = \begin{pmatrix} 0 \\ 3 \end{pmatrix}$, find

1) the linear Bezier segment $\vec{b}_{01}(t)$ from $V_0$ to $V_1$,

2) the quadratic Bezier segment $\vec{b}_{012}(t)$ from $V_0$ to $V_2$,

3) and eventually the cubic Bezier segment $\vec{b}_{0123}(t)$ from $V_0$ to $V_3$.

4) Eliminate the parameter $t$ from the parametric equation of the linear Bezier segment $\vec{b}_{01}(t)$, in order to find the linear function by its explicit recipe $y = ax + b$.

5) Find the intersection of the linear Bezier segment $\vec{b}_{01}(t)$ and the cubic Bezier segment $\vec{b}_{0123}(t)$. Hint: substitute the above step 3 in step 4.

# Chapter 13 · Transformations

Animating computer graphics is all about displaying transformations of line segments into line segments. **Linear transformations** are accordingly capable of mapping line segments onto line segments. Widespread examples of linear transformations are scalings, reflections, rotations and shearings, while translations are, strictly speaking, not linear. We apply each transformation using its matrix operator.

We define such a transformation or its operator $L$ as **linear** when it features both properties

$$L(\vec{a}+\vec{b}) = L(\vec{a}) + L(\vec{b}),$$

$$L(\lambda \vec{a}) = \lambda L(\vec{a}),$$

given the vectors $\vec{a}, \vec{b} \in \mathbb{R}^n$ and a scalar $\lambda \in \mathbb{R}$.

Linear transformations can be combined: a sequence of linear transformations can be replaced by one linear transformation. And the other way around: a linear transformation can be decomposed into a sequence of linear transformations. This decomposition is a powerful practical property, given the fact that a linear transformation and its decomposition yield the same effect.

Since linear transformations are key for all 2D and 3D applications, we outline them in detail in this chapter.

## 13.1 Translation

We start with the basic screen effect of moving in a certain direction along a straight line, from $A$ to $B$. We call these geometrical shifts **translations** which we typeset as $T_{\overrightarrow{AB}}$. We can realise such a translation from point $A$ to point $B$ either by a vector sum or by a matrix product.

Imagine an object located in a cartesian frame in the point $P(3,2)$, which we want to move one unit to the right and four units upwards. We achieve this effect simply by adding 1 to its $x$-label and 4 to its $y$-label. This adjusts its location from the point $P(3,2)$ to the point $P'(3+1,2+4) = P'(4,6)$.

This approach is straightforward in the case of one point or just a handful of points. In case of objects built up by a lot of points, we like to move them more systematically.

A translation is an effect which shifts each point of the plane (or space) by the same free vector $\overrightarrow{AB}$. The previous example translates one point horizontally over distance 1 and vertically over distance 4, which we typeset as

$$\begin{pmatrix} 3 \\ 2 \end{pmatrix} + \begin{pmatrix} 1 \\ 4 \end{pmatrix} = \begin{pmatrix} 4 \\ 6 \end{pmatrix}.$$

If $P(x,y)$ is an original point and $P'(x',y')$ the resulting image point reached by the displacement vector $\overrightarrow{AB} = (\Delta x, \Delta y)$, then we generalise the **two-dimensional translation** using vectors as

$$\begin{pmatrix} x' \\ y' \end{pmatrix} = \begin{pmatrix} x \\ y \end{pmatrix} + \begin{pmatrix} \Delta x \\ \Delta y \end{pmatrix}. \qquad (13.1)$$

The values of $\Delta x$ and of $\Delta y$ do not need to be positive. A negative $\Delta x$ causes a shift to the left. A negative $\Delta y$ causes a shift downwards.

By taking the $z$-label into account, we can translate objects in 3D space as well. We express a **three-dimensional translation** using vectors as

$$\begin{pmatrix} x' \\ y' \\ z' \end{pmatrix} = \begin{pmatrix} x \\ y \\ z \end{pmatrix} + \begin{pmatrix} \Delta x \\ \Delta y \\ \Delta z \end{pmatrix}, \qquad (13.2)$$

with offsets $\Delta x, \Delta y$ and $\Delta z$ describing the 3D translation vector $\overrightarrow{AB}$.

*Example*: 2D translation.

We give the general matrix expression to translate flat objects 20 pixels to the right and 10 pixels downwards on the screen, which we will use to shift the golden rectangle bordered by the vertices $A(10,10)$ and $B(20, 16.18)$. Assuming that the edges of this rectangle are parallel to the cartesian axes, then both vertices, $A$ at the lower left and $B$ at the upper right corner, determine the complete rectangle.

In order to translate 20 units to the right and 10 units downwards, we set $\Delta x = 20$ and $\Delta y = -10$ (negative because of the downwards effect).

$$\begin{pmatrix} x' \\ y' \end{pmatrix} = \begin{pmatrix} x \\ y \end{pmatrix} + \begin{pmatrix} 20 \\ -10 \end{pmatrix}$$

Evaluating each original vertex yields its corresponding image vertex. Vertex $A$ translates to the image vertex $A'$ via

$$\begin{pmatrix} x' \\ y' \end{pmatrix} = \begin{pmatrix} 10 \\ 10 \end{pmatrix} + \begin{pmatrix} 20 \\ -10 \end{pmatrix} = \begin{pmatrix} 30 \\ 0 \end{pmatrix}.$$

Similarly, vertex $B$ translates to the image vertex $B'(40,\ 6.18)$. Translating both corner vertices translates the complete golden rectangle with them.

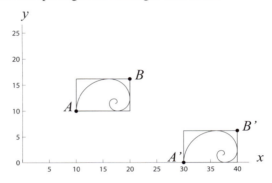

*Figure 13.1*: Translation of a golden rectangle

If we just need a single translation, we should perform it by vector addition, as it is most simple and fast. If we plan to scale or rotate after translation, we should choose to translate by matrix multiplication (see paragraph 13.6). In this latter case, we augment each vector in the translation's matrix expression with an extra component 1 for technical reasons. We call the components of these augmented vectors **homogeneous**.

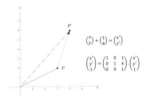

We express a **two-dimensional translation** using matrices by applying a dot product on homogeneous components

$$\begin{pmatrix} x' \\ y' \\ 1 \end{pmatrix} = \begin{pmatrix} 1 & 0 & \Delta x \\ 0 & 1 & \Delta y \\ 0 & 0 & 1 \end{pmatrix} \cdot \begin{pmatrix} x \\ y \\ 1 \end{pmatrix}. \tag{13.3}$$

We define the $3 \times 3$ matrix as the translation **operator** $T_{\overrightarrow{AB}}$ taking an original point $P$ as argument to return its image point $P'$, as shifted by the displacement vector $\overrightarrow{AB}$. Its matrix element $\Delta x$ is the horizontal displacement along the $x$-axis and the element $\Delta y$ is the vertical shift along the $y$-axis.

We emphasise that the extra component 1 and the extra operator row 0 0 1 are without physical meaning. These extra homogeneous components just allow us technically to perform translations by matrix multiplication. We calculate the image value $x'$ by taking its original value $x$ (multiplied by 1), dropping the original value $y$ (multiplied by 0) and adding offset $\Delta x$ (multiplied by 1). We calculate the image value $y'$ similarly by dropping the original value $x$, taking its original value $y$ and adding offset $\Delta y$ to it. In other words, this dot product achieves the former vector addition.

$$\begin{pmatrix} x' \\ y' \\ 1 \end{pmatrix} = \begin{pmatrix} 1 \cdot x + 0 \cdot y + \Delta x \cdot 1 \\ 0 \cdot x + 1 \cdot y + \Delta y \cdot 1 \\ 0 \cdot x + 0 \cdot y + 1 \cdot 1 \end{pmatrix} = \begin{pmatrix} x + \Delta x \\ y + \Delta y \\ 1 \end{pmatrix}$$

We straightforwardly extend this matrix product for translating 3D objects, which involves a $4 \times 4$ translation matrix operator.

We express a **three-dimensional translation** by the matrix product

$$\begin{pmatrix} x' \\ y' \\ z' \\ 1 \end{pmatrix} = \begin{pmatrix} 1 & 0 & 0 & \Delta x \\ 0 & 1 & 0 & \Delta y \\ 0 & 0 & 1 & \Delta z \\ 0 & 0 & 0 & 1 \end{pmatrix} \cdot \begin{pmatrix} x \\ y \\ z \\ 1 \end{pmatrix}. \tag{13.4}$$

Calculating the inverse translation matrix $T_{\overrightarrow{AB}}^{-1}$ (see exercise 128), we express the inverse operation as

$$\begin{pmatrix} x \\ y \\ z \\ 1 \end{pmatrix} = \begin{pmatrix} 1 & 0 & 0 & -\Delta x \\ 0 & 1 & 0 & -\Delta y \\ 0 & 0 & 1 & -\Delta z \\ 0 & 0 & 0 & 1 \end{pmatrix} \cdot \begin{pmatrix} x' \\ y' \\ z' \\ 1 \end{pmatrix}. \tag{13.5}$$

We recall $\Delta x$ as the horizontal and $\Delta y$ as the vertical displacement, and we interpret $\Delta z$ logically as the displacement along the $z$-axis.

*Example*: 3D translation via matrices

We use the matrix operator for the translation of 2 pixels to the left, 3 pixels upwards and 5 pixels to the front, to translate the cubic Bezier segment (see chapter 12) defined by its control points $A(0,0,0)$, $B(1,2,3)$, $C(2,-1,1)$ and $D(3,1,2)$.

To realise the above translation, we set $\Delta x = -2$, $\Delta y = 3$ and $\Delta z = 5$ in its matrix operator.

$$\begin{pmatrix} x' \\ y' \\ z' \\ 1 \end{pmatrix} = \begin{pmatrix} 1 & 0 & 0 & -2 \\ 0 & 1 & 0 & 3 \\ 0 & 0 & 1 & 5 \\ 0 & 0 & 0 & 1 \end{pmatrix} \cdot \begin{pmatrix} x \\ y \\ z \\ 1 \end{pmatrix}$$

Evaluating each original point yields its accordingly shifted image point. Control point $A(0,0,0)$ translates to the image point $A'$ via

$$\begin{pmatrix} x' \\ y' \\ z' \\ 1 \end{pmatrix} = \begin{pmatrix} 1 & 0 & 0 & -2 \\ 0 & 1 & 0 & 3 \\ 0 & 0 & 1 & 5 \\ 0 & 0 & 0 & 1 \end{pmatrix} \cdot \begin{pmatrix} 0 \\ 0 \\ 0 \\ 1 \end{pmatrix} = \begin{pmatrix} -2 \\ 3 \\ 5 \\ 1 \end{pmatrix}.$$

Similarly translating the control points $B, C$ and $D$ yields $B'(-1,5,8)$, $C'(0,2,6)$ and $D'(1,4,7)$. Translating all four control points translates the complete Bezier segment.

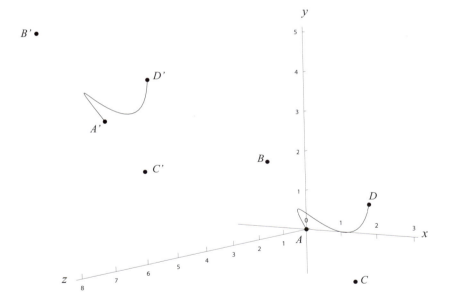

*Figure 13.2*: Translation of a cubic Bezier segment

## 13.2 Scaling

The dot product on matrices also suits the scaling of objects. Let us take the same approach as we took for translating objects: we scale a complete object by transforming each of its anchor points separately.

We express the **two-dimensional scaling** $S_O$ in matrix form as

$$\begin{pmatrix} x' \\ y' \\ 1 \end{pmatrix} = \begin{pmatrix} s_x & 0 & 0 \\ 0 & s_y & 0 \\ 0 & 0 & 1 \end{pmatrix} \cdot \begin{pmatrix} x \\ y \\ 1 \end{pmatrix}, \tag{13.6}$$

given $s_x > 0$ as the scale factor along the $x$-axis and $s_y > 0$ as the one along the $y$-axis.

For uniform scaling we set $s_x = s_y$. When we apply different nonzero values $s_x \neq s_y$, the matrix operator causes a non-uniform scaling. Positive scale factors $s_x$ and $s_y$ smaller than 1 will shrink objects, whereas those larger than 1 will enlarge objects.

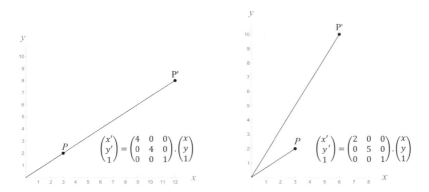

*Figure 13.3*: Uniform and non-uniform scaling

*Example*: A 2D standard scaling

We give the general matrix expression to enlarge objects three times, which we will use to scale the golden rectangle between the vertices $A(10,10)$ and $B(20, 16.18)$.

For this uniform scaling we set $s_x = s_y = 3$, leading to

$$\begin{pmatrix} x' \\ y' \\ 1 \end{pmatrix} = \begin{pmatrix} 3 & 0 & 0 \\ 0 & 3 & 0 \\ 0 & 0 & 1 \end{pmatrix} \cdot \begin{pmatrix} x \\ y \\ 1 \end{pmatrix}.$$

Evaluating each original vertex returns its corresponding scaled vertex. For instance vertex $A$ scales to $A'$ via

$$\begin{pmatrix} 3 & 0 & 0 \\ 0 & 3 & 0 \\ 0 & 0 & 1 \end{pmatrix} \cdot \begin{pmatrix} 10 \\ 10 \\ 1 \end{pmatrix} = \begin{pmatrix} 30 \\ 30 \\ 1 \end{pmatrix}.$$

Indeed, figure 13.4 confirms an image golden rectangle of three times its original size. But as a side effect, our enlarged rectangle also tripled its distance to the origin. This is due to the fact that the **standard transformation** $S_O$ is constructed with respect to the origin $O$, which pushes enlarged objects away from the origin and pulls shrinked objects towards the origin. To keep a scaled object at its location (for instance anchored at its centre point), requires a composite transformation (see paragraph 13.6). We straightforwardly extend

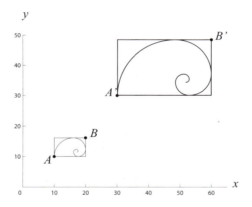

*Figure 13.4*: 2D uniform scaling

this matrix product for scaling 3D objects, which requires just one dimension extra.

We perform a **three-dimensional standard scaling** by applying the matrix product

$$\begin{pmatrix} x' \\ y' \\ z' \\ 1 \end{pmatrix} = \begin{pmatrix} s_x & 0 & 0 & 0 \\ 0 & s_y & 0 & 0 \\ 0 & 0 & s_z & 0 \\ 0 & 0 & 0 & 1 \end{pmatrix} \cdot \begin{pmatrix} x \\ y \\ z \\ 1 \end{pmatrix}. \tag{13.7}$$

Calculating the inverse standard scale operator $S_O^{-1}$ (see exercise 128), we express the inverse operator as

$$
\begin{pmatrix} x \\ y \\ z \\ 1 \end{pmatrix} = \begin{pmatrix} \frac{1}{s_x} & 0 & 0 & 0 \\ 0 & \frac{1}{s_y} & 0 & 0 \\ 0 & 0 & \frac{1}{s_z} & 0 \\ 0 & 0 & 0 & 1 \end{pmatrix} \cdot \begin{pmatrix} x' \\ y' \\ z' \\ 1 \end{pmatrix}
\tag{13.8}
$$

We recall the positive scale factors $s_x$ along the x-axis, $s_y$ along the y-axis and $s_z$ along the z-axis.

*Example*: 3D non-uniform scaling.

Let us find the scale operator which doubles the height of 3D objects and shrinks their depth to half the size, in order to rescale the cubic Bezier segment determined by its four control points $A(0,0,0)$, $B(1,2,3)$, $C(2,-1,1)$ and $D(3,1,2)$.

To double an objects' vertical size, we set scale factors $s_x = 1$ (no change) and $s_y = 2$. To shrink an object by 50% along the z-axis, we set $s_z = 0.5$.

$$
\begin{pmatrix} x' \\ y' \\ z' \\ 1 \end{pmatrix} = \begin{pmatrix} 1 & 0 & 0 & 0 \\ 0 & 2 & 0 & 0 \\ 0 & 0 & 0.5 & 0 \\ 0 & 0 & 0 & 1 \end{pmatrix} \cdot \begin{pmatrix} x \\ y \\ z \\ 1 \end{pmatrix}
$$

Making all four control points of this Bezier segment subject to this scale operator yields $A'(0,0,0)$, $B'(1, 4, 1.5)$, $C'(2, -2, 0.5)$ and $D'(3,2,1)$.

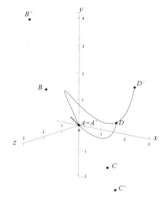

## 13.3 Rotation

There are several ways to rotate objects. In this book we consistently outline the method we have already used for translation $T_{\overrightarrow{AB}}$ and for standard scaling $S_O$, based on the matrix product.

### ROTATION IN 2D

Rotating a point $P$ on the unit circle, having $P$ labelled as $(x,y) = (\cos\alpha, \sin\alpha)$, over a positive angle $\theta$ returns its image point $P'$ labelled as

$$(x', y') = (\cos(\alpha + \theta),\ \sin(\alpha + \theta)).$$

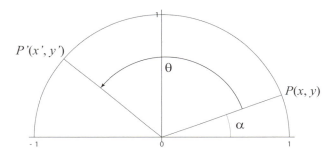

*Figure 13.5*: Standard rotation over a positive angle $\theta$

We reverse engineer the well known trigonometric Sum Identities of sine and cosine:

$$\begin{cases} \cos(\alpha + \theta) = \cos\alpha\cos\theta - \sin\alpha\sin\theta \\ \sin(\alpha + \theta) = \sin\alpha\cos\theta + \cos\alpha\sin\theta \end{cases}$$

$$\Leftrightarrow \begin{cases} \cos(\alpha + \theta) = x\cos\theta - y\sin\theta \\ \sin(\alpha + \theta) = y\cos\theta + x\sin\theta \end{cases}$$

$$\Leftrightarrow \begin{cases} x' = x\cos\theta - y\sin\theta \\ y' = y\cos\theta + x\sin\theta \end{cases}$$

$$\Leftrightarrow \begin{pmatrix} x' \\ y' \end{pmatrix} = \underbrace{\begin{pmatrix} \cos\theta & -\sin\theta \\ \sin\theta & \cos\theta \end{pmatrix}}_{\text{rotation operator}} \cdot \begin{pmatrix} x \\ y \end{pmatrix}.$$

In order to combine screen effects in the future, we augment this rotation operator to a matrix compatible with the translation and scale matrices. For this reason, we again use homogeneous coordinates, which lead to the matrix operator of the **two-dimensional rotation** $R_{O,\theta}$ around centre $O$ and over an angle $\theta$. Since its rotation centre lies in the origin $O$, this rotator is also a standard transformation.

$$\begin{pmatrix} x' \\ y' \\ 1 \end{pmatrix} = \begin{pmatrix} \cos\theta & -\sin\theta & 0 \\ \sin\theta & \cos\theta & 0 \\ 0 & 0 & 1 \end{pmatrix} \cdot \begin{pmatrix} x \\ y \\ 1 \end{pmatrix} \tag{13.9}$$

If we rotate over an angle $\theta$, then the inverse transformation rotates over the angle $-\theta$. We recall that for coterminal angles, their cosines are equal and their sines are opposite, which leads us to the inverse standard rotation operator $R_{O,\theta}^{-1}$ (see exercise 128) as

$$\begin{pmatrix} x \\ y \\ 1 \end{pmatrix} = \begin{pmatrix} \cos\theta & \sin\theta & 0 \\ -\sin\theta & \cos\theta & 0 \\ 0 & 0 & 1 \end{pmatrix} \cdot \begin{pmatrix} x' \\ y' \\ 1 \end{pmatrix}.$$

*Example*: Standard 2D rotation.

Let us find the standard matrix operator to rotate flat objects over a right angle, in order to apply it to the golden rectangle determined by its corner vertices $A(10,10)$ and $B(20, 16.18)$.

$$\begin{pmatrix} x' \\ y' \\ 1 \end{pmatrix} = \begin{pmatrix} \cos 90° & -\sin 90° & 0 \\ \sin 90° & \cos 90° & 0 \\ 0 & 0 & 1 \end{pmatrix} \cdot \begin{pmatrix} x \\ y \\ 1 \end{pmatrix} = \begin{pmatrix} 0 & -1 & 0 \\ 1 & 0 & 0 \\ 0 & 0 & 1 \end{pmatrix} \cdot \begin{pmatrix} x \\ y \\ 1 \end{pmatrix}$$

Evaluating both vertices $A$ and $B$ returns their rotated image points as $A'(-10,10)$ and $B'(-16.18, 20)$.

There again is a side effect to note, as our rotated rectangle also shifted to another location. This is due to the fact that the standard rotation operator $R_{O,\theta}$ is constructed with respect to the origin $O$, just like the standard scale operator $S_O$ was. The major consequence of it is the rotation of objects around the origin. As soon as we want to rotate objects around their centre or around an arbitrary point, we will need to combine standard transformations (see paragraph 13.6).

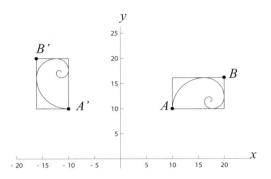

*Figure 13.6*: Standard 2D rotation

## ROTATION IN 3D

Let us study a 3D rotation in detail. As rotations in 2D operate in a plane (for instance in a screen), rotations in 3D allow some more degrees of freedom: they operate in three planes.

These are the three planes they rotate in:

▷ a vertical plane $xy$ in which the 'roll' rotates around the $z$-axis,

▷ a longitudinal plane $yz$ in which the 'pitch' rotates around the $x$-axis,

▷ a horizontal plane $xz$ in which the 'yaw' rotates around the $y$-axis.

Let us for a start study the 'roll' around the $z$-axis, since it is the equivalent of the rotation in 2D. The $z$-axis is perpendicular to the (flat) screen, so a rotation around it is actually the rotation around the origin $O$ in the $xy$-plane. Hence its 3D rotation operator $R_{z,\theta}$ is similar to the 2D rotator $R_{O,\theta}$ as it takes just one extra row and column. In this book, we call this rotation a **roll**, the term commonly used for it in aviation.

As the three-dimensional rotation around the $z$-axis $R_{z,\theta}$ only affects the $x$- and $y$-labels of points, we deduct its operator matrix as

$$\begin{pmatrix} x' \\ y' \\ z' \\ 1 \end{pmatrix} = \begin{pmatrix} \cos\theta & -\sin\theta & 0 & 0 \\ \sin\theta & \cos\theta & 0 & 0 \\ 0 & 0 & 1 & 0 \\ 0 & 0 & 0 & 1 \end{pmatrix} \cdot \begin{pmatrix} x \\ y \\ z \\ 1 \end{pmatrix}. \qquad (13.10)$$

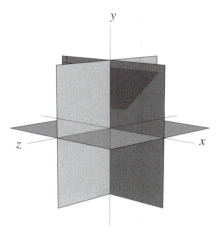

*Figure 13.7*: Standard 3D rotation planes

A second possible 3D rotation is the one around the $x$-axis, which we typeset as $R_{x,\theta}$ and in this book denote as **pitch**. Rotating around the $x$-axis does not affect the $x$-labels, so the first row and column of its operator matrix conserve the $x$-values.

Consequently we deduct the three-dimensional rotation $R_{x,\theta}$ around the $x$-axis over the angle $\theta$ as

$$\begin{pmatrix} x' \\ y' \\ z' \\ 1 \end{pmatrix} = \begin{pmatrix} 1 & 0 & 0 & 0 \\ 0 & \cos\theta & -\sin\theta & 0 \\ 0 & \sin\theta & \cos\theta & 0 \\ 0 & 0 & 0 & 1 \end{pmatrix} \cdot \begin{pmatrix} x \\ y \\ z \\ 1 \end{pmatrix}. \tag{13.11}$$

As a third and final type of 3D rotation, in this book we call **yaw** the rotation around the $y$-axis, typeset as $R_{y,\theta}$. Rotating around the $y$-axis does not affect the $y$-labels, so the second row and column of its operator matrix conserve the $y$-values. Consequently we express the three-dimensional rotation $R_{y,\theta}$ around the $y$-axis over the angle $\theta$ as

$$\begin{pmatrix} x' \\ y' \\ z' \\ 1 \end{pmatrix} = \begin{pmatrix} \cos\theta & 0 & \sin\theta & 0 \\ 0 & 1 & 0 & 0 \\ -\sin\theta & 0 & \cos\theta & 0 \\ 0 & 0 & 0 & 1 \end{pmatrix} \cdot \begin{pmatrix} x \\ y \\ z \\ 1 \end{pmatrix} \tag{13.12}$$

We warn for a technical detail: the negative sign in yaw $R_{y,\theta}$ resides at its lower left element. All three types of 3D rotation are standard transformations since they each take a cartesian axis to rotate around. By the way, take care with these definitions of roll, pitch and yaw when you consult other books and resources for their axes may differ. In a subsequent paragraph, we will discuss how to rotate around an arbitrary centre and how to combine standard 3D rotations to rotate simultaneously around the $x$-, $y$- and $z$-axis.

*Example*: Standard 3D rotation around the $y$-axis.

We construct the matrix operator to rotate 3D objects around the $y$-axis over a straight angle, in order to rotate the triangle with vertices $A(4,0,1), B(0,3,2)$ and $C(-1,2,-1)$.

$$\begin{pmatrix} x' \\ y' \\ z' \\ 1 \end{pmatrix} = \begin{pmatrix} \cos\pi & 0 & \sin\pi & 0 \\ 0 & 1 & 0 & 0 \\ -\sin\pi & 0 & \cos\pi & 0 \\ 0 & 0 & 0 & 1 \end{pmatrix} \cdot \begin{pmatrix} x \\ y \\ z \\ 1 \end{pmatrix}$$

$$= \begin{pmatrix} -1 & 0 & 0 & 0 \\ 0 & 1 & 0 & 0 \\ 0 & 0 & -1 & 0 \\ 0 & 0 & 0 & 1 \end{pmatrix} \cdot \begin{pmatrix} x \\ y \\ z \\ 1 \end{pmatrix}$$

Evaluating its vertices $A, B$ and $C$ returns the rotated triangle as $A'(-4,0,-1), B'(0,3,-2)$ and $C'(1,2,1)$.

## 13.4 Reflection

Also reflections can be described using matrices. Two-dimensionally we reflect over the $x$-axis, over the $y$-axis or over the origin $O$. We can describe alternatively the reflection over the origin as the standard rotation over the straight angle, or as the combined reflection over the $x$-axis and the $y$-axis simultaneously.

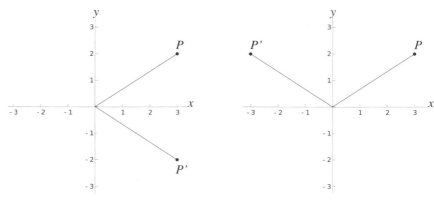

(a) reflection over the $x$-axis                    (b) reflection over the $y$-axis

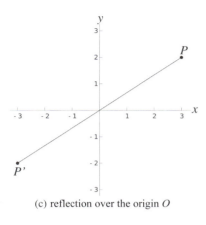

(a) $\begin{vmatrix} 1 & 0 & 0 \\ 0 & -1 & 0 \\ 0 & 0 & 1 \end{vmatrix} \cdot \begin{vmatrix} x \\ y \\ 1 \end{vmatrix} = \begin{vmatrix} x \\ -y \\ 1 \end{vmatrix}$

(b) $\begin{vmatrix} -1 & 0 & 0 \\ 0 & 1 & 0 \\ 0 & 0 & 1 \end{vmatrix} \cdot \begin{vmatrix} x \\ y \\ 1 \end{vmatrix} = \begin{vmatrix} -x \\ y \\ 1 \end{vmatrix}$

(c) $\begin{vmatrix} -1 & 0 & 0 \\ 0 & -1 & 0 \\ 0 & 0 & 1 \end{vmatrix} \cdot \begin{vmatrix} x \\ y \\ 1 \end{vmatrix} = \begin{vmatrix} -x \\ -y \\ 1 \end{vmatrix}$

(c) reflection over the origin $O$

The **two-dimensional reflection over the $x$-axis** leaves the $x$-labels invariant and affects just the sign of the $y$-labels. We typeset it as $M_x$ and express its matrix operator as

$$\begin{pmatrix} x' \\ y' \\ 1 \end{pmatrix} = \begin{pmatrix} 1 & 0 & 0 \\ 0 & -1 & 0 \\ 0 & 0 & 1 \end{pmatrix} \cdot \begin{pmatrix} x \\ y \\ 1 \end{pmatrix}. \tag{13.13}$$

The **two-dimensional reflection over the $y$-axis** leaves the $y$-labels invariant and affects just the sign of the $x$-labels. We typeset it as $M_y$ and express its matrix operator as

$$\begin{pmatrix} x' \\ y' \\ 1 \end{pmatrix} = \begin{pmatrix} -1 & 0 & 0 \\ 0 & 1 & 0 \\ 0 & 0 & 1 \end{pmatrix} \cdot \begin{pmatrix} x \\ y \\ 1 \end{pmatrix}. \tag{13.14}$$

The **two-dimensional reflection over the origin** affects the sign of both the $x$- and the $y$-labels. We typeset it as $M_O$ and calculate its matrix operator via $R_{O,180°}$ as

$$\begin{pmatrix} x' \\ y' \\ 1 \end{pmatrix} = \begin{pmatrix} -1 & 0 & 0 \\ 0 & -1 & 0 \\ 0 & 0 & 1 \end{pmatrix} \cdot \begin{pmatrix} x \\ y \\ 1 \end{pmatrix}. \tag{13.15}$$

## 13.5 Shearing

A **standard shearing** deforms an object in only one direction, proportionally to the distance to an invariable axis. We illustrate this effect via the *'false italic'* typesetting font of which the bottom pixels remain invariable and all other pixels are pushed to the right; the further up they are, the more they are pushed to the right.

The **standard shearing along the x-axis** by an angle called the **shear strain** $\sigma_x$ conserves the $y$-labels and affects the $x$-labels, but conserves the $x$-axis itself. We typeset this horizontal shearing as $S_{\sigma_x}$ and express its matrix operator as

$$\begin{pmatrix} x' \\ y' \\ 1 \end{pmatrix} = \begin{pmatrix} 1 & \tan\sigma_x & 0 \\ 0 & 1 & 0 \\ 0 & 0 & 1 \end{pmatrix} \cdot \begin{pmatrix} x \\ y \\ 1 \end{pmatrix}.$$

*Proof:*

The standard shearing of an original point $P(x,y)$ by $S_{\sigma_x}$ conserves its $y$-label and adds a proportional length $|PP'|$ to its $x$-label, typeset as

$$\begin{pmatrix} x' \\ y' \end{pmatrix} = \begin{pmatrix} x+|PP'| \\ y \end{pmatrix}.$$

We retrieve this proportional length $|PP'|$ from the right triangle $OPP'$ (see figure 13.8) as

$$\tan\sigma_x = \frac{|PP'|}{|OP|} = \frac{|PP'|}{y} \implies |PP'| = y\tan\sigma_x.$$

Replacing $|PP'|$ by $y\tan\sigma_x$, we reverse engineer the column matrix to a matrix product

$$\begin{pmatrix} x' \\ y' \end{pmatrix} = \begin{pmatrix} x+y\tan\sigma_x \\ y \end{pmatrix} = \begin{pmatrix} 1 & \tan\sigma_x \\ 0 & 1 \end{pmatrix} \cdot \begin{pmatrix} x \\ y \end{pmatrix}.$$

Adding the homogeneous component finalises the proof.                                                   ∎

The **standard shearing along the y-axis** by an angle called the **shear strain** $\sigma_y$ conserves the $x$-labels and affects the $y$-labels, but conserves the $y$-axis itself. We typeset this vertical shearing as $S_{\sigma_y}$ and express its matrix operator as

$$\begin{pmatrix} x' \\ y' \\ 1 \end{pmatrix} = \begin{pmatrix} 1 & 0 & 0 \\ \tan\sigma_y & 1 & 0 \\ 0 & 0 & 1 \end{pmatrix} \cdot \begin{pmatrix} x \\ y \\ 1 \end{pmatrix}.$$

We prove this matrix operator completely similar to the previous proof for the $S_{\sigma_x}$-operator.

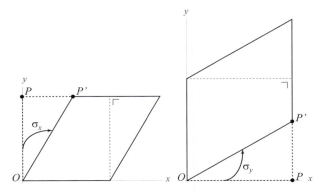

*Figure 13.8*: Standard shearing along the *x*-axis and along the *y*-axis

Eventually we are able to shear an object *simultaneously* in both the direction of the *x*-axis and the *y*-axis. We realise such a **combined standard shearing** by both shear strains $\sigma_x$ and $\sigma_y$ by applying its matrix operator

$$S_{\sigma_x,\sigma_y} = \begin{pmatrix} 1 & \tan\sigma_x & 0 \\ \tan\sigma_y & 1 & 0 \\ 0 & 0 & 1 \end{pmatrix}.$$

*Example:* We construct the matrix operator to shear flat objects simultaneously by $\sigma_x = 50°$ along the *x*-axis and by $\sigma_y = 20°$ along the *y*-axis. We then apply this operator to transform the golden rectangle bordered by the vertices $A(0,0)$ and $B(10,\ 6.18)$.

$$\begin{pmatrix} x' \\ y' \\ 1 \end{pmatrix} = \begin{pmatrix} 1 & \tan 50° & 0 \\ \tan 20° & 1 & 0 \\ 0 & 0 & 1 \end{pmatrix} \begin{pmatrix} x \\ y \\ 1 \end{pmatrix}.$$

*Properties:*

▷  The plane standard shearing conserves area.

▷  The inverse standard shearing takes the opposite angle:

$$S_{\sigma_x}^{-1} = S_{-\sigma_x} \qquad \text{and} \qquad S_{\sigma_y}^{-1} = S_{-\sigma_y}.$$

▷  The inverse combined standard shearing does not take the opposite angles:

$$S_{\sigma_x,\sigma_y}^{-1} \neq S_{-\sigma_x,-\sigma_y}.$$

▷  The combined standard shearing is *not* a combination of standard shearings:

$$S_{\sigma_x,\sigma_y} \neq S_{\sigma_x} \cdot S_{\sigma_y} \qquad \text{and} \qquad S_{\sigma_x,\sigma_y} \neq S_{\sigma_y} \cdot S_{\sigma_x}.$$

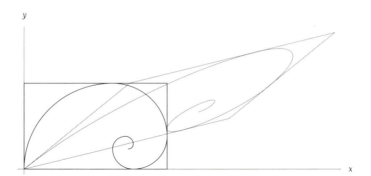

*Figure 13.9*: The combined standard shearing $S_{50°,20°}$ along both cartesian axes

## 13.6 Combining standard transformations

Let us now combine translations and the standard scaling, rotation, reflection and shearing into one transformation. This will allow us to tackle some former challenges such as scaling with respect to the object centre, rotation around an arbitrary point and combining 3D rotations.

We are aware that the matrix product is not commutative: altering the order of the matrices we multiply, changes the result in general. We visualise this non-commutative property by combining a translation (10 pixels upwards and 20 pixels to the right) and a standard reflection over the origin $O$. We apply this composite transformation to the golden rectangle bordered by the vertices $A(10, 10)$ and $B(20, 16.18)$. Firstly, we reflect after translation. Secondly, we will translate after reflection.

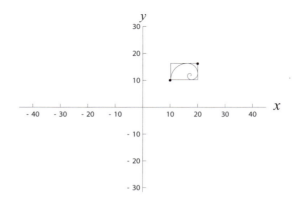

### 1. first translate

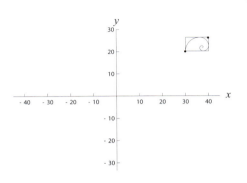

### 2. first reflect

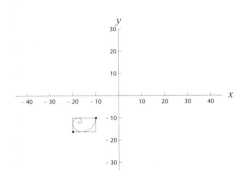

### then reflect

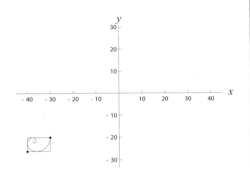

### then translate

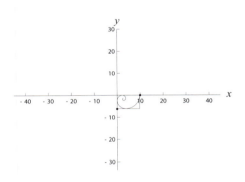

We set the translation operator $T$ and the standard reflection operator $M_O$ as

$$T = \begin{pmatrix} 1 & 0 & 20 \\ 0 & 1 & 10 \\ 0 & 0 & 1 \end{pmatrix} \text{ and } M_O = \begin{pmatrix} -1 & 0 & 0 \\ 0 & -1 & 0 \\ 0 & 0 & 1 \end{pmatrix}.$$

#### 1. first translate

$$\begin{pmatrix} 1 & 0 & 20 \\ 0 & 1 & 10 \\ 0 & 0 & 1 \end{pmatrix} \cdot \begin{pmatrix} 10 \\ 10 \\ 1 \end{pmatrix} = \begin{pmatrix} 30 \\ 20 \\ 1 \end{pmatrix}$$

#### then reflect

$$\begin{pmatrix} -1 & 0 & 0 \\ 0 & -1 & 0 \\ 0 & 0 & 1 \end{pmatrix} \cdot \begin{pmatrix} 30 \\ 20 \\ 1 \end{pmatrix} = \begin{pmatrix} -30 \\ -20 \\ 1 \end{pmatrix}$$

#### 2. first reflect

$$\begin{pmatrix} -1 & 0 & 0 \\ 0 & -1 & 0 \\ 0 & 0 & 1 \end{pmatrix} \cdot \begin{pmatrix} 10 \\ 10 \\ 1 \end{pmatrix} = \begin{pmatrix} -10 \\ -10 \\ 1 \end{pmatrix}$$

#### then translate

$$\begin{pmatrix} 1 & 0 & 20 \\ 0 & 1 & 10 \\ 0 & 0 & 1 \end{pmatrix} \cdot \begin{pmatrix} -10 \\ -10 \\ 1 \end{pmatrix} = \begin{pmatrix} 10 \\ 0 \\ 1 \end{pmatrix}$$

1. composite

$$
\begin{pmatrix} -1 & 0 & 0 \\ 0 & -1 & 0 \\ 0 & 0 & 1 \end{pmatrix} \cdot \begin{pmatrix} 1 & 0 & 20 \\ 0 & 1 & 10 \\ 0 & 0 & 1 \end{pmatrix} \cdot \begin{pmatrix} 10 \\ 10 \\ 1 \end{pmatrix}
$$

$$
= \begin{pmatrix} -1 & 0 & -20 \\ 0 & -1 & -10 \\ 0 & 0 & 1 \end{pmatrix} \cdot \begin{pmatrix} 10 \\ 10 \\ 1 \end{pmatrix}
$$

$$
= \begin{pmatrix} -30 \\ -20 \\ 1 \end{pmatrix}
$$

2. composite

$$
\begin{pmatrix} 1 & 0 & 20 \\ 0 & 1 & 10 \\ 0 & 0 & 1 \end{pmatrix} \cdot \begin{pmatrix} -1 & 0 & 0 \\ 0 & -1 & 0 \\ 0 & 0 & 1 \end{pmatrix} \cdot \begin{pmatrix} 10 \\ 10 \\ 1 \end{pmatrix}
$$

$$
= \begin{pmatrix} -1 & 0 & 20 \\ 0 & -1 & 10 \\ 0 & 0 & 1 \end{pmatrix} \cdot \begin{pmatrix} 10 \\ 10 \\ 1 \end{pmatrix}
$$

$$
= \begin{pmatrix} 10 \\ 0 \\ 1 \end{pmatrix}
$$

Altering the product order obviously changes the composite operator matrix. Aiming for a composite transformation, once we know its successive steps and in which order they have to be taken, we are able to express its composite operator matrix. Due to the non-commutativity of the matrix product, ordering its factors correctly is of major importance. The homogeneous column matrix containing the original location has to be placed as the outer right factor, then stacked with each successive operator, from right to left building up a matrix product.

We study a few examples of composite transformations to trigger our understanding of how to combine a translation and standard scaling, rotation, reflection or shearing.

## 2D ROTATION AROUND AN ARBITRARY CENTRE

We refer to the former example of the standard 2D rotation (see page 281) in which we rotated the golden rectangle over a right angle around the origin. If we aim to rotate this golden rectangle around its bottom left vertex $A(10, 10)$, we need to take three steps instead of one.

step 0                                            step 1: translate by $T_{\overrightarrow{AO}}$

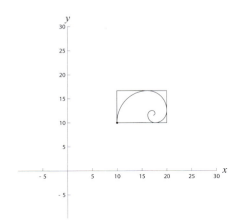

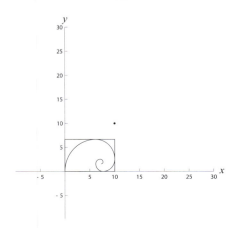

step 2: rotate by $R_{O,90°}$                     step 3: inversely translate by $T_{\overrightarrow{OA}}$

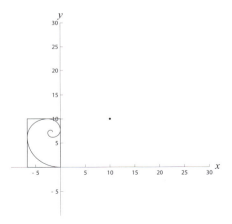

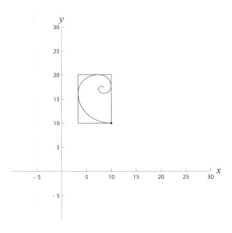

1)  A translation by $T_{\overrightarrow{AO}}$ in order to get the bottom left vertex $A(10,10)$ in the origin $O(0,0)$. This implies a shift of 10 pixels to the left and 10 pixels downwards, as dictated by $\overrightarrow{AO} = \vec{o} - \vec{a}$ (see formula (7.5) on page 131).

$$\begin{pmatrix} 1 & 0 & -10 \\ 0 & 1 & -10 \\ 0 & 0 & 1 \end{pmatrix}$$

2) A standard rotation over 90° around $O$.

$$\begin{pmatrix} \cos 90° & -\sin 90° & 0 \\ \sin 90° & \cos 90° & 0 \\ 0 & 0 & 1 \end{pmatrix}$$

3) The inverse translation by $T_{\overrightarrow{OA}}$ to restore the position of the bottom left vertex $A$. This implies a shift of 10 pixels to the right and 10 pixels upwards.

$$\begin{pmatrix} 1 & 0 & 10 \\ 0 & 1 & 10 \\ 0 & 0 & 1 \end{pmatrix}$$

We summarise the matrix expression of the composite operator $R_A(\theta)$ to calculate all image points for this example as

$$\begin{pmatrix} x' \\ y' \\ 1 \end{pmatrix} = \begin{pmatrix} 1 & 0 & 10 \\ 0 & 1 & 10 \\ 0 & 0 & 1 \end{pmatrix} \cdot \begin{pmatrix} \cos 90° & -\sin 90° & 0 \\ \sin 90° & \cos 90° & 0 \\ 0 & 0 & 1 \end{pmatrix} \cdot \begin{pmatrix} 1 & 0 & -10 \\ 0 & 1 & -10 \\ 0 & 0 & 1 \end{pmatrix} \cdot \begin{pmatrix} x \\ y \\ 1 \end{pmatrix}$$

$$= \begin{pmatrix} 0 & -1 & 20 \\ 1 & 0 & 0 \\ 0 & 0 & 1 \end{pmatrix} \cdot \begin{pmatrix} x \\ y \\ 1 \end{pmatrix} = R_A(\theta) \cdot \begin{pmatrix} x \\ y \\ 1 \end{pmatrix}.$$

Evaluating both bordering vertices $A$ and $B$ calculates the golden rectangle's new position as $A'(10, 10)$ and $B'(3.82, 20)$.

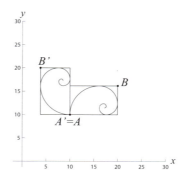

*Figure 13.10*: 2D rotation around an arbitrary point

## 3D SCALING ABOUT AN ARBITRARY CENTRE

Imagine we like to scale a 3D object, keeping it steady at its original location. Let us find the matrix operator to realise this, applying positive scale factors $s_x, s_y$ and $s_z$ with respect to its own centre $C(x_c, y_c, z_c)$. This transformation requires three steps:

1) translation by $T_{\overrightarrow{CO}}$ in order to get the object's centre $C(x_c, y_c, z_c)$ in the origin $O(0,0,0)$,

$$T_{\overrightarrow{CO}} = \begin{pmatrix} 1 & 0 & 0 & -x_c \\ 0 & 1 & 0 & -y_c \\ 0 & 0 & 1 & -z_c \\ 0 & 0 & 0 & 1 \end{pmatrix}$$

2) standard scaling $S_O$ applying positive scale factors $s_x, s_y$ and $s_z$,

$$S_O = \begin{pmatrix} s_x & 0 & 0 & 0 \\ 0 & s_y & 0 & 0 \\ 0 & 0 & s_z & 0 \\ 0 & 0 & 0 & 1 \end{pmatrix}$$

3) inverse translation by $T_{\overrightarrow{OC}}$ to restore the position of the object's centre $C$

$$T_{\overrightarrow{OC}} = \begin{pmatrix} 1 & 0 & 0 & x_c \\ 0 & 1 & 0 & y_c \\ 0 & 0 & 1 & z_c \\ 0 & 0 & 0 & 1 \end{pmatrix}.$$

We summarise the matrix expression of the composite operator $S_C$ as

$$\begin{pmatrix} x' \\ y' \\ z' \\ 1 \end{pmatrix} = \begin{pmatrix} 1 & 0 & 0 & x_c \\ 0 & 1 & 0 & y_c \\ 0 & 0 & 1 & z_c \\ 0 & 0 & 0 & 1 \end{pmatrix} \cdot \begin{pmatrix} s_x & 0 & 0 & 0 \\ 0 & s_y & 0 & 0 \\ 0 & 0 & s_z & 0 \\ 0 & 0 & 0 & 1 \end{pmatrix} \cdot \begin{pmatrix} 1 & 0 & 0 & -x_c \\ 0 & 1 & 0 & -y_c \\ 0 & 0 & 1 & -z_c \\ 0 & 0 & 0 & 1 \end{pmatrix} \cdot \begin{pmatrix} x \\ y \\ z \\ 1 \end{pmatrix}$$

$$= \begin{pmatrix} s_x & 0 & 0 & x_c - s_x x_c \\ 0 & s_y & 0 & y_c - s_y y_c \\ 0 & 0 & s_z & z_c - s_z z_c \\ 0 & 0 & 0 & 1 \end{pmatrix} \cdot \begin{pmatrix} x \\ y \\ z \\ 1 \end{pmatrix} = S_C \cdot \begin{pmatrix} x \\ y \\ z \\ 1 \end{pmatrix}.$$

## 2D REFLECTION OVER AN AXIS THROUGH THE ORIGIN

We determine the general matrix expression to reflect flat objects over a straight line $r$ described by $y = mx$, inclined to the $x$-axis by an angle $\varphi$. We recall determining this angle $\varphi$ via the slope of $r$ as $\tan\varphi = m$ (see formula (4.2)).

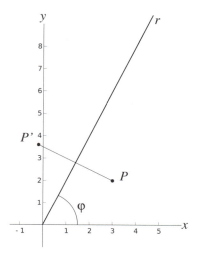

*Figure 13.11*: 2D reflection over a line $r$ through the origin $O$

Achieving this reflection requires combining three transformations in their appropriate order. Firstly, we need to rotate the line $r$ (as well as all points $P$ of the plane) clockwise around $O$ over an angle $\varphi$, to get the line $r$ onto the $x$-axis. Secondly, we perform the standard reflection over the $x$-axis. Finally, we inversely rotate the line $r$ (and the complete plane) counterclockwise over the angle $\varphi$, to restore the original orientation of the line $r$.

1) standard clockwise rotation over the angle $\varphi$ by

$$R_{O,-\varphi} = \begin{pmatrix} \cos(-\varphi) & -\sin(-\varphi) & 0 \\ \sin(-\varphi) & \cos(-\varphi) & 0 \\ 0 & 0 & 1 \end{pmatrix} = \begin{pmatrix} \cos\varphi & \sin\varphi & 0 \\ -\sin\varphi & \cos\varphi & 0 \\ 0 & 0 & 1 \end{pmatrix}$$

2) standard reflection over the $x$-axis by

$$M_x = \begin{pmatrix} 1 & 0 & 0 \\ 0 & -1 & 0 \\ 0 & 0 & 1 \end{pmatrix}$$

3) standard counterclockwise rotation over the angle $\varphi$ by

$$R_{O,\varphi} = \begin{pmatrix} \cos\varphi & -\sin\varphi & 0 \\ \sin\varphi & \cos\varphi & 0 \\ 0 & 0 & 1 \end{pmatrix}.$$

This leads to a composite matrix operator which is dependent of the line's inclination $\varphi$.

$$\begin{pmatrix} x' \\ y' \\ 1 \end{pmatrix} = \begin{pmatrix} \cos\varphi & -\sin\varphi & 0 \\ \sin\varphi & \cos\varphi & 0 \\ 0 & 0 & 1 \end{pmatrix} \cdot \begin{pmatrix} 1 & 0 & 0 \\ 0 & -1 & 0 \\ 0 & 0 & 1 \end{pmatrix} \cdot \begin{pmatrix} \cos\varphi & \sin\varphi & 0 \\ -\sin\varphi & \cos\varphi & 0 \\ 0 & 0 & 1 \end{pmatrix} \cdot \begin{pmatrix} x \\ y \\ 1 \end{pmatrix}$$

$$= \begin{pmatrix} \cos(2\varphi) & \sin(2\varphi) & 0 \\ \sin(2\varphi) & -\cos(2\varphi) & 0 \\ 0 & 0 & 1 \end{pmatrix} \cdot \begin{pmatrix} x \\ y \\ 1 \end{pmatrix}$$

## 2D REFLECTION OVER AN ARBITRARY AXIS

We extend the previous reflection in order to reflect over an arbitrary line not necessarily through the origin $O$. We assume the line $r$ described by $y = mx + c$ to intersect the $y$-axis at intercept $C(0,c)$ and to incline to the $x$-axis at an angle $\varphi$. We achieve reflecting a point $P$ over such a line $r$ by again combining standard transformations. This specific reflection we aim for, requires five standard transformations. Firstly, we shift all points vertically over a distance $c$ until the line $r$ runs through the origin. Secondly, we rotate around $O$ over an angle $-\varphi$, putting the line $r$ onto the $x$-axis. Essentially, we then reflect over the $x$-axis. Step four inversely rotates line $r$ over the angle $\varphi$. Finally, we inversely shift over the distance $c$, restoring the plane to its original position.

1) vertical translation over the distance $c$ to the origin $O$ by

$$T_{\overrightarrow{CO}} = \begin{pmatrix} 1 & 0 & 0 \\ 0 & 1 & -c \\ 0 & 0 & 1 \end{pmatrix}$$

2) standard clockwise rotation by

$$R_{O,-\varphi} = \begin{pmatrix} \cos(-\varphi) & -\sin(-\varphi) & 0 \\ \sin(-\varphi) & \cos(-\varphi) & 0 \\ 0 & 0 & 1 \end{pmatrix} = \begin{pmatrix} \cos\varphi & \sin\varphi & 0 \\ -\sin\varphi & \cos\varphi & 0 \\ 0 & 0 & 1 \end{pmatrix}$$

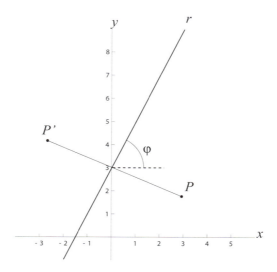

*Figure 13.12*: 2D arbitrary reflection line

3) standard reflection over the $x$-axis by

$$M_x = \begin{pmatrix} 1 & 0 & 0 \\ 0 & -1 & 0 \\ 0 & 0 & 1 \end{pmatrix}$$

4) standard counterclockwise rotation by

$$R_{O,\varphi} = \begin{pmatrix} \cos\varphi & -\sin\varphi & 0 \\ \sin\varphi & \cos\varphi & 0 \\ 0 & 0 & 1 \end{pmatrix}$$

5) vertical translation over the distance $c$ to the point $C$ by

$$T_{\overrightarrow{OC}} = \begin{pmatrix} 1 & 0 & 0 \\ 0 & 1 & c \\ 0 & 0 & 1 \end{pmatrix}.$$

This leads to a composite matrix operator which is dependent on the line's inclination $\varphi$ and its intercept $c$.

$$\begin{pmatrix} x' \\ y' \\ 1 \end{pmatrix} = \begin{pmatrix} \cos(2\varphi) & \sin(2\varphi) & -c\sin(2\varphi) \\ \sin(2\varphi) & -\cos(2\varphi) & c+c\cos(2\varphi) \\ 0 & 0 & 1 \end{pmatrix} \begin{pmatrix} x \\ y \\ 1 \end{pmatrix}.$$

*Example*:     We construct the matrix operator to reflect objects over the straight line $y = \frac{1}{2}x + 20$. We apply this transformation to reflect the golden rectangle with vertices $A(10,10)$ en $B(20,\ 16.18)$ over the above reflection line.

Constructing its matrix operator requires the reflection line's intercept $c$ and its inclination angle $\varphi$.

▷   The line $y = \frac{1}{2}x + 20$ intersects the $y$-axis in the intercept $C(0,20)$.

▷   We interpret the line's slope $m = \frac{1}{2}$ trigonometrically as $\frac{1}{2} = \tan\varphi$ which yields
$\varphi = \text{atan2}(1,2) \approx 26.57°$. Consequently, we calculate $\cos(2\varphi) \approx 0.6$ and
$\sin(2\varphi) \approx 0.8$.

Evaluating both $c = 20$ and $\varphi \approx 26.57°$ leads to the composite matrix operator which reflects points over the line $y = \frac{1}{2}x + 20$.

$$
\begin{pmatrix} x' \\ y' \\ 1 \end{pmatrix} = \begin{pmatrix} 0.6 & 0.8 & -16 \\ 0.8 & -0.6 & 32 \\ 0 & 0 & 1 \end{pmatrix} \cdot \begin{pmatrix} x \\ y \\ 1 \end{pmatrix}
$$

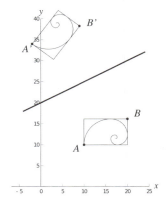

*Figure 13.13*: Reflection over the line $y = \frac{1}{2}x + 20$

We calculate both image vertices of the reflected golden rectangle as

$$
\begin{pmatrix} 0.6 & 0.8 & -16 \\ 0.8 & -0.6 & 32 \\ 0 & 0 & 1 \end{pmatrix} \cdot \begin{pmatrix} 10 \\ 10 \\ 1 \end{pmatrix} = \begin{pmatrix} -2 \\ 34 \\ 1 \end{pmatrix},
$$

$$
\begin{pmatrix} 0.6 & 0.8 & -16 \\ 0.8 & -0.6 & 32 \\ 0 & 0 & 1 \end{pmatrix} \cdot \begin{pmatrix} 20 \\ 16.18 \\ 1 \end{pmatrix} = \begin{pmatrix} 8.9 \\ 38.29 \\ 1 \end{pmatrix}.
$$

3D COMBINED ROTATION

Finally we aim for the composite matrix operator to rotate 3D objects, in this order by: a roll over 30°, a pitch over 180° and a yaw over 90°. We apply this composite transformation to rotate the triangle $ABC$ with vertices $A(220, 0, -30), B(0, 50, -150)$ and $C(40, 20, -100)$.

We need to combine three standard 3D rotations in their given order.

1) the standard rotation

$$R_{z,30°} = \begin{pmatrix} \cos 30° & -\sin 30° & 0 & 0 \\ \sin 30° & \cos 30° & 0 & 0 \\ 0 & 0 & 1 & 0 \\ 0 & 0 & 0 & 1 \end{pmatrix} = \begin{pmatrix} \frac{\sqrt{3}}{2} & -\frac{1}{2} & 0 & 0 \\ \frac{1}{2} & \frac{\sqrt{3}}{2} & 0 & 0 \\ 0 & 0 & 1 & 0 \\ 0 & 0 & 0 & 1 \end{pmatrix}$$

2) the standard rotation

$$R_{x,180°} = \begin{pmatrix} 1 & 0 & 0 & 0 \\ 0 & \cos 180° & -\sin 180° & 0 \\ 0 & \sin 180° & \cos 180° & 0 \\ 0 & 0 & 0 & 1 \end{pmatrix} = \begin{pmatrix} 1 & 0 & 0 & 0 \\ 0 & -1 & 0 & 0 \\ 0 & 0 & -1 & 0 \\ 0 & 0 & 0 & 1 \end{pmatrix}$$

3) the standard rotation

$$R_{y,90°} = \begin{pmatrix} \cos 90° & 0 & \sin 90° & 0 \\ 0 & 1 & 0 & 0 \\ -\sin 90° & 0 & \cos 90° & 0 \\ 0 & 0 & 0 & 1 \end{pmatrix} = \begin{pmatrix} 0 & 0 & 1 & 0 \\ 0 & 1 & 0 & 0 \\ -1 & 0 & 0 & 0 \\ 0 & 0 & 0 & 1 \end{pmatrix}$$

Combining these standard rotations yields the combined rotation matrix operator.

$$\begin{pmatrix} x' \\ y' \\ z' \\ 1 \end{pmatrix} = \begin{pmatrix} 0 & 0 & 1 & 0 \\ 0 & 1 & 0 & 0 \\ -1 & 0 & 0 & 0 \\ 0 & 0 & 0 & 1 \end{pmatrix} \cdot \begin{pmatrix} 1 & 0 & 0 & 0 \\ 0 & -1 & 0 & 0 \\ 0 & 0 & -1 & 0 \\ 0 & 0 & 0 & 1 \end{pmatrix} \cdot \begin{pmatrix} \frac{\sqrt{3}}{2} & -\frac{1}{2} & 0 & 0 \\ \frac{1}{2} & \frac{\sqrt{3}}{2} & 0 & 0 \\ 0 & 0 & 1 & 0 \\ 0 & 0 & 0 & 1 \end{pmatrix} \cdot \begin{pmatrix} x \\ y \\ z \\ 1 \end{pmatrix}$$

$$= \begin{pmatrix} 0 & 0 & -1 & 0 \\ \frac{-1}{2} & \frac{-\sqrt{3}}{2} & 0 & 0 \\ \frac{-\sqrt{3}}{2} & \frac{1}{2} & 0 & 0 \\ 0 & 0 & 0 & 1 \end{pmatrix} \cdot \begin{pmatrix} x \\ y \\ z \\ 1 \end{pmatrix}$$

Rotating the triangle $ABC$ with the vertices $A(220, 0, -30), B(0, 50, -150)$ and $C(40, 20, -100)$ returns their corresponding image vertices as $A'(30, \ -100, \ 173.2)$, $B'(150, \ -43.3, \ 25)$ and $C'(100, \ -37.32, \ -24.64)$.

## 13.7  Row-representation

Depending on the context, we express points $P$ either as column matrices or as row matrices. For instance, in the context of this book we typeset dot products $A \cdot P$ of the operator matrix $A$ and column matrices. The same convention is used in OpenGL, whereas DirectX® adopts row matrices to define points.

Transposing the dot product via

$$(A \cdot P)^T = P^T \cdot A^T,$$

implies we also need to transpose the operator matrix to $A^T$ in the latter case (conventionally using row matrices) and to reverse the product order.

All this leads to general two-dimensional matrix operators to

$$\left( \begin{pmatrix} a_{11} & a_{12} & \Delta x \\ a_{21} & a_{22} & \Delta y \\ 0 & 0 & 1 \end{pmatrix} \cdot \begin{pmatrix} x \\ y \\ 1 \end{pmatrix} \right)^T = \begin{pmatrix} x \\ y \\ 1 \end{pmatrix}^T \cdot \begin{pmatrix} a_{11} & a_{12} & \Delta x \\ a_{21} & a_{22} & \Delta y \\ 0 & 0 & 1 \end{pmatrix}^T$$

$$= \begin{pmatrix} x & y & 1 \end{pmatrix} \cdot \begin{pmatrix} a_{11} & a_{21} & 0 \\ a_{12} & a_{22} & 0 \\ \Delta x & \Delta y & 1 \end{pmatrix}.$$

## 13.8 Exercises

**Exercise 119** Determine the matrix operator to double the size of 3D objects along the x-axis and to halve their size along the z-axis. Scale the triangle $ABC$ spanned by the vertices $A(0,30,-100)$, $B(-50,100,-20)$ and $C(-20,0,-300)$ by this operator.

**Exercise 120** Draw the flat triangle $ABC$ defined by the vertices $A(0,0)$, $B(2,1)$ and $C(0,1)$. Rotate this triangle around $A$ counterclockwise over an angle of $45°$ and calculate all image vertices $A'$, $B'$ and $C'$. Draw this image triangle $A'B'C'$ in the same cartesian frame.

**Exercise 121** Determine the 2D transformation matrix to rotate objects counterclockwise around centre $(3,1)$ over an angle of $\frac{\pi}{4}$ radians. Apply this operator to rotate the triangle $ABC$ defined by the vertices $A(6,1)$, $B(8,2)$ and $C(7,3)$.

**Exercise 122** Prove $S_{\sigma_x}^{-1}$ and $S_{\sigma_y}^{-1}$ of the standard shearings as outlined on page 287.

**Exercise 123** Consider the 3D translation matrix given by $\Delta x = -1$, $\Delta y = -1$ and $\Delta z = -1$ for a start. Let this translation be followed by the 3D rotation around the x-axis over $+30°$. Another 3D rotation, around the y-axis over $+45°$, finalises the transformation. Combine all steps to transform the tetrahedron defined by the vertices $(0,0,0)$, $(1,0,0)$, $(0,1,0)$ and $(0,0,1)$.

**Exercise 124**

  ▷ Given the triangle $ABC$ defined by the vertices $A(-1,5)$, $B(9,0)$ and $C(-5,10)$, find its centroid $Z$ as the intersection point of its medians.

  ▷ Find the transformation operator to rotate this triangle clockwise around $Z$ over the right angle of $90°$.

  ▷ Calculate all returned image vertices $A',B',C'$ and draw both triangles, $ABC$ and its image $A'B'C'$, in different colours.

**Exercise 125** We want to resize the polyhedron defined by the vertices $A(2,2,1)$, $B(5,1,2)$, $C(5,1,-1)$, $D(2,2,-1)$, $E(2,5,1)$, $F(5,4,2)$, $G(5,1,4)$ and $H(2,5,4)$ into an image polyhedron by applying scale factor 2 along the x-axis, factor 4 along the y-axis and factor 3 along the z-axis, with respect to its corner vertex $A$. Determine the appropriate matrix operator and use it to calculate all image vertices of the returned cuboid.

**Exercise 126**   Given the hexagon defined by the vertices $A(2,2)$, $B(4,2)$, $C(5,3)$, $D(4,4)$, $E(2,4)$ and $F(1,3)$, determine the transformation matrix to rotate it clockwise around its centre $Z$ over an angle of $270°$. Calculate all returned image vertices and draw both the original and the image hexagon in different colours.

Hint: the centre $Z$ of the hexagon is the midpoint of the line segment $[AD]$.

**Exercise 127**   Reflect the trapezium $ABCD$ defined by the vertices $A(4,1)$, $B(5,2)$, $C(6,4)$ and $D(\frac{9}{2},3)$ over the line described by the parametric equation

$$\begin{cases} x = 2\lambda \\ y = -1 + 4\lambda. \end{cases}$$

Determine the standard matrix operators required at each step and in the correct order, to dot product them to the composite matrix operator. Calculate all returned image vertices and draw both the original and the image trapezium in different colours.

**Exercise 128**   Apply row reduction (see page 236) to calculate the 3D inverse of the

▷   translation matrix $T_{\overrightarrow{AB}}^{-1}$,

▷   standard scale operator $S_O^{-1}$,

▷   standard rotation operator $R_{z,\theta}^{-1}$.

**Exercise 129**

1)   Outline the required steps in their correct order to reflect over the line described by $y = x$.

2)   Dot product the above matrices to calculate the composite matrix operator which reflects over the line described by $y = x$.

3)   Apply this composite matrix operator to reflect the point $P(-3,1)$ over the line $y = x$.

4)   Draw the reflection line, the original point $P$ and its returned image point $P'$ in the same cartesian plane.

**Exercise 130**

1)   Outline the required steps in their correct order to reflect over the line $y = \frac{1}{\sqrt{3}}x$.

2)   Dot product the above matrices to calculate the composite matrix operator which reflects over the line $y = \frac{1}{\sqrt{3}}x$.

3)   Apply this composite matrix operator to reflect the circle centred on $M(3,4)$ and containing the point $P(4,5)$.

We credit this chapter to its inspirer **Koen Samyn** [14]. Even when already familiar with transformations, the subsequent chapters in this book require a deeper insight into the nature of such transformations. Apart from analysing the three standard transformations, we inevitably introduce the TRS-convention for creating composite transformations. For a better understanding, we also initiate three of such composite applications: pivoting, orbiting and look-at. While the pivot and orbit transformations tie up scene graphs (see chapter 15), the latter look-at transformation bridges us towards the so-called **steering behaviours** pioneered by **Craig Reynolds** (1953), for which we kindly refer you to the specialised literature.

TYPESETTING

For the scope of all subsequent chapters in this book, we define the **world** $(x,y)$−**frame** as the reference system originating in the standard origin $O(0,0)$ and spanned by the unit base vectors $\hat{e}_1 = \begin{pmatrix} 1 \\ 0 \end{pmatrix}$ and $\hat{e}_2 = \begin{pmatrix} 0 \\ 1 \end{pmatrix}$. We will refer all involved vectors systematically with respect to the world $(x,y)$−frame throughout the subsequent chapters, unless explicitly mentioned otherwise. However, we restrict all analysis here to the 2D space; the acquired insights extend straightforwardly to the 3D space.

## 14.1 Translation analysis

We hereby revisit the **two-dimensional translation** (see formula (13.3))

$$\begin{pmatrix} x' \\ y' \\ 1 \end{pmatrix} = \begin{pmatrix} 1 & 0 & \Delta x \\ 0 & 1 & \Delta y \\ 0 & 0 & 1 \end{pmatrix} \cdot \begin{pmatrix} x \\ y \\ 1 \end{pmatrix}.$$

For the scope of all subsequent chapters in this book, we typeset its matrix operator $T_{\overrightarrow{AB}}$ in short as $T_{\vec{t}}$ and therefore replace its displacement components by

$$\begin{cases} \Delta x = t_1 \\ \Delta y = t_2 \end{cases}$$

leading to

$$\begin{pmatrix} x' \\ y' \\ 1 \end{pmatrix} = \underbrace{\begin{pmatrix} 1 & 0 & t_1 \\ 0 & 1 & t_2 \\ 0 & 0 & 1 \end{pmatrix}}_{T_{\begin{pmatrix} t_1 \\ t_2 \end{pmatrix}}} \cdot \begin{pmatrix} x \\ y \\ 1 \end{pmatrix}. \tag{14.1}$$

Let us analyse the translation through an example. For instance, given the horizontal shift $t_1 = 5$ and the vertical offset $t_2 = 4$, we evaluate the former matrix formula as

$$\begin{pmatrix} x' \\ y' \\ 1 \end{pmatrix} = \begin{pmatrix} 1 & 0 & 5 \\ 0 & 1 & 4 \\ 0 & 0 & 1 \end{pmatrix} \cdot \begin{pmatrix} x \\ y \\ 1 \end{pmatrix}.$$

Consequently we draw its corresponding

▷  displacement vector $\vec{t} = \begin{pmatrix} 5 \\ 4 \end{pmatrix}$ and

▷  unit vectors $\hat{v}_1 = \begin{pmatrix} 1 \\ 0 \end{pmatrix}$ and $\hat{v}_2 = \begin{pmatrix} 0 \\ 1 \end{pmatrix}$,

as mentioned with respect to the underlying world $(x, y)$−frame (see figure 14.1).

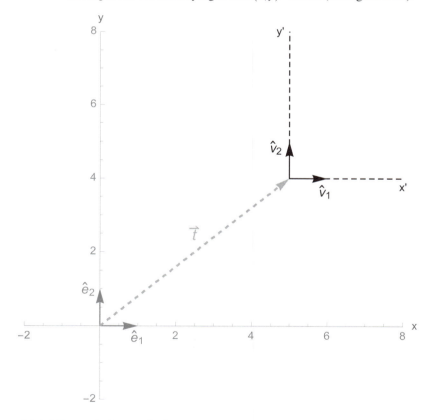

*Figure 14.1*: 2D translation analysis

The latter couple of vectors $\hat{v}_1$ and $\hat{v}_2$ are not location vectors but *free* vectors, being the local base vectors of a new **image** $(x',y')$–**frame** and are actually displacements of the world base vectors $\hat{e}_1$ and $\hat{e}_2$ because

$$T_{\binom{5}{4}}\hat{e}_1 = \hat{v}_1 \iff \begin{pmatrix} 1 & 0 & 5 \\ 0 & 1 & 4 \\ 0 & 0 & 1 \end{pmatrix} \cdot \begin{pmatrix} 1 \\ 0 \\ 0 \end{pmatrix} = \begin{pmatrix} 1 \\ 0 \\ 0 \end{pmatrix}$$

and

$$T_{\binom{5}{4}}\hat{e}_2 = \hat{v}_2 \iff \begin{pmatrix} 1 & 0 & 5 \\ 0 & 1 & 4 \\ 0 & 0 & 1 \end{pmatrix} \cdot \begin{pmatrix} 0 \\ 1 \\ 0 \end{pmatrix} = \begin{pmatrix} 0 \\ 1 \\ 0 \end{pmatrix}.$$

The image origin $O'(5,4)$ is the translation image of the default origin $O(0,0)$ given

$$T_{\binom{5}{4}}\vec{o} = \vec{o}\,' \iff \begin{pmatrix} 1 & 0 & 5 \\ 0 & 1 & 4 \\ 0 & 0 & 1 \end{pmatrix} \cdot \begin{pmatrix} 0 \\ 0 \\ 1 \end{pmatrix} = \begin{pmatrix} 5 \\ 4 \\ 1 \end{pmatrix}.$$

Note how the first two columns of the translator matrix contain $\hat{v}_1$ and $\hat{v}_2$ respectively, each bearing the **homogeneous component** $0$ corresponding to *free* vectors. The last column contains the displacement vector $\vec{t}$, bearing the homogeneous component $1$ corresponding to a *location* vector.

*Example*: 2D translation of a polygon

Translate the polygon $PQR$ by the previous $T_{\binom{5}{4}}$ given its vertices $P(0,0)$, $Q(2,1)$ and $R(1,-2)$. We calculate each image vertex similarly as

$$\begin{pmatrix} x' \\ y' \\ 1 \end{pmatrix} = \begin{pmatrix} 1 & 0 & 5 \\ 0 & 1 & 4 \\ 0 & 0 & 1 \end{pmatrix} \cdot \begin{pmatrix} 0 \\ 0 \\ 1 \end{pmatrix} = \begin{pmatrix} 5 \\ 4 \\ 1 \end{pmatrix},$$

$$\begin{pmatrix} x' \\ y' \\ 1 \end{pmatrix} = \begin{pmatrix} 1 & 0 & 5 \\ 0 & 1 & 4 \\ 0 & 0 & 1 \end{pmatrix} \cdot \begin{pmatrix} 2 \\ 1 \\ 1 \end{pmatrix} = \begin{pmatrix} 7 \\ 5 \\ 1 \end{pmatrix},$$

$$\begin{pmatrix} x' \\ y' \\ 1 \end{pmatrix} = \begin{pmatrix} 1 & 0 & 5 \\ 0 & 1 & 4 \\ 0 & 0 & 1 \end{pmatrix} \cdot \begin{pmatrix} 1 \\ -2 \\ 1 \end{pmatrix} = \begin{pmatrix} 6 \\ 2 \\ 1 \end{pmatrix}.$$

Resuming the previous $P'(5,4)$, $Q'(7,5)$ and $R'(6,2)$, we draw both the original polygon $PQR$ and the translated polygon $P'Q'R'$ in the world $(x,y)$–frame (see figure 14.2).

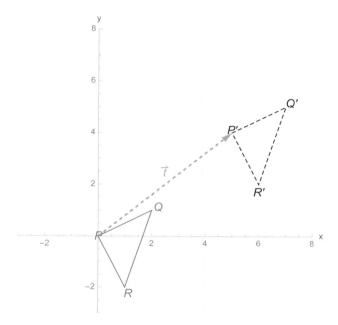

*Figure 14.2*: 2D translation of a polygon

Finally, we prove the former 2D translation matrix (14.1) to be equivalent to the linear combination of its columns. Given we translate the location vector $\vec{v} = x \cdot \hat{e}_1 + y \cdot \hat{e}_2$ by the displacement vector $\vec{t} = \begin{pmatrix} t_1 \\ t_2 \end{pmatrix}$ up to $\vec{v}\,'$ we start from their vector addition.

*Proof:*

$$\vec{v}\,' = \vec{v} + \vec{t}$$
$$\Longleftrightarrow \vec{v}\,' = x \cdot \hat{e}_1 + y \cdot \hat{e}_2 + 1 \cdot \vec{t}$$
$$\Leftrightarrow \begin{pmatrix} x' \\ y' \end{pmatrix} = x \begin{pmatrix} 1 \\ 0 \end{pmatrix} + y \begin{pmatrix} 0 \\ 1 \end{pmatrix} + 1 \begin{pmatrix} t_1 \\ t_2 \end{pmatrix}$$
$$\Leftrightarrow \begin{pmatrix} x' \\ y' \end{pmatrix} = \begin{pmatrix} 1 \cdot x + 0 \cdot y + t_1 \cdot 1 \\ 0 \cdot x + 1 \cdot y + t_2 \cdot 1 \end{pmatrix}$$
$$\Leftrightarrow \begin{pmatrix} x' \\ y' \\ 1 \end{pmatrix} = \begin{pmatrix} 1 & 0 & t_1 \\ 0 & 1 & t_2 \\ 0 & 0 & 1 \end{pmatrix} \cdot \begin{pmatrix} x \\ y \\ 1 \end{pmatrix} \qquad \blacksquare$$

Appending the homogeneous row $\boxed{0 \quad 0 \quad 1}$ completes the above proof.

## 14.2 Scaling analysis

We hereby revisit the **two-dimensional standard scaling** (see formula (13.6))

$$\begin{pmatrix} x' \\ y' \\ 1 \end{pmatrix} = \begin{pmatrix} s_x & 0 & 0 \\ 0 & s_y & 0 \\ 0 & 0 & 1 \end{pmatrix} \cdot \begin{pmatrix} x \\ y \\ 1 \end{pmatrix}.$$

For the scope of all subsequent chapters in this book, we typeset its matrix operator $S_O$ alternatively with its positive scale factors $s_x$ and $s_y$ leading to

$$\begin{pmatrix} x' \\ y' \\ 1 \end{pmatrix} = \underbrace{\begin{pmatrix} s_x & 0 & 0 \\ 0 & s_y & 0 \\ 0 & 0 & 1 \end{pmatrix}}_{S_{\begin{pmatrix} s_x \\ s_y \end{pmatrix}}} \cdot \begin{pmatrix} x \\ y \\ 1 \end{pmatrix}. \qquad (14.2)$$

Let us analyse the scaling transformation through an example. Given the horizontal scale factor $s_x = 2$ and vertical scale factor $s_y = 3$ for a **non-uniform scaling**, we evaluate the former matrix formula as

$$\begin{pmatrix} x' \\ y' \\ 1 \end{pmatrix} = \begin{pmatrix} 2 & 0 & 0 \\ 0 & 3 & 0 \\ 0 & 0 & 1 \end{pmatrix} \cdot \begin{pmatrix} x \\ y \\ 1 \end{pmatrix}.$$

Consequently we draw its corresponding

▷ displacement vector $\vec{t} = \begin{pmatrix} 0 \\ 0 \end{pmatrix}$ for a *pure* scaling and

▷ non-unit vectors $\vec{v}_1 = \begin{pmatrix} 2 \\ 0 \end{pmatrix}$ and $\vec{v}_2 = \begin{pmatrix} 0 \\ 3 \end{pmatrix}$,

as mentioned with respect to the underlying world $(x, y)$–frame (see figure 14.3).

The latter couple of vectors $\vec{v}_1$ and $\vec{v}_2$ are not location vectors but free vectors that jointly span the new image image $(x', y')$–frame as local base vectors.

Its first base vector $\vec{v}_1 = \begin{pmatrix} 2 \\ 0 \end{pmatrix}$ is the scaled image of the world's base vector $\hat{e}_1 = \begin{pmatrix} 1 \\ 0 \end{pmatrix}$

$$\begin{pmatrix} 2 & 0 & 0 \\ 0 & 3 & 0 \\ 0 & 0 & 1 \end{pmatrix} \cdot \begin{pmatrix} 1 \\ 0 \\ 0 \end{pmatrix} = \begin{pmatrix} 2 \\ 0 \\ 0 \end{pmatrix}.$$

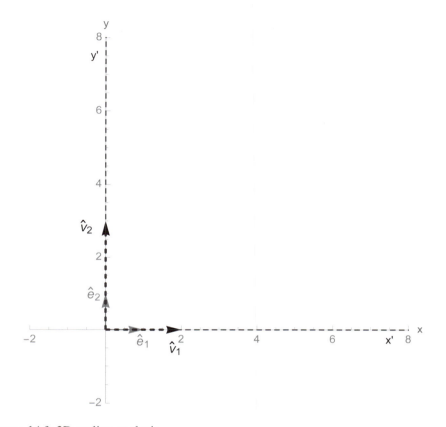

*Figure 14.3*: 2D scaling analysis

We see both *free* vector $\hat{e}_1$ and its image $\vec{v}_1$ featuring the homogeneous component 0.

Similarly, the second image base vector $\vec{v}_2 = \begin{pmatrix} 0 \\ 3 \end{pmatrix}$ is the the scaling of $\hat{e}_2 = \begin{pmatrix} 0 \\ 1 \end{pmatrix}$.

$$\begin{pmatrix} 2 & 0 & 0 \\ 0 & 3 & 0 \\ 0 & 0 & 1 \end{pmatrix} \cdot \begin{pmatrix} 0 \\ 1 \\ 0 \end{pmatrix} = \begin{pmatrix} 0 \\ 3 \\ 0 \end{pmatrix}.$$

We realise both *free* vector $\hat{e}_2$ and its image $\vec{v}_2$ keep the homogeneous component 0. We notice how the first two columns of the standard scaling matrix contain $\vec{v}_1$ and $\vec{v}_2$ respectively, each bearing the homogeneous component 0. The last column contains in the case of a pure scaling the displacement vector set to $\vec{o}$ with homogeneous component 1 for this *location* vector.

*Example*: 2D standard scaling of a polygon

Scale the polygon *PQR* by the previous $S_{\binom{2}{3}}$ given its vertices $P(0,0)$, $Q(2,1)$ and $R(1,-2)$. We calculate each image vertex similarly as

$$\begin{pmatrix} x' \\ y' \\ 1 \end{pmatrix} = \begin{pmatrix} 2 & 0 & 0 \\ 0 & 3 & 0 \\ 0 & 0 & 1 \end{pmatrix} \cdot \begin{pmatrix} 0 \\ 0 \\ 1 \end{pmatrix} = \begin{pmatrix} 0 \\ 0 \\ 1 \end{pmatrix},$$

$$\begin{pmatrix} x' \\ y' \\ 1 \end{pmatrix} = \begin{pmatrix} 2 & 0 & 0 \\ 0 & 3 & 0 \\ 0 & 0 & 1 \end{pmatrix} \cdot \begin{pmatrix} 2 \\ 1 \\ 1 \end{pmatrix} = \begin{pmatrix} 4 \\ 3 \\ 1 \end{pmatrix},$$

$$\begin{pmatrix} x' \\ y' \\ 1 \end{pmatrix} = \begin{pmatrix} 2 & 0 & 0 \\ 0 & 3 & 0 \\ 0 & 0 & 1 \end{pmatrix} \cdot \begin{pmatrix} 1 \\ -2 \\ 1 \end{pmatrix} = \begin{pmatrix} 2 \\ -6 \\ 1 \end{pmatrix}.$$

Resuming the previous $P'(0,0)$, $Q'(4,3)$ and $R'(2,-6)$, we draw both the original polygon *PQR* and the scaled polygon $P'Q'R'$ in the world $(x,y)$-frame (see figure 14.4).

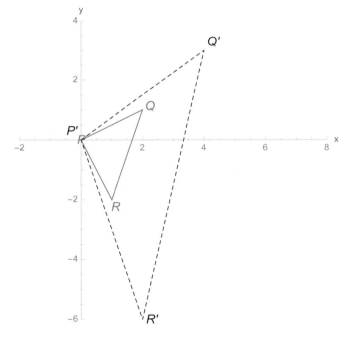

*Figure 14.4*: 2D standard scaling of a polygon with respect to the world's origin

Finally, we prove the former 2D standard scaling matrix (14.2) to be equivalent to the linear combination of its columns. Considering a location vector $\vec{v} = x \cdot \hat{e}_1 + y \cdot \hat{e}_2$ we investigate the effect of replacing its world base unit vectors

$$\hat{e}_1 = \begin{pmatrix} 1 \\ 0 \end{pmatrix} \text{ and } \hat{e}_2 = \begin{pmatrix} 0 \\ 1 \end{pmatrix} \text{ by the scaled base vectors } \vec{v}_1 = \begin{pmatrix} s_x \\ 0 \end{pmatrix} \text{ and } \vec{v}_2 = \begin{pmatrix} 0 \\ s_y \end{pmatrix},$$

spanning the image location vector $\vec{v}\,'$ purely with the displacement vector $\vec{t} = \vec{o}$.

*Proof:*

$$\vec{v}\,' = x \cdot \vec{v}_1 + y \cdot \vec{v}_2 + 1 \cdot \vec{t}$$

$$\Longleftrightarrow \begin{pmatrix} x' \\ y' \end{pmatrix} = x \begin{pmatrix} s_x \\ 0 \end{pmatrix} + y \begin{pmatrix} 0 \\ s_y \end{pmatrix} + 1 \begin{pmatrix} 0 \\ 0 \end{pmatrix}$$

$$\Longleftrightarrow \begin{pmatrix} x' \\ y' \end{pmatrix} = \begin{pmatrix} s_x \cdot x + 0 \cdot y + 0 \cdot 1 \\ 0 \cdot x + s_y \cdot y + 0 \cdot 1 \end{pmatrix}$$

$$\Longleftrightarrow \begin{pmatrix} x' \\ y' \\ 1 \end{pmatrix} = \begin{pmatrix} s_x & 0 & 0 \\ 0 & s_y & 0 \\ 0 & 0 & 1 \end{pmatrix} \cdot \begin{pmatrix} x \\ y \\ 1 \end{pmatrix} \qquad \blacksquare$$

Appending the homogeneous row $\boxed{0 \quad 0 \quad 1}$ completes the above proof. We keep in mind the condition set $s_x > 0$, $s_y > 0$ and $\vec{t} = \vec{0}$ being required for pure scaling effects.

## 14.3 Rotation analysis

We hereby revisit the **two-dimensional standard rotation** (see formula (13.9))

$$\begin{pmatrix} x' \\ y' \\ 1 \end{pmatrix} = \begin{pmatrix} \cos\theta & -\sin\theta & 0 \\ \sin\theta & \cos\theta & 0 \\ 0 & 0 & 1 \end{pmatrix} \cdot \begin{pmatrix} x \\ y \\ 1 \end{pmatrix}.$$

For the scope of all subsequent chapters in this book, we typeset its matrix operator $R_{O,\theta}$ emphasising its angle rotation angle $\theta$ as $R_O(\theta)$

$$\begin{pmatrix} x' \\ y' \\ 1 \end{pmatrix} = \underbrace{\begin{pmatrix} \cos\theta & -\sin\theta & 0 \\ \sin\theta & \cos\theta & 0 \\ 0 & 0 & 1 \end{pmatrix}}_{R_O(\theta)} \cdot \begin{pmatrix} x \\ y \\ 1 \end{pmatrix}. \tag{14.3}$$

Let us analyse the rotation through an example like $R_O(30°)$ for which we evaluate the former matrix formula as

$$\begin{pmatrix} x' \\ y' \\ 1 \end{pmatrix} = \begin{pmatrix} \cos 30° & -\sin 30° & 0 \\ \sin 30° & \cos 30° & 0 \\ 0 & 0 & 1 \end{pmatrix} \cdot \begin{pmatrix} x \\ y \\ 1 \end{pmatrix}$$

$$= \begin{pmatrix} \frac{\sqrt{3}}{2} & -\frac{1}{2} & 0 \\ \frac{1}{2} & \frac{\sqrt{3}}{2} & 0 \\ 0 & 0 & 1 \end{pmatrix} \cdot \begin{pmatrix} x \\ y \\ 1 \end{pmatrix}$$

$$\approx \begin{pmatrix} 0.87 & -0.50 & 0 \\ 0.50 & 0.87 & 0 \\ 0 & 0 & 1 \end{pmatrix} \cdot \begin{pmatrix} x \\ y \\ 1 \end{pmatrix}.$$

Consequently we draw its corresponding

▷ displacement vector $\vec{t} = \begin{pmatrix} 0 \\ 0 \end{pmatrix}$ and given we set $s_x = s_y = 1$,

▷ unit vectors $\hat{v}_1 \approx \begin{pmatrix} 0.87 \\ 0.50 \end{pmatrix}$ and $\hat{v}_2 \approx \begin{pmatrix} -0.50 \\ 0.87 \end{pmatrix}$,

of this **standard** rotation with respect to the world $(x, y)$−frame (see figure 14.5).

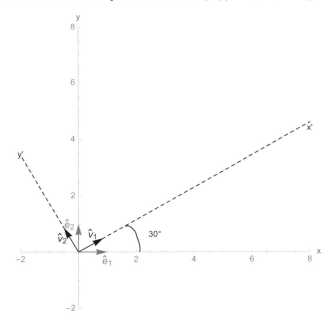

*Figure 14.5*: 2D standard rotation analysis

The latter couple of vectors $\hat{v}_1$ and $\hat{v}_2$ are not location vectors but free vectors that jointly span as new base vectors the new image $(x', y')-$frame. The first new base vector $\hat{v}_1$ is the rotation of the world's base vector $\hat{e}_1$ and the second vector $\hat{v}_2$ is the rotation of $\hat{e}_2$.

$$\begin{pmatrix} 0.87 & -0.50 & 0 \\ 0.50 & 0.87 & 0 \\ 0 & 0 & 1 \end{pmatrix} \cdot \begin{pmatrix} 1 \\ 0 \\ 0 \end{pmatrix} = \begin{pmatrix} 0.87 \\ 0.50 \\ 0 \end{pmatrix}$$

$$\begin{pmatrix} 0.87 & -0.50 & 0 \\ 0.50 & 0.87 & 0 \\ 0 & 0 & 1 \end{pmatrix} \cdot \begin{pmatrix} 0 \\ 1 \\ 0 \end{pmatrix} = \begin{pmatrix} -0.50 \\ 0.87 \\ 0 \end{pmatrix}$$

We see how all *free* vectors feature the homogeneous component 0 in the matrix formula. We notice how the first two columns of the standard rotation matrix $R_O(30°)$ contain $\hat{v}_1$ and $\hat{v}_2$, while the last column contains the displacement vector $\vec{o}$ with its homogeneous component 1 (corresponding to a *location* vector).

*Example*: 2D standard rotation of a polygon

Rotate the polygon $PQR$ by the previous $R_O(30°)$ given its vertices $P(0,0)$, $Q(2,1)$ and $R(1,-2)$. We calculate each image vertex similarly as

$$\begin{pmatrix} x' \\ y' \\ 1 \end{pmatrix} = \begin{pmatrix} 0.87 & -0.50 & 0 \\ 0.50 & 0.87 & 0 \\ 0 & 0 & 1 \end{pmatrix} \cdot \begin{pmatrix} 0 \\ 0 \\ 1 \end{pmatrix} = \begin{pmatrix} 0.00 \\ 0.00 \\ 1 \end{pmatrix},$$

$$\begin{pmatrix} x' \\ y' \\ 1 \end{pmatrix} = \begin{pmatrix} 0.87 & -0.50 & 0 \\ 0.50 & 0.87 & 0 \\ 0 & 0 & 1 \end{pmatrix} \cdot \begin{pmatrix} 2 \\ 1 \\ 1 \end{pmatrix} = \begin{pmatrix} 1.23 \\ 1.87 \\ 1 \end{pmatrix},$$

$$\begin{pmatrix} x' \\ y' \\ 1 \end{pmatrix} = \begin{pmatrix} 0.87 & -0.50 & 0 \\ 0.50 & 0.87 & 0 \\ 0 & 0 & 1 \end{pmatrix} \cdot \begin{pmatrix} 1 \\ -2 \\ 1 \end{pmatrix} = \begin{pmatrix} 1.87 \\ -1.23 \\ 1 \end{pmatrix}.$$

Resuming the previous $P'(0.00, 0.00)$, $Q'(1.23, 1.87)$ and $R'(1.87, -1.23)$, we draw both the original polygon $PQR$ and the scaled polygon $P'Q'R'$ in the world $(x, y)-$frame (see figure 14.6).

Finally, we prove the former 2D standard rotation matrix (14.3) to be equivalent to the linear combination of its columns. Considering a location vector $\vec{v} = x \cdot \hat{e}_1 + y \cdot \hat{e}_2$ we investigate the effect of replacing its world base vectors $\hat{e}_1 = \begin{pmatrix} 1 \\ 0 \end{pmatrix} \perp \hat{e}_2 = \begin{pmatrix} 0 \\ 1 \end{pmatrix}$ by the rotated base vectors $\hat{v}_1 = \begin{pmatrix} \cos\theta \\ \sin\theta \end{pmatrix} \perp \hat{v}_2 = \begin{pmatrix} -\sin\theta \\ \cos\theta \end{pmatrix}$, which leads us into the rotated location vector $\vec{v}\,'$.

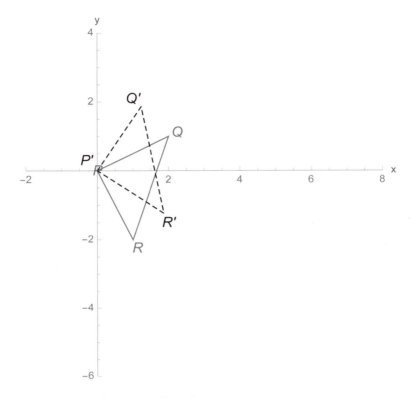

*Figure 14.6*: 2D standard rotation of a polygon

*Proof:*

$$\vec{v}\,' \quad = \quad x \cdot \hat{v}_1 \quad + \quad y \cdot \hat{v}_2 \quad + \quad 1 \cdot \vec{t}$$

$$\Longleftrightarrow \begin{pmatrix} x' \\ y' \end{pmatrix} = x \begin{pmatrix} \cos\theta \\ \sin\theta \end{pmatrix} + y \begin{pmatrix} -\sin\theta \\ \cos\theta \end{pmatrix} + 1 \begin{pmatrix} 0 \\ 0 \end{pmatrix}$$

$$\Longleftrightarrow \begin{pmatrix} x' \\ y' \end{pmatrix} = \begin{pmatrix} \cos\theta \cdot x - \sin\theta \cdot y + 0 \cdot 1 \\ \sin\theta \cdot x + \cos\theta \cdot y + 0 \cdot 1 \end{pmatrix}$$

$$\Longleftrightarrow \begin{pmatrix} x' \\ y' \\ 1 \end{pmatrix} = \begin{pmatrix} \cos\theta & -\sin\theta & 0 \\ \sin\theta & \cos\theta & 0 \\ 0 & 0 & 1 \end{pmatrix} \cdot \begin{pmatrix} x \\ y \\ 1 \end{pmatrix} \qquad\blacksquare$$

Appending the homogeneous row $\boxed{0}\;\boxed{0}\;\boxed{1}$ completes the above proof. We keep in mind the condition set $s_x = s_y = 1$ and $\vec{t} = \vec{0}$ is required for pure rotations.

## 14.4 Composite transformation analysis

We hereby revisit the **two-dimensional composite transformation** of the translation, the standard scaling and the standard rotation (see paragraph 13.6) but anchored by its local base vectors $\hat{v}_1 \perp \hat{v}_2$

$$
\begin{pmatrix} x' \\ y' \\ h \end{pmatrix} = \underbrace{\begin{pmatrix} v_{1_x} & v_{2_x} & t_1 \\ v_{1_y} & v_{2_y} & t_2 \\ 0 & 0 & 1 \end{pmatrix}}_{A} \cdot \begin{pmatrix} x \\ y \\ h \end{pmatrix}, \tag{14.4}
$$

whereby the **homogeneous component** is set

$$
h = \begin{cases} 0 & \text{for a free vector} \\ 1 & \text{for a location vector.} \end{cases} \tag{14.5}
$$

In order to immediately interprete the last column of this composite transformation matrix as its translation-part $\vec{t} = \begin{pmatrix} t_1 \\ t_2 \end{pmatrix}$ a well-ordered composition is required. We define the popular **TRS-convention** as this specific order to produce the action matrix $A = T \cdot R \cdot S$ which in this column-based book reads stackwise *'First scale, then rotate and finally translate (your object)'*. Let us analyse the general action matrix $A$ in terms of these well-ordered standard steps. We therefore calculate $A$ by precisely this underlying TRS-product.

$$
\begin{aligned}
A &= T_{\begin{pmatrix} t_1 \\ t_2 \end{pmatrix}} \cdot R_O(\theta) \cdot S_{\begin{pmatrix} s_x \\ s_y \end{pmatrix}} \\[2mm]
&= \begin{pmatrix} 1 & 0 & t_1 \\ 0 & 1 & t_2 \\ 0 & 0 & 1 \end{pmatrix} \cdot \left( \begin{pmatrix} \cos\theta & -\sin\theta & 0 \\ \sin\theta & \cos\theta & 0 \\ 0 & 0 & 1 \end{pmatrix} \cdot \begin{pmatrix} s_x & 0 & 0 \\ 0 & s_y & 0 \\ 0 & 0 & 1 \end{pmatrix} \right) \\[2mm]
&= \begin{pmatrix} 1 & 0 & t_1 \\ 0 & 1 & t_2 \\ 0 & 0 & 1 \end{pmatrix} \cdot \begin{pmatrix} s_x\cos\theta & s_y(-\sin\theta) & 0 \\ s_x\sin\theta & s_y\cos\theta & 0 \\ 0 & 0 & 1 \end{pmatrix} \\[2mm]
&= \begin{pmatrix} s_x\cos\theta & -s_y\sin\theta & t_1 \\ s_x\sin\theta & s_y\cos\theta & t_2 \\ 0 & 0 & 1 \end{pmatrix}
\end{aligned}
$$

Identifying the corresponding elements of either of the above expressions for A

$$
\begin{pmatrix} v_{1_x} & v_{2_x} & t_1 \\ v_{1_y} & v_{2_y} & t_2 \\ 0 & 0 & 1 \end{pmatrix} = \begin{pmatrix} s_x\cos\theta & -s_y\sin\theta & t_1 \\ s_x\sin\theta & s_y\cos\theta & t_2 \\ 0 & 0 & 1 \end{pmatrix}, \tag{14.6}
$$

reveals how we

▷ read the $T_{\binom{t_1}{t_2}}$ displacement vector directly in the last matrix column,

▷ calculate column wise the $S_{\binom{s_x}{s_y}}$ scale factors respectively as

$$
\begin{aligned}
(s_x\cos\theta)^2 + (s_x\sin\theta)^2 &= (v_{1_x})^2 + (v_{1_y})^2 \\
\Longleftrightarrow \qquad (s_x)^2 &= \|\vec{v}_1\|^2 \\
\Longrightarrow \qquad s_x &= \|\vec{v}_1\| = \sqrt{(v_{1_x})^2 + (v_{1_y})^2} \qquad (14.7)
\end{aligned}
$$

$$
\begin{aligned}
(s_y(-\sin\theta))^2 + (s_y\cos\theta)^2 &= (v_{2_x})^2 + (v_{2_y})^2 \\
\Longleftrightarrow \qquad (s_y)^2 &= \|\vec{v}_2\|^2 \\
\Longrightarrow \qquad s_y &= \|\vec{v}_2\| = \sqrt{(v_{2_x})^2 + (v_{2_y})^2} \qquad (14.8)
\end{aligned}
$$

▷ and finally retrieve the $R_O(\theta)$ rotation angle via the inverse tangent-with-quadrant function atan2 which takes two arguments (see page 85), given $s_x > 0$

$$
\begin{aligned}
\frac{s_x\sin\theta}{s_x\cos\theta} &= \frac{v_{1_y}}{v_{1_x}} \\
\Longleftrightarrow \qquad \tan\theta &= \frac{v_{1_y}}{v_{1_x}} \\
\Longrightarrow \qquad \theta &= \text{atan2}\left(v_{1_y}, v_{1_x}\right) \qquad (14.9)
\end{aligned}
$$

*Example*: 2D general transformation of a polygon

Transform the polygon $PQR$ by the composite TRS-action $A = T_{\binom{5}{4}} \cdot R_O(30°) \cdot S_{\binom{2}{3}}$ given its vertices $P(0,0)$, $Q(2,1)$ and $R(1,-2)$.

Let us first evaluate this TRS-action as

$$
\begin{aligned}
A &= \begin{pmatrix} s_x\cos\theta & s_y(-\sin\theta) & t_1 \\ s_x\sin\theta & s_y\cos\theta & t_2 \\ 0 & 0 & 1 \end{pmatrix} = \begin{pmatrix} 2\cos 30° & 3(-\sin 30°) & 5 \\ 2\sin 30° & 3\cos 30° & 4 \\ 0 & 0 & 1 \end{pmatrix} \\
&= \begin{pmatrix} \sqrt{3} & -\frac{3}{2} & 5 \\ 1 & \frac{3\sqrt{3}}{2} & 4 \\ 0 & 0 & 1 \end{pmatrix} \approx \begin{pmatrix} 1.73 & -1.50 & 5.00 \\ 1.00 & 2.60 & 4.00 \\ 0 & 0 & 1 \end{pmatrix}.
\end{aligned}
$$

We then calculate each image vertex similarly as

$$
\begin{pmatrix} x' \\ y' \\ 1 \end{pmatrix} = \begin{pmatrix} 1.73 & -1.50 & 5.00 \\ 1.00 & 2.60 & 4.00 \\ 0 & 0 & 1 \end{pmatrix} \cdot \begin{pmatrix} 0 \\ 0 \\ 1 \end{pmatrix} = \begin{pmatrix} 5.00 \\ 4.00 \\ 1 \end{pmatrix},
$$

$$\begin{pmatrix} x' \\ y' \\ 1 \end{pmatrix} = \begin{pmatrix} 1.73 & -1.50 & 5.00 \\ 1.00 & 2.60 & 4.00 \\ 0 & 0 & 1 \end{pmatrix} \cdot \begin{pmatrix} 2 \\ 1 \\ 1 \end{pmatrix} = \begin{pmatrix} 6.96 \\ 8.60 \\ 1 \end{pmatrix},$$

$$\begin{pmatrix} x' \\ y' \\ 1 \end{pmatrix} = \begin{pmatrix} 1.73 & -1.50 & 5.00 \\ 1.00 & 2.60 & 4.00 \\ 0 & 0 & 1 \end{pmatrix} \cdot \begin{pmatrix} 1 \\ -2 \\ 1 \end{pmatrix} = \begin{pmatrix} 9.73 \\ -0.20 \\ 1 \end{pmatrix}.$$

Resuming the previous $P'(5.00, 4.00)$, $Q'(6.96, 8.60)$ and $R'(9.73, -0.20)$, we draw both the original polygon $PQR$ and the generally transformed polygon $P'Q'R'$ in the world $(x, y)$−frame (see figure 14.7).

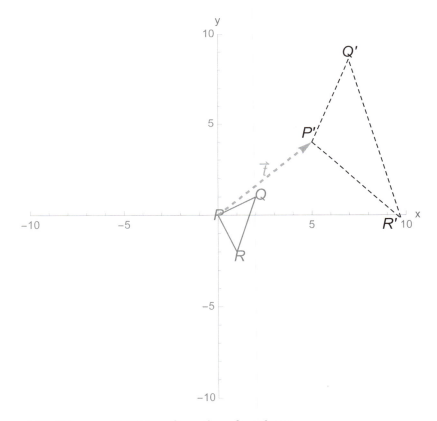

*Figure 14.7*: 2D general TRS-transformation of a polygon

Generally, we prove the former 2D general transformation matrix (14.4) to be equivalent to the linear combination of its columns. We apply the general homogeneous component $h$ (see formula (14.5)). Considering any vector $\vec{v} = x \cdot \hat{e}_1 + y \cdot \hat{e}_2$ we investigate the effect of having replaced its world base unit vectors $\hat{e}_1 \perp \hat{e}_2$ by the RS-transformed base vectors $\vec{v}_1 = \begin{pmatrix} v_{1x} \\ v_{1y} \end{pmatrix} \perp \vec{v}_2 = \begin{pmatrix} v_{2x} \\ v_{2y} \end{pmatrix}$ and translated by the displacement vector $\vec{t} = \begin{pmatrix} t_1 \\ t_2 \end{pmatrix}$ thus resulting in the image vector $\vec{v}\,'$.

*Proof:*

$$\vec{v}\,' = x \cdot \vec{v}_1 + y \cdot \vec{v}_2 + h \cdot \vec{t} \tag{14.10}$$

$$\Longleftrightarrow \begin{pmatrix} x' \\ y' \end{pmatrix} = x \begin{pmatrix} v_{1x} \\ v_{1y} \end{pmatrix} + y \begin{pmatrix} v_{2x} \\ v_{2y} \end{pmatrix} + h \begin{pmatrix} t_1 \\ t_2 \end{pmatrix}$$

$$\Longleftrightarrow \begin{pmatrix} x' \\ y' \end{pmatrix} = \begin{pmatrix} x \cdot v_{1x} + y \cdot v_{2x} + h \cdot t_1 \\ x \cdot v_{1y} + y \cdot v_{2y} + h \cdot t_2 \end{pmatrix}$$

$$\Longleftrightarrow \begin{pmatrix} x' \\ y' \\ h \end{pmatrix} = \begin{pmatrix} v_{1x} & v_{2x} & t_1 \\ v_{1y} & v_{2y} & t_2 \\ 0 & 0 & 1 \end{pmatrix} \cdot \begin{pmatrix} x \\ y \\ h \end{pmatrix}. \qquad\blacksquare$$

Appending the homogeneous row $\boxed{0} \ \boxed{0} \ \boxed{1}$ completes the above proof.

## 14.5 Conventions

Depending on the context, we express points $P$ either as column matrices or as row matrices. For instance, throughout this book we typeset dot products $A \cdot P$ of the operator matrix $A$ and column matrices. The same convention is used in OpenGL, whereas DirectX® adopts row matrices to define points.

Transposing the dot product via

$$(A \cdot P)^T = P^T \cdot A^T,$$

consequently implies we need to transpose the operator matrix to $A^T$ in the latter case (conventionally using row matrices) and to reverse the product order.

For general two-dimensional matrix operators, all this leads to

$$\left( \begin{pmatrix} v_{1x} & v_{2x} & t_1 \\ v_{1y} & v_{2y} & t_2 \\ 0 & 0 & 1 \end{pmatrix} \cdot \begin{pmatrix} x \\ y \\ 1 \end{pmatrix} \right)^T = \begin{pmatrix} x \\ y \\ 1 \end{pmatrix}^T \cdot \begin{pmatrix} v_{1x} & v_{2x} & t_1 \\ v_{1y} & v_{2y} & t_2 \\ 0 & 0 & 1 \end{pmatrix}^T$$

$$= \begin{pmatrix} x & y & 1 \end{pmatrix} \cdot \begin{pmatrix} v_{1x} & v_{1y} & 0 \\ v_{2x} & v_{2y} & 0 \\ t_1 & t_2 & 1 \end{pmatrix}.$$

## 14.6 Applications

P IVOT  TRANSFORMATION

Neutralising the former scaling operator by setting $s_x = s_y = 1$, the TRS-composite transformation simplifies to

$$P_B(\theta) \quad = \quad T_{\overrightarrow{OB}} \cdot R_O(\theta) \cdot S_{\binom{1}{1}} \quad = \quad T_{\overrightarrow{OB}} \cdot R_O(\theta) \cdot I_3 \quad = \quad T_{\overrightarrow{OB}} \cdot R_O(\theta),$$

which pivots an object around the local centre $B(b_1, b_2)$.

Alternatively, we call this an **off-centre rotation**, **non-standard rotation** or a **spinning** around the point $B$. We calculate this pivoting operator as

$$P_{(b_1,b_2)}(\theta) \quad = \quad T_{\binom{b_1}{b_2}} \cdot R_O(\theta) \tag{14.11}$$

$$= \quad \begin{pmatrix} 1 & 0 & b_1 \\ 0 & 1 & b_2 \\ 0 & 0 & 1 \end{pmatrix} \cdot \begin{pmatrix} \cos\theta & -\sin\theta & 0 \\ \sin\theta & \cos\theta & 0 \\ 0 & 0 & 1 \end{pmatrix} = \begin{pmatrix} \cos\theta & -\sin\theta & b_1 \\ \sin\theta & \cos\theta & b_2 \\ 0 & 0 & 1 \end{pmatrix}$$

*Example*:  2D pivot transformation

Pivot the polygon $PQR$ around the local centre $B(5,4)$ by a $30°$ angle, given its vertices $P(0,0)$, $Q(2,1)$ and $R(1,-2)$.

Let us first evaluate this pivot transformation $P_B(\theta)$ as

$$P_{(5,4)}(30°) \quad = \quad \begin{pmatrix} \cos 30° & -\sin 30° & 5 \\ \sin 30° & \cos 30° & 4 \\ 0 & 0 & 1 \end{pmatrix} \approx \begin{pmatrix} 0.87 & -0.50 & 5 \\ 0.50 & 0.87 & 4 \\ 0 & 0 & 1 \end{pmatrix}.$$

We then calculate each image vertex similarly as

$$\begin{pmatrix} x' \\ y' \\ 1 \end{pmatrix} = \begin{pmatrix} 0.87 & -0.50 & 5 \\ 0.50 & 0.87 & 4 \\ 0 & 0 & 1 \end{pmatrix} \cdot \begin{pmatrix} 0 \\ 0 \\ 1 \end{pmatrix} = \begin{pmatrix} 5 \\ 4 \\ 1 \end{pmatrix},$$

$$\begin{pmatrix} x' \\ y' \\ 1 \end{pmatrix} = \begin{pmatrix} 0.87 & -0.50 & 5 \\ 0.50 & 0.87 & 4 \\ 0 & 0 & 1 \end{pmatrix} \cdot \begin{pmatrix} 2 \\ 1 \\ 1 \end{pmatrix} = \begin{pmatrix} 6.24 \\ 5.87 \\ 1 \end{pmatrix},$$

$$\begin{pmatrix} x' \\ y' \\ 1 \end{pmatrix} = \begin{pmatrix} 0.87 & -0.50 & 5 \\ 0.50 & 0.87 & 4 \\ 0 & 0 & 1 \end{pmatrix} \cdot \begin{pmatrix} 1 \\ -2 \\ 1 \end{pmatrix} = \begin{pmatrix} 6.87 \\ 2.76 \\ 1 \end{pmatrix}.$$

Resuming the previous $P'(5,4)$, $Q'(6.24,\ 5.87)$ and $R'(6.87,\ 2.76)$, we draw both the original polygon $PQR$ and the pivoted polygon $P'Q'R'$ in the world $(x,y)-$frame (see figure 14.8).

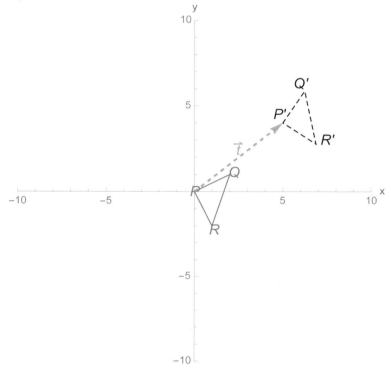

*Figure 14.8*: 2D pivoting of a polygon

O RBIT   TRANSFORMATION

Neutralising the former scaling operator by setting $s_x = s_y = 1$ and reversing the TRS-ordered transformation into an SRT-composition simplifies to

$$O_B(\theta) \;\;=\;\; S_{\binom{1}{1}} \cdot R_O(\theta) \cdot T_{\overrightarrow{OB}} \;\;=\;\; I_3 \cdot R_O(\theta) \cdot T_{\overrightarrow{OB}} \;\;=\;\; R_O(\theta) \cdot T_{\overrightarrow{OB}},$$

which orbits an object anchored in $B(b_1,b_2)$ around the world's origin $O(0,0)$.

We calculate this orbiting operator as

$$O_{(b_1,b_2)}(\theta) \quad = \quad R_O(\theta) \cdot T_{\left(\begin{smallmatrix} b_1 \\ b_2 \end{smallmatrix}\right)} \tag{14.12}$$

$$= \quad \begin{pmatrix} \cos\theta & -\sin\theta & 0 \\ \sin\theta & \cos\theta & 0 \\ 0 & 0 & 1 \end{pmatrix} \cdot \begin{pmatrix} 1 & 0 & b_1 \\ 0 & 1 & b_2 \\ 0 & 0 & 1 \end{pmatrix}$$

$$= \quad \begin{pmatrix} \cos\theta & -\sin\theta & b_1\cos\theta - b_2\sin\theta \\ \sin\theta & \cos\theta & b_1\sin\theta + b_2\cos\theta \\ 0 & 0 & 1 \end{pmatrix}$$

*Example*: 2D orbit transformation

Orbit the polygon $PQR$ anchored in $B(5,4)$ around the world's origin $O$ by a $30°$ angle, given its vertices $P(0,0)$, $Q(2,1)$ and $R(1,-2)$.

Let us first evaluate this orbit transformation $O_B(\theta)$ as

$$O_{(5,4)}(30°) \quad = \quad \begin{pmatrix} \cos 30° & -\sin 30° & 5\cos 30° - 4\sin 30° \\ \sin 30° & \cos 30° & 5\sin 30° + 4\cos 30° \\ 0 & 0 & 1 \end{pmatrix}$$

$$\approx \quad \begin{pmatrix} 0.87 & -0.50 & 5(0.87) - 4(0.50) \\ 0.50 & 0.87 & 5(0.50) + 4(0.87) \\ 0 & 0 & 1 \end{pmatrix} = \begin{pmatrix} 0.87 & -0.50 & 2.35 \\ 0.50 & 0.87 & 5.98 \\ 0 & 0 & 1 \end{pmatrix}.$$

We then calculate each image vertex similarly as

$$\begin{pmatrix} x' \\ y' \\ 1 \end{pmatrix} = \begin{pmatrix} 0.87 & -0.50 & 2.35 \\ 0.50 & 0.87 & 5.98 \\ 0 & 0 & 1 \end{pmatrix} \cdot \begin{pmatrix} 0 \\ 0 \\ 1 \end{pmatrix} = \begin{pmatrix} 2.35 \\ 5.98 \\ 1 \end{pmatrix},$$

$$\begin{pmatrix} x' \\ y' \\ 1 \end{pmatrix} = \begin{pmatrix} 0.87 & -0.50 & 2.35 \\ 0.50 & 0.87 & 5.98 \\ 0 & 0 & 1 \end{pmatrix} \cdot \begin{pmatrix} 2 \\ 1 \\ 1 \end{pmatrix} = \begin{pmatrix} 3.59 \\ 7.85 \\ 1 \end{pmatrix},$$

$$\begin{pmatrix} x' \\ y' \\ 1 \end{pmatrix} = \begin{pmatrix} 0.87 & -0.50 & 2.35 \\ 0.50 & 0.87 & 5.98 \\ 0 & 0 & 1 \end{pmatrix} \cdot \begin{pmatrix} 1 \\ -2 \\ 1 \end{pmatrix} = \begin{pmatrix} 4.22 \\ 4.74 \\ 1 \end{pmatrix}.$$

Resuming the previous $P'(2.35, 5.98)$, $Q'(3.59, 7.85)$ and $R'(4.22, 4.74)$, we draw both the original polygon $PQR$ and the orbited polygon $P'Q'R'$ in the world $(x,y)$−frame (see figure 14.9).

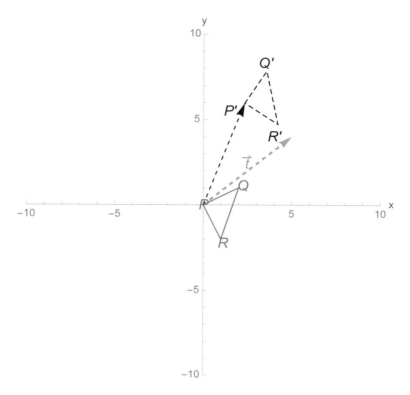

*Figure 14.9*: 2D orbiting of a polygon

LOOK-AT TRANSFORMATION

We neutralise the scaling operator by setting $s_x = s_y = 1$ for scaling is excluded here. We then define **look-at** as to keep pointing an object – from any location – towards a positioned target. We commonly depict this 'looking' object by an isosceles triangle because it automatically shows its orientation. We present such an isosceles shape $KLM$ centred in the world's origin, given its vertices $K$, $M$ and apex $L$ (see figure 14.10). In this figure, we want the shape $KLM$ to firstly translate to the location $B$ and to then aim itself to the target $T$.

▷   We may verify the given shape's centroid. We define the **centroid** $C$ of a polygon $KLM$ as the averaged location vector of its vertices

$$\vec{c} = \frac{\vec{k} + \vec{l} + \vec{m}}{3} \tag{14.13}$$

▷ We realise the translation straightforwardly by the translator $T_{\overrightarrow{OB}}$.

▷ For rotating the shape's apex $L$ towards the target in $T(t_1, t_2)$, we need to set up the unit vectors $\hat{v}_1$ and $\hat{v}_2$. We firstly determine the shape's desired direction vector by

$$\hat{v}_1 = \frac{\overrightarrow{BT}}{\|\overrightarrow{BT}\|} = \widehat{BT}.$$

▷ We secondly set up the second unit vector $\hat{v}_2$ at a $+90°$ angle to $\hat{v}_1$ in order to span a right-handed shape's local $(\hat{v}_1, \hat{v}_2)$−frame. Therefore we apply the standard rotation $R_O(90°)$ on the direction vector $\hat{v}_1$ which is a *free* vector, hence its homogeneous component is set $h = 0$ (see formula (14.3)).

$$\hat{v}_2 = R_O(90°)(\hat{v}_1)$$

$$\Downarrow$$

$$\begin{pmatrix} v_{2_x} \\ v_{2_y} \\ 0 \end{pmatrix} = \begin{pmatrix} \cos 90° & -\sin 90° & 0 \\ \sin 90° & \cos 90° & 0 \\ 0 & 0 & 1 \end{pmatrix} \cdot \begin{pmatrix} v_{1_x} \\ v_{1_y} \\ 0 \end{pmatrix}$$

$$= \begin{pmatrix} 0 & -1 & 0 \\ 1 & 0 & 0 \\ 0 & 0 & 1 \end{pmatrix} \cdot \begin{pmatrix} v_{1_x} \\ v_{1_y} \\ 0 \end{pmatrix} = \begin{pmatrix} -v_{1_y} \\ v_{1_x} \\ 0 \end{pmatrix}$$

Given the above, we now TRS-calculate the look-at operator as

$$A_{\overrightarrow{BT}} = T_{\overrightarrow{OB}} \cdot R_{\widehat{BT}} \cdot S_{\binom{1}{1}} = T_{\overrightarrow{OB}} \cdot R_{\widehat{BT}} \cdot I_3 = T_{\overrightarrow{OB}} \cdot R_{\widehat{BT}},$$

which points an isosceles triangle situated in $B$ towards a target positioned in $T$.

We calculate this look-at operator as

$$A_{\overrightarrow{BT}} = T_{\binom{b_1}{b_2}} \cdot R_{\widehat{BT}} \tag{14.14}$$

$$= \begin{pmatrix} 1 & 0 & b_1 \\ 0 & 1 & b_2 \\ 0 & 0 & 1 \end{pmatrix} \cdot \begin{pmatrix} v_{1_x} & -v_{1_y} & 0 \\ v_{1_y} & v_{1_x} & 0 \\ 0 & 0 & 1 \end{pmatrix}$$

$$= \begin{pmatrix} v_{1_x} & -v_{1_y} & b_1 \\ v_{1_y} & v_{1_x} & b_2 \\ 0 & 0 & 1 \end{pmatrix}, \qquad \text{given } \hat{v}_1 = \frac{\overrightarrow{BT}}{\|\overrightarrow{BT}\|} = \widehat{BT}.$$

*Example*: 2D look-at transformation

Reposition the isosceles shape $KLM$, given its vertices $K(-1, 1)$, $L(2, 0)$ and $M(-1, -1)$ centred on the world's origin $O(0,0)$, having it pointing at a target in $T(10,6)$ from location $B(5,4)$.

Firstly, we check whether the centroid $C$ of the given shape $KLM$ indeed equals the world's origin $O$ (see formula (14.13 )).

$$\vec{c} = \frac{\vec{k}+\vec{l}+\vec{m}}{3}$$

$$\begin{pmatrix} c_1 \\ c_2 \end{pmatrix} = \frac{1}{3}\left( \begin{pmatrix} -1 \\ 1 \end{pmatrix} + \begin{pmatrix} 2 \\ 0 \end{pmatrix} + \begin{pmatrix} -1 \\ -1 \end{pmatrix} \right) = \begin{pmatrix} 0 \\ 0 \end{pmatrix}$$

Secondly, we evaluate this look-at transformation $A_{\overrightarrow{BT}}$ as

$$A_{\overrightarrow{BT}} = \begin{pmatrix} v_{1_x} & -v_{1_y} & 5 \\ v_{1_y} & v_{1_x} & 4 \\ 0 & 0 & 1 \end{pmatrix}, \quad \text{given } \hat{v}_1 = \frac{\overrightarrow{BT}}{\|\overrightarrow{BT}\|} = \widehat{BT}$$

We therefore need to determine the shape's desired direction vector as

$$\hat{v}_1 = \frac{\overrightarrow{BT}}{\|\overrightarrow{BT}\|} = \frac{1}{\|\overrightarrow{BT}\|}(\vec{t}-\vec{b})$$

$$= \frac{1}{\sqrt{29}}\left( \begin{pmatrix} 10 \\ 6 \end{pmatrix} - \begin{pmatrix} 5 \\ 4 \end{pmatrix} \right) = \frac{1}{\sqrt{29}} \begin{pmatrix} 5 \\ 2 \end{pmatrix}$$

$$A_{\overrightarrow{BT}} \approx \begin{pmatrix} 0.93 & -0.37 & 5.00 \\ 0.37 & 0.93 & 4.00 \\ 0 & 0 & 1 \end{pmatrix}, \quad \text{given } \hat{v}_1 \approx \begin{pmatrix} 0.93 \\ 0.37 \end{pmatrix} \quad (14.15)$$

Thirdly, we calculate each image vertex of $KLM$ similarly as

$$\begin{pmatrix} x' \\ y' \\ 1 \end{pmatrix} = \begin{pmatrix} 0.93 & -0.37 & 5 \\ 0.37 & 0.93 & 4 \\ 0 & 0 & 1 \end{pmatrix} \cdot \begin{pmatrix} -1 \\ 1 \\ 1 \end{pmatrix} = \begin{pmatrix} 3.70 \\ 4.56 \\ 1 \end{pmatrix},$$

$$\begin{pmatrix} x' \\ y' \\ 1 \end{pmatrix} = \begin{pmatrix} 0.93 & -0.37 & 5 \\ 0.37 & 0.93 & 4 \\ 0 & 0 & 1 \end{pmatrix} \cdot \begin{pmatrix} 2 \\ 0 \\ 1 \end{pmatrix} = \begin{pmatrix} 6.86 \\ 4.74 \\ 1 \end{pmatrix},$$

$$\begin{pmatrix} x' \\ y' \\ 1 \end{pmatrix} = \begin{pmatrix} 0.93 & -0.37 & 5 \\ 0.37 & 0.93 & 4 \\ 0 & 0 & 1 \end{pmatrix} \cdot \begin{pmatrix} -1 \\ -1 \\ 1 \end{pmatrix} = \begin{pmatrix} 4.44 \\ 2.70 \\ 1 \end{pmatrix}.$$

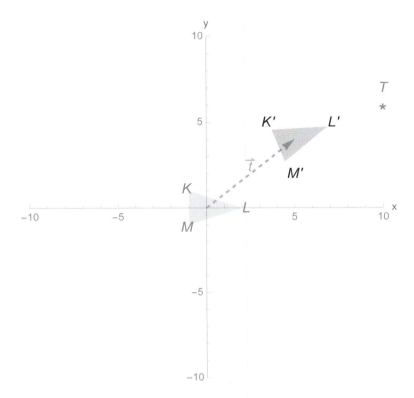

*Figure 14.10*: Look-at of an isosceles shape *KLM*

Resuming the previous results $K'(3.70, \ 4.56)$, $L'(6.86, \ 4.74)$ and $M'(4.44, \ 2.70)$, we draw both the original isosceles shape *KLM* and its repositioned image $K'L'M'$ in the world $(x, y)$−frame (see figure 14.10).

## 14.7 Exercises

**Exercise 131**  Translate the unit circle by the displacement $\vec{t} = \begin{pmatrix} -3 \\ 4 \end{pmatrix}$ applying its matrix operator (see formula (14.1)). Calculate and draw both the image shape and its original circle in different colours within the same $(x,y)$−frame.

**Exercise 132**  Standardly scale the square $ABCD$ given its vertices $A(1,1), B(1,2), C(2,2)$ and $D(2,1)$ by the scale factors $s_x = 5$ and $s_y = 2$ applying its matrix operator (see formula (14.2)). Calculate and draw both the image shape and the original square in different colours within the same $(x,y)$−frame.

**Exercise 133**  Standardly rotate the triangle $UVW$ given its vertices $U(2,0), V(4,1)$ and $W(4,-1)$ by the angle $-23°$ applying its matrix operator (see formula (14.3)). Calculate and draw both the image shape and the original triangle in different colours.

**Exercise 134**  Given the translation $\text{T}_{\begin{pmatrix} -3 \\ 4 \end{pmatrix}}$, the standard scaling $\text{S}_{\begin{pmatrix} 5 \\ 2 \end{pmatrix}}$ and the standard rotation $\text{R}_O(23°)$ deliver these composite transformation matrices:

1) the TRS-compliant action matrix $A = \text{T}_{\begin{pmatrix} -3 \\ 4 \end{pmatrix}} \cdot \text{R}_O(23°) \cdot \text{S}_{\begin{pmatrix} 5 \\ 2 \end{pmatrix}}$

2) the loosely assembled action matrix $B = \text{R}_O(23°) \cdot \text{S}_{\begin{pmatrix} 5 \\ 2 \end{pmatrix}} \cdot \text{T}_{\begin{pmatrix} -3 \\ 4 \end{pmatrix}}$

**Exercise 135**  Given the polygon $UVW$ spanned by the vertices $U(2,0), V(4,1)$ and $W(4,-1)$, calculate and draw in different colours, referring to the previously created transformations (see exercise 134)

1) its image shape $U'V'W'$ by the action transformation A, within the same $(x,y)$−frame,

2) its image shape $U''V''W''$ by the action transformation B, within the same $(x,y)$−frame.

**Exercise 136**  Applying the TRS-based analysis (see formulas (14.7), (14.8) and (14.9)) and referring to the previously created transformations (see exercise 134), attempt to retrieve the translation T (displacement), the standard R (angle) and standard S (scale factors) ingredients of

1) the TRS-based action transformation A

2) the loosely composite action transformation B.

**Exercise 137**  Apply the TRS-based analysis (see formulas (14.7), (14.8) and (14.9)) to retrieve the translation T (displacement), the standard R (angle) and standard S (scale factors) ingredients of the TRS-based action transformation

$$A = \begin{pmatrix} 1.73 & -1.50 & 5.00 \\ 1.00 & 2.60 & 4.00 \\ 0 & 0 & 1 \end{pmatrix}.$$

**Exercise 138**  Pivot the square $ABCD$ from exercise 132 around its centroid $Z$ by a $45°$ angle. Calculate and draw both the original square and its pivoted shape $A'B'C'D'$ in different colours within the same $(x,y)$–frame.

**Exercise 139**  Orbit the square $ABCD$ from exercise 132 anchored in its centroid $Z$ around the origin over a $60°$ angle. Calculate and draw both the original square and its orbited shape $A''B''C''D''$ in different colours within the same $(x,y)$–frame.

**Exercise 140**  Referring to exercises 138 and 139 by firstly applying $O_Z(60°)$ and then $P_Z(45°)$ realise their composite transformation $G = P_Z(45°) \cdot O_Z(60°)$. Calculate and draw both the original square and its composite transformation $G$ image in different colours, within the same $(x,y)$–frame.

**Exercise 141**  Referring to exercises 138 and 139 by firstly applying $P_Z(45°)$ and then $O_Z(60°)$ realise their composite transformation $H = O_Z(60°) \cdot P_Z(45°)$. Calculate and draw both the original square and its composite transformation $H$ image in different colours, within the same $(x,y)$–frame. Compare this outcome to the result of exercise 140

**Exercise 142**  Given the isosceles shape $KLM$ with apex $L(4,0)$ and base vertices $K(-2,1)$ and $M(-2,-1)$, make this pointer $KLM$ look from its centre location $B(1,7)$ at target $T(-9,6)$ by imposing the look-at transformation to it, following the steps below (see formula (14.14)).

   1)  Compute the look-at action matrix $A_{\overrightarrow{BT}}$

   2)  Calculate and draw the image pointer $K'L'M'$ from centre $B$ looking at target $T$.

**Exercise 143**  Apply the TRS-based analysis (see formulas (14.7), (14.8) and (14.9)) to retrieve the translation T (displacement), the standard R (angle) and standard S (scale factors) ingredients of the the look-at action matrix $A_{\overrightarrow{BT}}$ as computed in exercise 142.

**Exercise 144**  Apply the TRS-based analysis (see formulas (14.7), (14.8) and (14.9)) to retrieve the translation T (displacement), the standard R (angle) and standard S (scale factors) ingredients of the the look-at action matrix computed on page 324:

$$A_{\overrightarrow{BT}} \approx \begin{pmatrix} 0.93 & -0.37 & 5.00 \\ 0.37 & 0.93 & 4.00 \\ 0 & 0 & 1 \end{pmatrix}.$$

# Chapter 15 · Scene Graphs

We credit this chapter to its inspirer **Koen Samyn** [14]. When it comes to programming challenging structures like complex vehicles, articulated bodies or solar systems then applying scene graphs is the better way forward. Thanks to the parent-child designed rigging of such a structure, animating them will be all the easier later on. This chapter introduces scene graphs and illustrates this concept by bone structures as well as solar systems, respectively tied up by the former pivot and orbit transformations (see chapter 14).

## 15.1 Concept of a scene graph

We define a **scene graph** as a parent-child tied up composite layered object. We structure a scene graph by spanning its object tree and we may visualise (parts of it) graphically.

Using scene graphs to express multilayered objects such as characters, vehicles, solar systems or (parts of) backgrounds, offers an intrinsic transparency and ease of transforming them. When we move the parent object, all of its children should move with it. We introduce all key aspects of a scene graph through an elementary example.

*Example*: Elementary Scene Graph

We therefore revisit the former shape *KLM* (see figure 14.10) in its **model space**, considering it as for instance a jet flying in our scene and hence having it equipped with two jet engines.

▷ We design a scene graph from the **object tree**. This object tree rooted in the **world** $(\hat{e}_1, \hat{e}_2)$–**space**, reveals the parent (jet) in its jet **local** $(\hat{v}_1, \hat{v}_2)$–**space** and its childs (jet engines) residing respectively in the left engine **local** $(\hat{l}_1, \hat{l}_2)$–**space** and the right engine **local** $(\hat{r}_1, \hat{r}_2)$–**space**.

▷ This object tree is tied up by the **embedding transformations** in between its successive local spaces. We typeset these matrix operators as $E_i$ specifying the layer by the index $i$.

The embedding transformation $E_1$ from the world space to the jet local space was previously calculated as (see formula (14.15))

$$E_1 = A_{\overrightarrow{BT}} \approx \begin{pmatrix} 0.93 & -0.37 & 5.00 \\ 0.37 & 0.93 & 4.00 \\ 0 & 0 & 1 \end{pmatrix}.$$

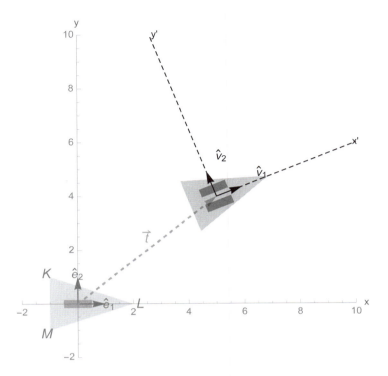

*Figure 15.1*: Scene graph example based on the shape *KLM*

The embedding transformations from the jet local space to the left and right engine local spaces are pure translations along the $y-$axis of the jet local space

$$E_{2,l} \;=\; \begin{pmatrix} 1 & 0 & 0.00 \\ 0 & 1 & +0.30 \\ 0 & 0 & 1 \end{pmatrix},$$

$$E_{2,r} \;=\; \begin{pmatrix} 1 & 0 & 0.00 \\ 0 & 1 & -0.30 \\ 0 & 0 & 1 \end{pmatrix}.$$

▷ Apart from offering the best head start to scene graph designs, an object tree completed by its tying embedding transformations serves as a road map to derive any partial expression from it. If, for instance, we need the right engine's **world expression**, then we append all the embedded transformations tying the path from the said engine down to the root. We typeset such chained matrix operator as $E^{(W)}$ referring to the world space, and denoted by the upper index $^{(W)}$.

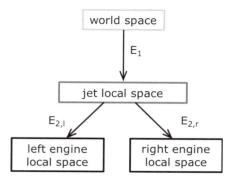

*Figure 15.2*: Object tree for the *KLM*-shape scene graph

$$
\begin{aligned}
E_r^{(W)} &= E_1 \cdot E_{2,r} \\
&\approx \begin{pmatrix} 0.93 & -0.37 & 5.00 \\ 0.37 & 0.93 & 4.00 \\ 0 & 0 & 1 \end{pmatrix} \cdot \begin{pmatrix} 1 & 0 & 0.00 \\ 0 & 1 & -0.30 \\ 0 & 0 & 1 \end{pmatrix} \\
&= \begin{pmatrix} 0.93 & -0.37 & 5.11 \\ 0.37 & 0.93 & 3.72 \\ 0 & 0 & 1 \end{pmatrix}
\end{aligned}
$$

## 15.2 Bone structures

We define a **blueprint bone** as a diamond shape $B_0$ modelled in its **model** $(\hat{e}_1,\ \hat{e}_2)$–**space**.

We define a **bone structure** as a scene graph comprising clones of the blueprint bone.

We learn the aspects of bone structures through this instructive example below.

*Example*:  Sample of a Bone Structure

&#9655;  We firstly create its blueprint running from $O(0,0)$ to end **joint** $J_0(1,0)$ stocked in

$$
B_0 = \begin{pmatrix} 0 & 0.50 & 1 & 0.50 \\ 0 & -0.25 & 0 & 0.25 \\ 1 & 1 & 1 & 1 \end{pmatrix}.
$$

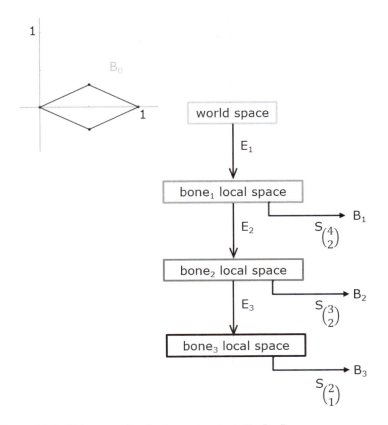

*Figure 15.3*: Object tree for the bone structure $B_1$-$B_2$-$B_3$

▷ Secondly, we design an arm-like bone structure based on $B_0$ and the object tree
for it. We initially model a blueprint in its model space which we firstly scale to
some desired proportions, then orient it before positioning it in its local space. For
a better understanding of bone structures, let us also draw such an arm-like scene
graph in its layered 2D space (see figure 15.4). We notice this bone structure's
resemblance to a human arm with its upper arm $B_1$, forearm $B_2$ and the hand $B_3$,
each of them jointed parent-to-child wise. Hence, when the upper arm pivots up,
the forearm plus the hand will obey likewise, as one.

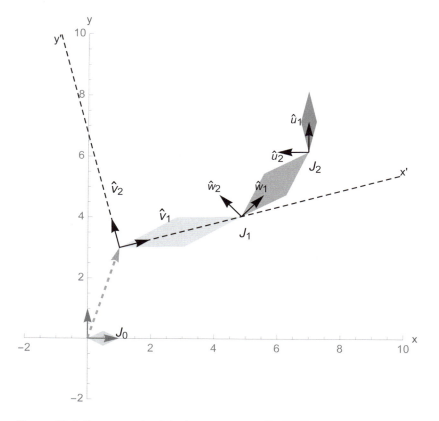

*Figure 15.4*: Scene graph of the bone structure $B_1$-$B_2$-$B_3$

Moreover, the object tree of this arm (see figure 15.3) indeed shows how each limb is jointed to its parent by pivot transformations as outlined in paragraph 14.6:

$$E_1 \;=\; P_{(1,3)}(15°) \;=\; T_{\binom{1}{3}} \cdot R_O(15°),$$

$$E_2 \;=\; P_{(4,0)}(30°) \;=\; T_{\binom{4}{0}} \cdot R_O(30°),$$

$$E_3 \;=\; P_{(3,0)}(45°) \;=\; T_{\binom{3}{0}} \cdot R_O(45°).$$

These embedding tranformations are each expressed within their local spaces.

▷ To also illustrate the object tree's usefulness as a roadmap, for instance we retrieve the world expression of the hand's joint $J_2$ (see figure 15.4). The easiest way to get there is to trace the forearm's world expression $B_2^{(W)}$ and then select its third column for it contains the connection point to the hand. Therefore, we need to append all the embedding transformations running the path from the forearm $B_2$ down to the root:

$$
\begin{aligned}
B_2^{(W)} &= E_1 \cdot E_2 \cdot S_{\binom{3}{2}} \cdot B_0 \\
&= \left(T_{\binom{1}{3}} \cdot R_O(15°)\right) \cdot \left(T_{\binom{4}{0}} \cdot R_O(30°)\right) \cdot S_{\binom{3}{2}} \cdot B_0
\end{aligned}
$$

In order to minimise our calculation load, we hereby restrict the forearm simply to its third column which will yield:

$$
\begin{aligned}
J_2^{(W)} &= \left(T_{\binom{1}{3}} \cdot R_O(15°)\right) \cdot \left(T_{\binom{4}{0}} \cdot R_O(30°)\right) \cdot S_{\binom{3}{2}} \cdot \begin{pmatrix} 1 \\ 0 \\ 1 \end{pmatrix} \\
&= \left(T_{\binom{1}{3}} \cdot R_O(15°)\right) \cdot \left(T_{\binom{4}{0}} \cdot R_O(30°)\right) \cdot \begin{pmatrix} 3 & 0 & 0 \\ 0 & 2 & 0 \\ 0 & 0 & 1 \end{pmatrix} \cdot \begin{pmatrix} 1 \\ 0 \\ 1 \end{pmatrix} \\
&= \left(T_{\binom{1}{3}} \cdot R_O(15°)\right) \cdot \left( \begin{pmatrix} 1 & 0 & 4 \\ 0 & 1 & 0 \\ 0 & 0 & 1 \end{pmatrix} \cdot \begin{pmatrix} \cos 30° & -\sin 30° & 0 \\ \sin 30° & \cos 30° & 0 \\ 0 & 0 & 1 \end{pmatrix} \right) \cdot \begin{pmatrix} 3 \\ 0 \\ 1 \end{pmatrix} \\
&\approx \left(T_{\binom{1}{3}} \cdot R_O(15°)\right) \cdot \begin{pmatrix} 0.87 & -0.50 & 4 \\ 0.50 & 0.87 & 0 \\ 0 & 0 & 1 \end{pmatrix} \cdot \begin{pmatrix} 3 \\ 0 \\ 1 \end{pmatrix} \\
&= \left( \begin{pmatrix} 1 & 0 & 1 \\ 0 & 1 & 3 \\ 0 & 0 & 1 \end{pmatrix} \cdot \begin{pmatrix} \cos 15° & -\sin 15° & 0 \\ \sin 15° & \cos 15° & 0 \\ 0 & 0 & 1 \end{pmatrix} \right) \cdot \begin{pmatrix} 6.61 \\ 1.50 \\ 1 \end{pmatrix} \\
&\approx \begin{pmatrix} 0.97 & -0.26 & 1 \\ 0.26 & 0.97 & 3 \\ 0 & 0 & 1 \end{pmatrix} \cdot \begin{pmatrix} 6.61 \\ 1.50 \\ 1 \end{pmatrix} = \begin{pmatrix} 7.02 \\ 6.17 \\ 1 \end{pmatrix}
\end{aligned}
$$

## 15.3 Solar systems

We previously applied the paramount TRS-convention which reigns the stackwise order of standard transformations operating on location vectors $\vec{p}$ throughout column based as $\vec{p}' = T(R(S(\vec{p})))$. However, it goes without saying that designing a solar system will take the orbit transformations as outlined in paragraph 14.6. Orbit transformations do reverse the said $TR$-order into $RT$-composites. Again, let us learn the aspects of solar systems that feature relative motions by example.

*Example*: Sample of a Solar System

▷ We firstly consider the **blueprint unit disc** in its **model** $(\hat{e}_1, \hat{e}_2)$−**space**.

$$C_0 \quad = \quad C((0,0),\ 1).$$

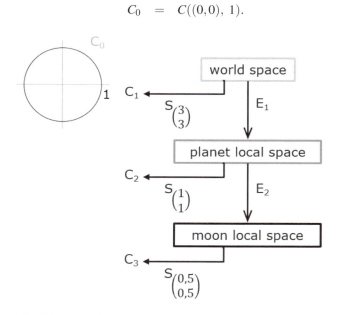

*Figure 15.5*: Object tree for the solar system $C_1$-$C_2$-$C_3$

▷ Secondly, we design a prototype Sun-Earth-Moon system based on $C_0$ and the object tree for it. We initially model a blueprint in its model space which we scale to its desired size, *then firstly* position it in its local space *before* we impose the standard rotation to it. For better insight into solar system modelling, let us also draw this prototype Sun-Earth-Moon system in its layered 2D space (see figure 15.6).

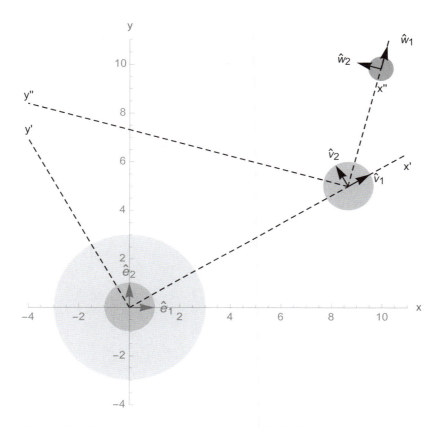

*Figure 15.6*: Scene graph of the solar system $C_1$-$C_2$-$C_3$

Moreover, the object tree of this simple solar system (see figure 15.5) reveals how each body moves relative to its parent by orbit transformations outlined in paragraph 14.6:

$$E_1 = O_{(10,0)}(30°) = R_O(30°) \cdot T_{\binom{10}{0}},$$

$$E_2 = O_{(5,0)}(45°) = R_O(45°) \cdot T_{\binom{5}{0}}.$$

These embedding tranformations are each expressed within their local spaces.

▷ Illustrating once more the object tree's usefulness as a roadmap, we for instance retrieve the world expression of the Moon $C_3$ (see figure 15.6). The easiest way to trace this world expression $C_3^{(W)}$ is to append all the embedding transformations running the path from the Moon $C_3$ down to the root:

$$
\begin{aligned}
C_3^{(W)} &= E_1 \cdot E_2 \cdot S_{\left(\begin{smallmatrix} 0.5 \\ 0.5 \end{smallmatrix}\right)} \cdot C_0 \\
&= \left(R_O(30°) \cdot T_{\left(\begin{smallmatrix} 10 \\ 0 \end{smallmatrix}\right)}\right) \cdot \left(R_O(45°) \cdot T_{\left(\begin{smallmatrix} 5 \\ 0 \end{smallmatrix}\right)}\right) \cdot S_{\left(\begin{smallmatrix} 0.5 \\ 0.5 \end{smallmatrix}\right)} \cdot C_0
\end{aligned}
$$

Eventually we stepwise perform each matrix product of the above chain:

$$
\begin{aligned}
C_3^{(W)} &= \left(R_O(30°) \cdot T_{\left(\begin{smallmatrix} 10 \\ 0 \end{smallmatrix}\right)}\right) \cdot \left(\left(\begin{pmatrix} \cos 45° & -\sin 45° & 0 \\ \sin 45° & \cos 45° & 0 \\ 0 & 0 & 1 \end{pmatrix} \cdot \begin{pmatrix} 1 & 0 & 5 \\ 0 & 1 & 0 \\ 0 & 0 & 1 \end{pmatrix}\right)\right) \cdot S(C_0) \\
&\approx \begin{pmatrix} \cos 30° & -\sin 30° & 0 \\ \sin 30° & \cos 30° & 0 \\ 0 & 0 & 1 \end{pmatrix} \cdot \begin{pmatrix} 1 & 0 & 10 \\ 0 & 1 & 0 \\ 0 & 0 & 1 \end{pmatrix} \cdot \begin{pmatrix} 0.71 & -0.71 & 3.54 \\ 0.71 & 0.71 & 3.54 \\ 0 & 0 & 1 \end{pmatrix} \cdot S(C_0 \\
&\approx \begin{pmatrix} 0.87 & -0.50 & 8.66 \\ 0.50 & 0.87 & 5.00 \\ 0 & 0 & 1 \end{pmatrix} \cdot \begin{pmatrix} 0.71 & -0.71 & 3.54 \\ 0.71 & 0.71 & 3.54 \\ 0 & 0 & 1 \end{pmatrix} \cdot S_{\left(\begin{smallmatrix} 0.5 \\ 0.5 \end{smallmatrix}\right)} \cdot C_0 \\
&= \begin{pmatrix} 0.26 & -0.97 & 9.97 \\ 0.97 & 0.26 & 9.85 \\ 0 & 0 & 1 \end{pmatrix} \cdot \begin{pmatrix} 0.5 & 0 & 0 \\ 0 & 0.5 & 0 \\ 0 & 0 & 1 \end{pmatrix} \cdot C_0 \\
&= \begin{pmatrix} 0.13 & -0.49 & 9.97 \\ 0.49 & 0.13 & 9.85 \\ 0 & 0 & 1 \end{pmatrix} \cdot C_0
\end{aligned}
$$

We therefore conclude the world expression of the Moon as

$$
C_3^{(W)} = C\left((9.97, 9.85), \frac{1}{2}\right).
$$

## 15.4 Exercises

**Exercise 145**  Below you see a representation of three bones $B_1, B_2$ and $B_3$ that model a robot arm. These bones $B_1, B_2$ and $B_3$ are of lengths $4, 3$ and $5$ respectively. The first bone's position is in point $(1, 0)$. In this way the orthonormal $(\hat{e}_1, \hat{e}_2)$−world space contains three embedded so-called local spaces.

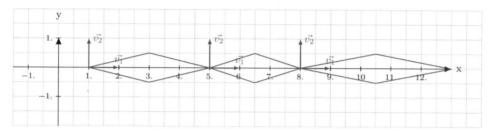

We model the bones themselves by scaling the blueprint diamond $B_0$ according to their required sizes.

$$B_0 \;=\; \begin{pmatrix} 0 & 0.5 & 1 & 0.5 \\ 0 & 0.5 & 0 & -0.5 \\ 1 & 1 & 1 & 1 \end{pmatrix}$$

1) Firstly, design the parent-to-child object tree for this robot arm.

2) Secondly, determine the embedding tranformations $E_i$ that link each successive local space of the scene graph. Do this with pen and paper for $B_1$ constantly inclined by $20°$ within the world space, then $B_2$ variably inclined by $\alpha°$ within its $B_1$−local space and eventually $B_3$ variably inclined by $\beta°$ within its $B_2$−local space.

3) Finally, implement all the previous in **GeoGebra** [48], swiftly realising a simulation of this robot arm by means of two angular sliders respectively for $\alpha$ and $\beta$ clipped between $0°$ and $90°$.

**Exercise 146**  Below you see a representation of a solar system featuring a central star with a radius of 30. A planet with a radius of 15 is circling the central star at a distance of 100 between the star centre and the planet centre. The entire planetary orbit counts 350 days. A moon with radius 5 is circling the planet at a distance of 40 and the lunar orbit counts 50 days. An isosceles space craft is circling the moon at a distance of 15 and completes its orbit in only 5 days. In this way the orthonormal $(\hat{e}_1, \hat{e}_2)$−world space contains three embedded subsequent local spaces.

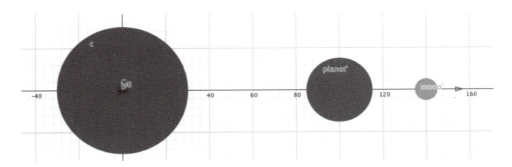

We model the said isosceles space craft by its vertices.

$$\text{craft} \quad = \quad \begin{pmatrix} 1.5 & -1.5 & -1.5 \\ 0.0 & 1.0 & -1.0 \\ 1 & 1 & 1 \end{pmatrix}$$

1) Firstly, design the parent-to-child object tree for this solar system with space craft.

2) Secondly, determine the embedding tranformations $E_i$ that link each successive local space of the scene graph. Do this with pen and paper making use of the variable *day* where appropriate.

3) Implement this in GeoGebra [48], realising a simulation of this solar system by means of a linear slider for the introduced variable *day* clipped between 0 and 350.

**Exercise 147**   A tank patrolling a circular trajectory around its headquarters keeps pointing its turret with barrel towards the South. The rectangular blueprint $T_0$ becomes the *tank*. The blueprint $T_0$ scales non-uniformly by scale factor $s_y = 0.5$ to its square *turret*. Also $T_0$ scales non-uniformly by scale factor $s_x = 0.25$ to a stretched *barrel*. The centroid of the *tank* rides in a circle around the headquarters in the origin. The pivot centroid of its *turret* is located in the centroid of the *tank*. Then the *barrel* is fixed immovably midway to the *turret*.

$$T_0 \quad = \quad \begin{pmatrix} -1 & 1 & 1 & -1 \\ 2 & 2 & -2 & -2 \\ 1 & 1 & 1 & 1 \end{pmatrix}$$

1) Design the parent-to-child object tree for this layered object *tank-turret-barrel*.

2) Determine the embedding tranformations $E_i$ that link each successive local space of this scene graph. Do this with pen and paper making use of two angular variables $\alpha$ and $\beta$ in order to keep the turret-with-barrel aiming to the south.

3) Implement this in GeoGebra [48] to reality check your solution by means of two angular variables $\alpha$ and $\beta$ clipped between 0 and $2\pi$ radian.

We credit this chapter to its inspirer **Koen Samyn** [14]. In an initial step in the propagation through the grapical pipeline, the computer selects the meaningful target from our entire scene: scanning a landscape, following the player's car on the race circuit, zooming in on a flying jet, …. Given the immobility of the computer's virtual camera, it cannot move around in the scene. Hence the so-called view transformation brings the camera captures to the game window, which we cover here.

## 16.1 The Rendering Pipeline

The final aim of computer graphics is to capture a meaningful part of the digital 3D world and to draw one single 2D image out of it. In order to perform this huge task in a minimum slice of time, computer graphics engineering opts for pipelining this task.

### THE CONCEPT OF THE PIPELINE

Given the default frame rate for video games equal to 60 *fps*, the computer needs to perform all the above mathematical workload in every delta time of only $\frac{1}{60}s \approx 17ms$. This is only feasible by dividing the former huge task into separate sub tasks, performed in the most appropriate order given their dependencies. We hereby define the **rendering pipeline** as the sequence of real time sub tasks which eventually draw one single 2D image out of a digital 3D scene. In the best case, the graphical pipeline streamlines the huge task, allowing for its simultaneously performed sub tasks. However, the entire pipeline is as a chain only as fast as its slowest sub task and may slow down with lingering processes.

### THE STAGES OF THE PIPELINE

Generally, we hereby distinguish four subsequent stages in the rendering pipeline from its 3D feed, all the way down to its 2D digest (see figure 16.1)

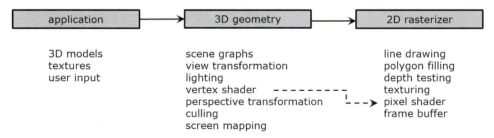

*Figure 16.1*: Subsequent stages of the rendering pipeline

Application stage

The application stage is flexible in its information loading: it takes all of your **3D models** with their **textures** up to the runtime **user input**. We easily imagine worked 3D models by extending our former 2D scene graphs by one dimension (see chapter 15). For more information and understanding of textures, we kindly refer you to the dedicated literature. User input may comprise runtime mouse clicks and varying view angles to keep track of target. This application stage finally feeds the entirety of this information into the geometry stage.

Geometry stage

This 3D stage captures the modelled **scene graphs** – present in the digital 3D scenery – by means of the famous **view transformation** (see paragraph 16.3). Subsequently illumination gets added through vector modelled **lighting** for which we refer you to the specialised literature. Meanwhile the GPU runs a **vertex shader** which is a script program running in 3D space on each vertex of your viewed scene and possibly – apart from shading per vertex – used for alternative tasks like skinning, animation, interpolations, ….. Now in order to attain 'what the camera sees', we firstly define a **frustum** as a truncated view pyramid for which everything outside it gets excluded from vision. Based on this frustum, either an orthographic (boxy frustum) or a realistic **perspective transformation** (ziggurat frustum) will squash the scenery onto a view plane. For more insight into these perspective transformations, we need to kindly refer you to the specialised literature (see figure 16.2).

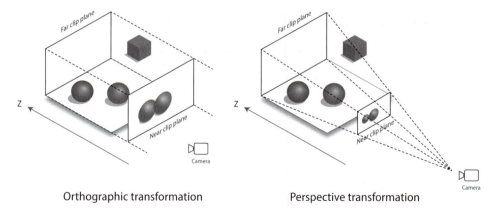

Orthographic transformation               Perspective transformation

*Figure 16.2*: Two different perspective transformations

We define **culling** or clipping as the exclusion of everything outside of the view frustum. However straightforward culling may sound, mathematically it is far from a walk in the park. Some 3D models may intersect some frustum sides which requires determining intersections, increasing the calculation load. This geometry stage ends by **screen mapping** all of the above acquired 3D coordinates onto the pixel coordinates of the rasterisation stage.

Rasterisation stage

We define rasterisation as the **scan conversion** whereby we scan line wise, meaning visualising the pixel coordinates row by row. Scan conversion covers **line drawing** and **polygon filling** which are both mathematical algorithms. But scan conversion also takes **depth testing** which was kept on the $Z-$axis (see figure 16.2) known as the **z-buffer**, and **texturing** for both of which we refer you to the specialised literature. Meanwhile the GPU runs a **pixel shader** which is a script program running in 2D space on each rasterised pixel and possibly – whilst shading per pixel – using extra information from the vertex shader to do so (see the former Geometry stage). This rasterisation stage finally feeds pixels into the **frame buffer** of the display stage.

Display stage

This stage refreshes image after image from the frame buffer to the screen, according to the frame rate.

## 16.2  Camera transformation

We define the fixed **game window** as the world origin-centred default view port of the size (*width* × *height*). Such a game window measures in pixels the casual $(640 \times 480)$ or $(1920 \times 1080)$ or any other size. We define the **camera transformation** as a TRS-ordered forward transformation. We may compare the forward camera transformation to a photographer roaming around in the real world in order to capture meaningful targets.

We define the **camera window** as the frustum's clipped plane (see figure 16.2) positioned (with its bottom left corner) in the origin $\vec{c}$ of the camera local space and inclined by an angle $\theta$ with respect to the world horizon. We hence typeset the forward camera transformations as $F_{\vec{c}}(\theta)$ given the camera's location $\vec{c}$ and the view window inclination $\theta$.

*Example*: Spotting a jet

In this simplified scenery featuring a jet aircraft, the camera origin is located in $C(400, 100)$ where it cuts out an $21.8°$ inclined $(640 \times 480)$−sized camera window (see figure 16.3).

▷  We summarise the according camera transformation as

$$
F_{\binom{400}{100}}(21.8°) = T_{\binom{400}{100}} \cdot R_O(21.8°) \cdot S_{\binom{1}{1}}
$$

$$
= \begin{pmatrix} \cos 21.8° & -\sin 21.8° & 400 \\ \sin 21.8° & \cos 21.8° & 100 \\ 0 & 0 & 1 \end{pmatrix}
$$

$$
\approx \begin{pmatrix} 0.93 & -0.37 & 400 \\ 0.37 & 0.93 & 100 \\ 0 & 0 & 1 \end{pmatrix}. \tag{16.1}
$$

Since this camera transformation does not feature zooming yet, we have set $s_x = s_y = 1$. We will elaborate on zoom cameras later in paragraph 16.5.

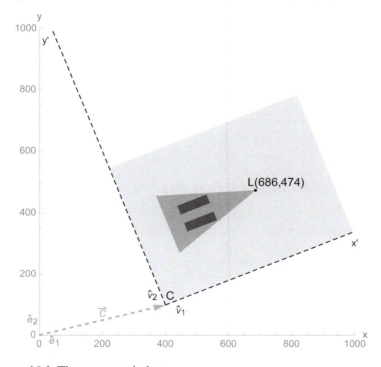

*Figure 16.3*: The camera window

▷ We generalize such a camera transformation, equipping it with a zoom function

$$F_{\vec{c}}(\theta) \;=\; T_{\binom{c_1}{c_2}} \cdot R_O(\theta) \cdot S_{\binom{s_x}{s_y}}$$

$$= \begin{pmatrix} 1 & 0 & c_1 \\ 0 & 1 & c_2 \\ 0 & 0 & 1 \end{pmatrix} \cdot \begin{pmatrix} \cos\theta & -\sin\theta & 0 \\ \sin\theta & \cos\theta & 0 \\ 0 & 0 & 1 \end{pmatrix} \cdot \begin{pmatrix} s_x & 0 & 0 \\ 0 & s_y & 0 \\ 0 & 0 & 1 \end{pmatrix}$$

$$= \begin{pmatrix} s_x\cos\theta & -s_y\sin\theta & c_1 \\ s_x\sin\theta & s_y\cos\theta & c_2 \\ 0 & 0 & 1 \end{pmatrix}. \tag{16.2}$$

Since this matrix obeys the TRS-order, we equivalently represent it by its local base vectors $(\vec{v}_1, \vec{v}_2)$ based on the general formula (14.6) column wise as

$$F_{\vec{c}}(\theta) \;=\; \begin{pmatrix} v_{1x} & v_{2x} & c_1 \\ v_{1y} & v_{2y} & c_2 \\ 0 & 0 & 1 \end{pmatrix}. \tag{16.3}$$

## 16.3 View transformation

We define the **view transformation** as the inverse of the camera transformation for it returns the camera window onto the game window. We hereby typeset the view transformation as $V_{\vec{c}}(\theta)$ given a local camera origin $\vec{c}$ and inclination $\theta$. The view transformation moves the capture of the camera window to the game window, from where its rendering starts. The view transformation continuously feeds the geometry stage of the rendering pipeline. When we imagine our jet (see figure 16.3) 'flying' around in our scene, then the view transformation may track this jet by anchoring the local camera to this plane.

Where we previously compared the forward camera transformation to a roaming photographer, literally the opposite is the case for the backwards view transformation. The view transformations actually re-position the complete world space in order to get the meaningful target in front of the immobile 'photographer'. Read the former sentence again for it is awkward but true.

We now calculate both the matrix representation and its local base vectors equivalent of the general view transformation $V_{\vec{c}}(\theta)$, similarly to the previous $F_{\vec{c}}(\theta)$. We start from the view transformation's definition, applying the 'Socks-and-Boots' rule (see page 234).

$$
\begin{aligned}
V_{\vec{c}}(\theta) \;&=\; F_{\vec{c}}(\theta)^{-1} \\[2mm]
&=\; \left(T_{\binom{c_1}{c_2}} \cdot R_O(\theta) \cdot S_{\binom{s_x}{s_y}}\right)^{-1} \\[2mm]
&=\; S^{-1}_{\binom{s_x}{s_y}} \cdot R_O^{-1}(\theta) \cdot T^{-1}_{\binom{c_1}{c_2}} \\[2mm]
&=\; S_{\binom{1/s_x}{1/s_y}} \cdot R_O(-\theta) \cdot T_{\binom{-c_1}{-c_2}} \\[2mm]
&=\; \begin{pmatrix} \frac{1}{s_x} & 0 & 0 \\ 0 & \frac{1}{s_y} & 0 \\ 0 & 0 & 1 \end{pmatrix} \cdot \begin{pmatrix} \cos(-\theta) & -\sin(-\theta) & 0 \\ \sin(-\theta) & \cos(-\theta) & 0 \\ 0 & 0 & 1 \end{pmatrix} \cdot \begin{pmatrix} 1 & 0 & -c_1 \\ 0 & 1 & -c_2 \\ 0 & 0 & 1 \end{pmatrix} \\[2mm]
&=\; \begin{pmatrix} \frac{1}{s_x} & 0 & 0 \\ 0 & \frac{1}{s_y} & 0 \\ 0 & 0 & 1 \end{pmatrix} \cdot \begin{pmatrix} \cos\theta & \sin\theta & 0 \\ -\sin\theta & \cos\theta & 0 \\ 0 & 0 & 1 \end{pmatrix} \cdot \begin{pmatrix} 1 & 0 & -c_1 \\ 0 & 1 & -c_2 \\ 0 & 0 & 1 \end{pmatrix} \\[2mm]
&=\; \begin{pmatrix} \frac{1}{s_x}\cos\theta & \frac{1}{s_x}\sin\theta & 0 \\ -\frac{1}{s_y}\sin\theta & \frac{1}{s_y}\cos\theta & 0 \\ 0 & 0 & 1 \end{pmatrix} \cdot \begin{pmatrix} 1 & 0 & -c_1 \\ 0 & 1 & -c_2 \\ 0 & 0 & 1 \end{pmatrix} \\[2mm]
&=\; \begin{pmatrix} \frac{1}{s_x}\cos\theta & \frac{1}{s_x}\sin\theta & -\frac{1}{s_x}(c_1\cos\theta + c_2\sin\theta) \\ -\frac{1}{s_y}\sin\theta & \frac{1}{s_y}\cos\theta & -\frac{1}{s_y}(c_1(-\sin\theta) + c_2\cos\theta) \\ 0 & 0 & 1 \end{pmatrix} \\[2mm]
&=\; \begin{pmatrix} \dfrac{\cos\theta}{s_x} & \dfrac{\sin\theta}{s_x} & \dfrac{-\hat{v}_1 \cdot \vec{c}}{s_x} \\[2mm] -\dfrac{\sin\theta}{s_y} & \dfrac{\cos\theta}{s_y} & \dfrac{-\hat{v}_2 \cdot \vec{c}}{s_y} \\[2mm] 0 & 0 & 1 \end{pmatrix} \quad\quad (16.4)
\end{aligned}
$$

Given $\hat{v}_1 = (\cos\theta,\ \sin\theta)$ and $\hat{v}_2 = (-\sin\theta,\ \cos\theta)$, we may equivalently again represent this view transformation by its underlying local base vectors $(\vec{v}_1,\ \vec{v}_2)$ as

$$
V_{\vec{c}}(\theta) \;=\; \begin{pmatrix} \dfrac{\hat{v}_{1x}}{\|\vec{v}_1\|} & \dfrac{\hat{v}_{1y}}{\|\vec{v}_1\|} & -\dfrac{\hat{v}_1 \cdot \vec{c}}{\|\vec{v}_1\|} \\[3mm] \dfrac{\hat{v}_{2x}}{\|\vec{v}_2\|} & \dfrac{\hat{v}_{2y}}{\|\vec{v}_2\|} & -\dfrac{\hat{v}_2 \cdot \vec{c}}{\|\vec{v}_2\|} \\[3mm] 0 & 0 & 1 \end{pmatrix} . \quad\quad (16.5)
$$

While we interpret the usual forward transformations throughout this book column wise, we need to change this into row wise interpretation for the inverted *backwards* transformation.

*Example*:  Calculating a view transformation

In this simplified scenery featuring a jet, the camera origin is located in $C(400, 100)$ where it captures a $21.8°$ inclined $(640 \times 480)$−sized camera window (see figure 16.3) to return to the default game window.

▷  We evaluate the according view transformation (see formula (16.4)) as

$$V_{\binom{400}{100}}(21.8°) = \begin{pmatrix} \dfrac{\cos 21.8°}{1} & \dfrac{\sin 21.8°}{1} & \dfrac{-\hat{v}_1 \cdot \vec{c}}{1} \\[2mm] -\dfrac{\sin 21.8°}{1} & \dfrac{\cos 21.8°}{1} & \dfrac{-\hat{v}_2 \cdot \vec{c}}{1} \\[2mm] 0 & 0 & 1 \end{pmatrix}$$

$$= \begin{pmatrix} \cos 21.8° & \sin 21.8° & -\hat{v}_1 \cdot \vec{c} \\ -\sin 21.8° & \cos 21.8° & -\hat{v}_2 \cdot \vec{c} \\ 0 & 0 & 1 \end{pmatrix}$$

Since in this view transformation zooming does not occur, we have set $s_x = s_y = 1$. Therefore both local base vectors $\vec{v}_1$ and $\vec{v}_2$ can immediately be interpreted row wise as the two *unit* vectors:

$$\hat{v}_1 = \begin{pmatrix} \cos 21.8° \\ \sin 21.8° \end{pmatrix} \approx \begin{pmatrix} 0.93 \\ 0.37 \end{pmatrix} \in \mathbb{R}^2 \text{ and } \hat{v}_2 = \begin{pmatrix} -\sin 21.8° \\ \cos 21.8° \end{pmatrix} \approx \begin{pmatrix} -0.37 \\ 0.93 \end{pmatrix} \in \mathbb{R}^2$$

whilst for the camera's position we insert  $\vec{c} = \begin{pmatrix} 400 \\ 100 \end{pmatrix} \in \mathbb{R}^2$

in order to calculate their dot products for the translational part as

$\hat{v}_1 \cdot \vec{c} = 0.93 \cdot 400 + 0.37 \cdot 100 = 409.00 \in \mathbb{R}$

$\hat{v}_2 \cdot \vec{c} = -0.37 \cdot 400 + 0.93 \cdot 100 = -55.70 \in \mathbb{R}.$

▷  Given all the above we assemble this example view transformation matrix as

$$V_{\binom{400}{100}}(21.8°) \approx \begin{pmatrix} 0.93 & 0.37 & -409.00 \\ -0.37 & 0.93 & 55.70 \\ 0 & 0 & 1 \end{pmatrix}. \tag{16.6}$$

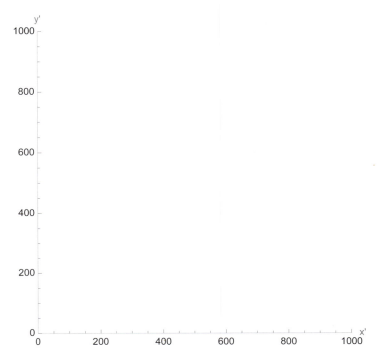

*Figure 16.4*: The game window

## 16.4 View operator

Starting from the transformation as the linear combination with $h = 1$ set for locations (see formula (14.10)) comes

$$\vec{v}^{\,(W)} \;=\; x' \cdot \vec{v}_1 \;+\; y' \cdot \vec{v}_2 \;+\; 1 \cdot \vec{c}$$

$$\Longleftrightarrow \begin{pmatrix} x \\ y \end{pmatrix}^{(W)} = x' \begin{pmatrix} v_{1_x} \\ v_{1_y} \end{pmatrix} + y' \begin{pmatrix} v_{2_x} \\ v_{2_y} \end{pmatrix} + 1 \begin{pmatrix} c_1 \\ c_2 \end{pmatrix}$$

$$\Longleftrightarrow \begin{pmatrix} x \\ y \\ 1 \end{pmatrix}^{(W)} = \begin{pmatrix} v_{1_x} \\ v_{1_y} \\ 1 \end{pmatrix} x' + \begin{pmatrix} v_{2_x} \\ v_{2_y} \\ 1 \end{pmatrix} y' + \begin{pmatrix} c_1 \\ c_2 \\ 1 \end{pmatrix} 1$$

because we refer the meaningful target points $P$ within the camera local $(x', y')$–window.

This linear combination is equivalent to the matrix expression (see page 318):

$$\begin{pmatrix} x \\ y \\ 1 \end{pmatrix}^{(W)} = \begin{pmatrix} v_{1_x} & v_{2_x} & c_1 \\ v_{1_y} & v_{2_y} & c_2 \\ 0 & 0 & 1 \end{pmatrix} \cdot \begin{pmatrix} x' \\ y' \\ 1 \end{pmatrix}.$$

Because this is a forward transformation we need to continue it as

$$\begin{pmatrix} x \\ y \\ 1 \end{pmatrix}^{(W)} = F_{\vec{c}}(\theta) \cdot \begin{pmatrix} x' \\ y' \\ 1 \end{pmatrix}.$$

Left multiplying this matrix equation with $F_{\vec{c}}(\theta)^{-1}$ simplifies to

$$F_{\vec{c}}(\theta)^{-1} \cdot \begin{pmatrix} x \\ y \\ 1 \end{pmatrix}^{(W)} = F_{\vec{c}}(\theta)^{-1} \cdot F_{\vec{c}}(\theta) \cdot \begin{pmatrix} x' \\ y' \\ 1 \end{pmatrix} \Longleftrightarrow$$

$$F_{\vec{c}}(\theta)^{-1} \cdot \begin{pmatrix} x \\ y \\ 1 \end{pmatrix}^{(W)} = I_3 \cdot \begin{pmatrix} x' \\ y' \\ 1 \end{pmatrix}.$$

In conclusion, we prove $F_{\vec{c}}(\theta)^{-1} = V_{\vec{c}}(\theta)$ an operator which indeed fetches the targeted points $P^{(W)}$ from anywhere in the $(x,y)$−world to present them in front of the camera's eye captured in the $(x',y')$−camera window as $P^{(C)}$.

$$\begin{pmatrix} x' \\ y' \\ 1 \end{pmatrix}^{(C)} = V_{\vec{c}}(\theta) \cdot \begin{pmatrix} x \\ y \\ 1 \end{pmatrix}^{(W)}.$$

*Example*:  Viewing a jet

In the simplified scenery, the camera (located in $C(400, 100)$ cuts out an $21.8°$ inclined $(640 \times 480)$−sized camera window (see figure 16.3)) capturing for instance the jet's nose point L, expressed in world coordinates as $L^{(W)}(686, 474)$.

▷  We previously assembled this view transformation matrix as

$$V_{\binom{400}{100}}(21.8°) \approx \begin{pmatrix} 0.93 & 0.37 & -409.00 \\ -0.37 & 0.93 & 55.70 \\ 0 & 0 & 1 \end{pmatrix}.$$

▷ This transformation views the jet's nose point $L^{(W)}(686, 474)$ therefore as

$$
\begin{pmatrix} x' \\ y' \\ 1 \end{pmatrix} = V_{\binom{400}{100}}(21.8°) \cdot \begin{pmatrix} 686 \\ 474 \\ 1 \end{pmatrix}^{(W)}
$$

$$
\approx \begin{pmatrix} 0.93 & 0.37 & -409.00 \\ -0.37 & 0.93 & 55.70 \\ 0 & 0 & 1 \end{pmatrix} \cdot \begin{pmatrix} 686 \\ 474 \\ 1 \end{pmatrix}^{(W)} = \begin{pmatrix} 404.36 \\ 242.70 \\ 1 \end{pmatrix}^{(C)}
$$

▷ Transformation $V_{\binom{400}{100}}(21.8°)$ views the jet's nose point captured in the $(x', y')-$camera window as $L^{(C)}(404.36, 242.70)$

## 16.5 Camera with zoom

We may also set up a view transformation which zooms in or out of the scenery (see paragraph 16.3). When we uniformly set $s_x = s_y = 2$, we double the length of the camera local base vectors $(\vec{v}_1, \vec{v}_2)$ and therefore **zoom out** on the scene contained in the camera window. Learning by example, we apply this on the jet's nose point $L^{(W)}(686, 474)$ to experience its effect.

*Example 1*: Zooming out of the jet

In the scene, the camera (located in $C(400, 100)$ where it spans a $21.8°$ inclined ($640 \times 480$)$-$sized camera window (see figure 16.3) zooms out by factor 2 on the complete captured part of the scene. As an example calculation, we track this effect on the jet's nose point $L$ expressed in world coordinates as $L^{(W)}(686, 474)$.

▷ We firstly update the view transformation matrix featuring $s_x = s_y = 2$ as

$$
V_{\binom{400}{100}}(21.8°)\binom{2}{2} \approx \begin{pmatrix} \frac{0.93}{2} & \frac{0.37}{2} & \frac{-409.00}{2} \\ \frac{-0.37}{2} & \frac{0.93}{2} & \frac{55.70}{2} \\ 0 & 0 & 1 \end{pmatrix}.
$$

▷ This zoom transformation views the jet's nose point $L^{(W)}(686, 474)$ therefore as

$$
\begin{pmatrix} x' \\ y' \\ 1 \end{pmatrix} \approx \begin{pmatrix} \frac{0.93}{2} & \frac{0.37}{2} & \frac{-409.00}{2} \\ \frac{-0.37}{2} & \frac{0.93}{2} & \frac{55.70}{2} \\ 0 & 0 & 1 \end{pmatrix} \cdot \begin{pmatrix} 686 \\ 474 \\ 1 \end{pmatrix}^{(W)} = \begin{pmatrix} 202.18 \\ 121.35 \\ 1 \end{pmatrix}^{(C)}
$$

▷ This latter transformation views the jet's nose point within the zoomed camera window as $L^{(C)}(202.18,\ 167.04)$ with respect to the local $(x',y')$–frame.

Doubling the local camera window halves the relative size of the jet captured inside it, so when brought into the game window, the jet's points $K$, $L$ and $M$ keep all of their halved coordinates.

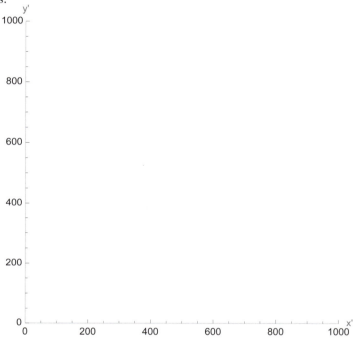

*Figure 16.5*: Zooming-out view transformation

*Example 2*:  Zooming in on the jet

In this simplified scenery, the camera located in $C(400,100)$ where it captures an $21.8°$ inclined $(640 \times 480)$–sized camera window (see figure 16.3) **zooms in** applying factor $0.5$ on the jet's nose point $L$ expressed in world coordinates as $L^{(W)}(686,474)$.

▷ We firstly update this view transformation matrix featuring $s_x = s_y = 0.5$ as

$$V_{\binom{400}{100}}^{(21.8°)}\binom{0.5}{0.5} \approx \begin{pmatrix} \frac{0.93}{0.5} & \frac{0.37}{0.5} & \frac{-409.00}{0.5} \\ \frac{-0.37}{0.5} & \frac{0.93}{0.5} & \frac{55.70}{0.5} \\ 0 & 0 & 1 \end{pmatrix}.$$

▷ This zoom transformation views the plane's nose point $L^{(W)}(686, 474)$ therefore as

$$
\begin{pmatrix} x' \\ y' \\ 1 \end{pmatrix} \approx \begin{pmatrix} \frac{0.93}{0.5} & \frac{0.37}{0.5} & \frac{-409.00}{0.5} \\ \frac{-0.37}{0.5} & \frac{0.93}{0.5} & \frac{55.70}{0.5} \\ 0 & 0 & 1 \end{pmatrix} \cdot \begin{pmatrix} 686 \\ 474 \\ 1 \end{pmatrix}^{(W)} = \begin{pmatrix} 807.94 \\ 484.01 \\ 1 \end{pmatrix}^{(C)}
$$

▷ This latter transformation views the jet's nose point within the zoomed camera window as $L^{(C)}(807.94,\ 484.01)$ with respect to the local $(x', y')$−frame.

So, if we uniformly set $s_x = s_y = 50\%$, this shrinks the camera local base vectors $(\vec{v}_1, \vec{v}_2)$ spanning the local window, and therefore zooms in on the scene captured in the camera window. Zooming in too excessively therefore risks (partially) leaving the meaningful target out of the camera window. Hence we would (partially) lose the meaningful target from sight in our game window.

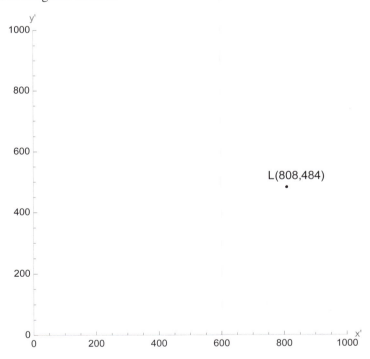

*Figure 16.6*: Zooming-in view transformation

## 16.6 Exercises

**Exercise 148**  One of the checks in game design is verifying that the local camera window stays within the world boundaries, which we define as a **boundary check**. When the camera is aligned to the world this is easy, but when we are introducing scaling or rotation it is better using matrices to verify how far our camera is in a valid position. Imagine a world space is 100 pixels wide and 50 high and the camera window is 15 wide and 15 high, and is rotated over a $15°$ angle around its bottom left vertex $C$ in position $(87, 10)$. Addressing this problem takes

1) constructing the camera transformation,

2) determining the vertices of the camera window,

3) transforming these camera vertices by our camera matrix and

4) verifying if the camera window image stays within the world space boundaries.

**Exercise 149**  A rectangular camera window has a width of 10 by a height of 8 units. If the left bottom vertex $C$ of the rectangle is located at the point $(3, 4)$ and it is rotated over an angle of $30°$ around $C$ then determine the

1) camera transformation to put the camera window at this position,

2) view transformation to return the camera capture horizontally filling our fixed game window measuring 800 by 600 pixels.

It is good practice to reality check and visualise your results in GeoGebra [48]

**Exercise 150**

▷ Verify that the matrix product of the matrices (16.1) and (16.6) returns matrix $I_3$.

▷ Prove that the matrix product of the matrices (16.2) and (16.4) yields matrix $I_3$.

▷ Prove that the matrix product of the matrices (16.3) and (16.5) produces matrix $I_3$.

**Exercise 151**  Construct a closed arrow shape spanning edges between vertices, in their order, collected as $\{(0, 20), (20, 0), (10, 0), (10, -20), (-10, -20), (-20, 0), (0, 20)\}$. We modelled this arrow around the origin $O$ which we adopt as its pivot point.

1) Positioning this arrow in the world space implies we need to construct its embedding transformation $E$. Embed the arrow with its pivot point in $(17, 19)$, then rotate it by an angle of $170°$ around the pivot point and finally scale it to $\frac{1}{10}$ of its original size. Calculate and visualise the according arrow image positioned in the world space in GeoGebra [48].

2) Return the camera captured arrow from world space to the fixed game window, which is requiring a view transformation. Establish the view transformation corresponding to a camera with its lower left pivot point in (12, 14), rotated by an angle of 30° around this pivot and uniformly scaled by factor $\frac{1}{1.13}$ with a camera window in pixels of 15 width and 15 height.

3) Stack the former view transformation on top of the latter embedding transformation $E$, to matrix multiply these into one action matrix. Now calculate and visualise the subsequent arrow image as brought into the game window in GeoGebra [48].

**Exercise 152** The Belgian UFO wave was a series of sightings of triangular UFOs over Belgium which lasted from November 1989 until April 1990. At a certain time, the pivot point $P$ of such a triangular UFO was located in $(299, 99)$ when its nose point $N$ was in $(300, 100)$. The military camera window measured 15 width by 10 height with the bottom left vertex $C$ of the camera's rectangle situated in $(297, 97)$ on the UFO's right wing tip as portrayed. This isosceles UFO is only principally portrayed, meaning its orientation and proportions may differ from this picture. The picture is solely provided for a better understanding of the above listed points $P$, $N$ and $C$.

 N

C

1) Given these conditions, answer the matrix product as such to deliver the view transformation which brings this UFO horizontally (meaning $\overrightarrow{PN}$ horizontally) into our game window.

2) Compute the former matrix product using GeoGebra [48] to answer the UFO view transformation matrix.

3) Also append a window-filling scaling for a ground based view port of 225 width by 150 height. This scaling is uniform. Recompute the former matrix product now including the latter scaling using GeoGebra [48] to answer the zoomed UFO view transformation matrix.

Here as he walked by
on the 16th of October 1843
Sir William Rowan Hamilton
in a flash of genius discovered
the fundamental formula for
quaternion multiplication
$$i^2 = j^2 = k^2 = ijk = -1$$
& cut it on a stone of this bridge

Rotator matrices are able to perform any desired rotation in 3D space but may suffer numerical instabilities. As an alternative to matrices, we can realise 3D rotations using quaternions. This chapter initiates us stepwise into the fascinating world of these extended complex numbers.

## 17.1 Complex numbers

Complex numbers were discovered already a few centuries ago. Their renaming from 'impossible' over 'imaginary' to 'complex' numbers illustrates to what extent they gave their users an uneasy feeling. Complex numbers were accidentally discovered in the $16^{\text{th}}$ century as a side-effect of solving cubic equations. Sometimes square roots of negative numbers appeared, which could be used temporarily in an algebraic way. By the end of the $18^{\text{th}}$ century, the Swiss mathematician **Leonhard Euler** (1707–1783) concluded the quadratic equation $x^2 = -1$ is impossible to solve for real numbers. Hence, 'impossible' numbers were born.

All natural, integer, rational and real numbers are represented on the **number line**. In other words, each number corresponds to a point on this straight line and vice versa. This allows us to interpret number addition and multiplication as linear transformations. Addition of numbers corresponds to translation of points (see paragraph 13.1). For instance, adding 3 to a number corresponds to a shift of 3 to the right on the number line. Similarly, subtracting 5 corresponds to a translation of 5 to the left. Multiplication of a number with a positive factor corresponds to a standard scaling of a point (see paragraph 13.2). For instance, the multiplication with factor 2 of the point labelled 1 on the line returns the scaled point to the right labelled 2. Multiplication of a number with a negative factor corresponds to the standard scaling combined with the point reflection over the origin (see paragraph 13.4). For instance, the multiplication with factor $-2$ of the point labelled 1 on the line returns the point to the left labelled $-2$, as the image of the standard scaling $S_O$ with the scale factor $s_x = 2$ combined with the point reflection $M_O$. Repeating this transformation for a second time returns the image point labelled 4. We confirm both successive multiplications arithmetically as $(-2) \cdot (-2) = +4$. Multiplication of a number with factor $-1$ corresponds to the point reflection $M_O$, reflecting points labelled $x$ over the origin to their image point $-x$. Repeating this transformation for a second time returns the initial point labelled $x$ as image. We confirm both successive multiplications arithmetically as $(-1) \cdot (-1) = +1$ or $(-1)^2 = +1$.

We realise there is no transformation on the number line which returns $-1$ after two successive operations on $+1$. In other words, there is no real number that squares to $-1$. As a consequence, it is impossible to take the square root of the number $-1$ in $\mathbb{R}$.

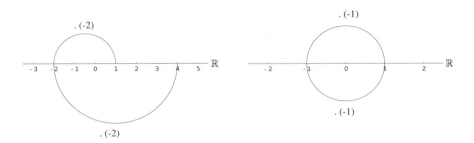

*Figure 17.1*: The real number line

Seeking for an 'impossible number' $i$ which squares to $-1$ corresponds to seeking for a transformation which returns $-1$ after two successive operations on the point labelled $+1$. The point reflection $M_O$ equals the standard rotation around the origin over the straight angle $R_{O,180°}$ (see paragraph 13.3). Two successive standard rotations over the right angle $R_{O,90°}$ yield the reflection over the origin $M_O$:

$$R_{O,90°} \cdot R_{O,90°} = R_{O,180°}.$$

Hence, the Swiss mathematician and politician **Jean-Robert Argand** (1768–1822) interpreted the 'impossible number' $i$ as a standard rotation over $90°$.

As a consequence, its image point after one operation on the point labelled $+1$ no longer lies on the (horizontal) number line. Applying the standard rotation $R_{O,90°}$ on the point labelled $+1$ yields its image point labelled $i$ on the vertical line, perpendicular to the real number line. The horizontal 'real' number line inspired Argand to name his vertical number line, containing the unity $i$, the 'imaginary' number line. We define $i$ as the imaginary unity via

$$i^2 = -1. \tag{17.1}$$

We define a **complex number** $z$ as an expression

$$z = a + bi,$$

given $a$ and $b$ real numbers, for instance $2 + 3i$. We call this the **algebraic representation** of $z$. We call $a = \text{Re } z$ the **real** part and $b = \text{Im } z$ the **imaginary part** of the number $z$. For instance, in $2 + 3i$ the real part is 2 and the imaginary part is 3. We typeset the set of all complex numbers as $\mathbb{C}$.

The former definition $z = a + bi$ is based on two real numbers $a$ and $b$. This allows us to interpret a plane point labelled $(a, b)$ as a complex number $z$. And the other way round, to each complex number $z$ a unique point $(a,b)$ corresponds geometrically in the plane. We call this plane the complex **number plane** or **Gauss plane**, named after the famous German mathematician.

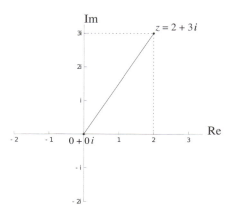

*Figure 17.2*: The complex number plane

Based on the correspondence of complex numbers to plane points, we can represent them alternatively by polar coordinates (see paragraph 6.3).

We define the absolute value or **modulus** $|z|$ of a complex number $z = a + bi$ as the distance from the origin to the point labelled $(a, b)$ in the complex plane. The Pythagorean theorem therefore yields

$$|z| = \sqrt{a^2 + b^2}.$$

We define the **argument** $\arg z$ of a complex number as the counterclockwise angle $\theta$, made by the segment from the origin to the point labelled $(a,b)$ and the positive real number line. Since $\theta$ corresponds to the polar angle, we determine it as

$$\arg z = \theta = \operatorname{atan2}(b,\ a),$$

given this atan2-function returns to the appropriate quadrant (see page 85). We realise that the complex number $z = 0 + 0i$ lacks a unique trigonometrical representation.

Analogously to polar coordinates, we determine the real part $a$ and the imaginary part $b$, given the known values of the modulus $|z|$ and the argument $\theta$, straightforwardly as

$$a = Re(z) = |z| \cos \theta$$
$$b = Im(z) = |z| \sin \theta.$$

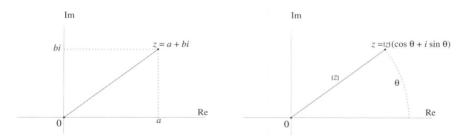

*Figure 17.3*: Algebraic versus trigonometric representation

Substituting the above parts in $z = a + bi$ yields $z = |z|\cos\theta + i|z|\sin\theta$, leading to the **trigonometrical representation** of a complex number as

$$z = |z|(\cos\theta + i\sin\theta).$$

*Example*:  We represent the complex number $z = 2 + 3i$ trigonometrically. We calculate its modulus as $|z| = \sqrt{2^2 + 3^2} = \sqrt{13}$. We retrieve its argument as $\arg z = \text{atan2}(3,\ 2) = 0.98\ \text{rad} = 56.31°$. We summarise this trigonometrical representation as

$$z = \sqrt{13}(\cos 56.31° + i\sin 56.31°).$$

We abbreviate the latter factor of this representation as $\text{cis}\theta = \cos\theta + i\sin\theta$.

But we *prove* the natural exponential function relating to $\text{cis}\theta$ as

$$\exp(i\theta) = \cos\theta + i\sin\theta. \tag{17.2}$$

*Proof:* **Euler's formula**

$$e^{\,i\theta} = \cos\theta + i\sin\theta$$

Relying on complex calculus, we extend the natural exponential function to complex arguments, making use of its maclaurin expansion which converges for all complex numbers:

$$\exp(z) \;=\; 1 + z + \frac{z^2}{2!} + \frac{z^3}{3!} + \frac{z^4}{4!} + \frac{z^5}{5!} + \frac{z^6}{6!} + \frac{z^7}{7!} + \dots \quad \text{valid for all } z \in \mathbb{C}.$$

Inserting the pure imaginary number $z = i\theta$ splits $\exp(z) = e^z$ into two maclaurin series:

$$
\begin{aligned}
e^{i\theta} &= 1 + i\theta + \frac{(i\theta)^2}{2!} + \frac{(i\theta)^3}{3!} + \frac{(i\theta)^4}{4!} + \frac{(i\theta)^5}{5!} + \frac{(i\theta)^6}{6!} + \frac{(i\theta)^7}{7!} + \dots \\
&= 1 + i\theta + \frac{i^2\theta^2}{2!} + \frac{i^3\theta^3}{3!} + \frac{i^4\theta^4}{4!} + \frac{i^5\theta^5}{5!} + \frac{i^6\theta^6}{6!} + \frac{i^7\theta^7}{7!} + \dots \\
&= 1 + i\theta - \frac{\theta^2}{2!} - \frac{i\theta^3}{3!} + \frac{\theta^4}{4!} + \frac{i\theta^5}{5!} - \frac{\theta^6}{6!} - \frac{i\theta^7}{7!} + \dots \qquad \text{see formula (17.1)}
\end{aligned}
$$

$$e^{i\theta} = 1 - \frac{\theta^2}{2!} + \frac{\theta^4}{4!} - \frac{\theta^6}{6!} + \ldots \quad +i\left(\theta - \frac{\theta^3}{3!} + \frac{\theta^5}{5!} - \frac{\theta^7}{7!} + \ldots\right)$$

$$= \qquad \cos\theta \qquad\qquad + i\sin\theta \qquad\qquad (4.6) \text{ and } (4.7) \quad \blacksquare$$

Geometrically the Euler formula means that $e^{i\theta}$ lies on the unit circle in the complex plane $\mathbb{C}$. The previous leads to the polar form or the **euler representation** of a complex number

$$z = |z|e^{i\theta}.$$

*Example 1*:  We euler represent the complex number $z = 2 + 3i$ as $z = \sqrt{13}\, e^{\,i56.31°}$.

*Example 2*:  We simplify in radian **Euler's identity** as $e^{\,i\pi} = \cos\pi + i\sin\pi = -1$.

## 17.2 Complex number arithmetic

### COMPLEX CONJUGATE

We define the **complex conjugate** number of $z = a + bi$ as $z^* = a - bi$. For instance, the complex conjugate number of $2 + 3i$ is $(2 + 3i)^* = 2 - 3i$. Consequently, successively complex conjugating the complex conjugate $z^*$ yields $(z^*)^* = z$ again.

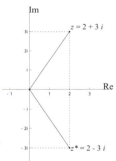

*Figure 17.4*: Complex conjugate number

Hence, for the complex conjugation in the euler representation this holds

$$z = |z|e^{\,i\theta} \implies z^* = |z|e^{\,-i\theta}.$$

### ADDITION AND SUBTRACTION

Each complex number corresponds geometrically to a point in the plane. We recall that each point in the plane can be interpreted as the head of a location vector. Addition of complex numbers hence corresponds to the addition of vectors (see paragraph 7.2).

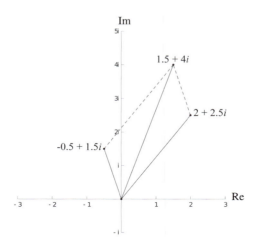

*Figure 17.5*: Addition of complex numbers

*Example:*

$$(2+2.5i)+(-0.5+1.5i) = (2+(-0.5))+(2.5+1.5)i = 1.5+4i$$

shows us that we have to add the real parts and the imaginary parts, similarly to the component-wise addition of vectors. The complex addition summarises as

$$(a+bi)+(c+di) = (a+c)+(b+d)i. \tag{17.3}$$

Consequently, the **opposite** of a complex number $z = a+bi$ is $-z = -a - bi$ because $(a+bi)+(-a-bi) = 0+0i$. We define the complex subtraction as the addition with the opposite number:

$$(a+bi)-(c+di) = (a+bi)+(-c-di) = (a-c)+(b-d)i. \tag{17.4}$$

## MULTIPLICATION

We explore complex multiplication, applying real arithmetic combined with the extra rule $i^2 = -1$. Applying the distributive property leads to

$$\begin{aligned}
(a+bi)\cdot(c+di) &= a(c+di)+bi(c+di) \\
&= ac+adi+bci+bidi \\
&= ac+(ad+bc)i+bd\ i^2 \\
&= ac+(ad+bc)i-bd \\
&= (ac-bd)+(ad+bc)i. \tag{17.5}
\end{aligned}$$

*Example*:
$$(2+3i) \cdot (-1+2i) = 2(-1+2i) + 3i(-1+2i) = -2+4i-3i+6\,i^2 = -8+i$$

Note that multiplying a complex number with its conjugate number returns a real number.

$$z \cdot z^* = (a+bi) \cdot (a-bi) = a^2 - abi + abi - b^2\,i^2$$
$$= a^2 + b^2$$

The complex multiplication can be performed more efficiently in the trigonometrical representation. Given two complex numbers $z = |z|(\cos\alpha + i\sin\alpha)$ and $w = |w|(\cos\beta + i\sin\beta)$, we simplify the complex multiplication, applying the trigonometrical Sum Identities (see paragraph 3.7).

$$\begin{aligned}
z \cdot w &= |z|(\cos\alpha + i\sin\alpha)|w|(\cos\beta + i\sin\beta) \\
&= |z||w|(\cos\alpha + i\sin\alpha)(\cos\beta + i\sin\beta) \\
&= |z||w|(\cos\alpha\cos\beta + i\cos\alpha\sin\beta + i\sin\alpha\cos\beta + i^2\sin\alpha\sin\beta) \\
&= |z||w|((\cos\alpha\cos\beta - \sin\alpha\sin\beta) + i(\cos\alpha\sin\beta + \sin\alpha\cos\beta)) \\
&= \underbrace{|z||w|}_{\text{product}}(\cos\underbrace{(\alpha+\beta)}_{\text{sum}} + i\sin\underbrace{(\alpha+\beta)}_{\text{sum}})
\end{aligned}$$

The complex multiplication in trigonometric representation summarises as:

▷   the modulus of a complex product $z \cdot w$ equals the product of the moduli of $z$ and $w$,

$$|z \cdot w| = |z| \cdot |w|$$

▷   the argument of a complex product $z \cdot w$ equals the sum of arguments of $z$ and $w$,

$$\arg(z \cdot w) = \arg z + \arg w.$$

The complex multiplication in euler representation summarises as

$$z \cdot w = |z|e^{\,i\alpha}|w|e^{\,i\beta} = |z||w|e^{\,i(\alpha+\beta)}.$$

*Example*:   We recalculate the previous complex product of $z = 2+3i$ and $w = -1+2i$. Firstly, we convert both complex numbers $z$ and $w$ into their trigonometrical representation. We determine their moduli as $|z| = \sqrt{13}$ and $|w| = \sqrt{5}$. We retrieve their arguments as $\arg z = \text{atan2}\,(3,\ 2)$ and $\arg w = \text{atan2}\,(2,\ -1)$.

The modulus of the product $z \cdot w$ equals

$$|z \cdot w| = |z| \cdot |w| = \sqrt{13} \cdot \sqrt{5} = \sqrt{65}.$$

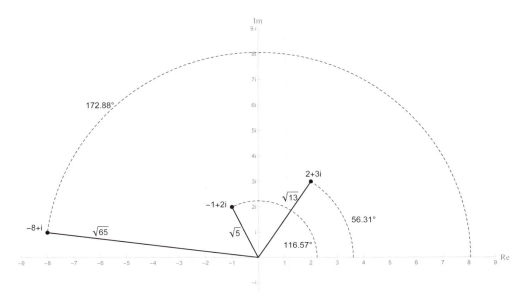

*Figure 17.6*: Complex multiplication.

The argument of the product $z \cdot w$ equals

$$\arg(z \cdot w) = \arg z + \arg w = \text{atan2}\,(3,\ 2) + \text{atan2}\,(2,\ -1) \approx 172.88°.$$

We approximate the complex product of $z$ and $w$ as

$$z \cdot w \approx \sqrt{65}(\cos 172.88° + i\sin 172.88°) \approx \sqrt{65}(-0.99 + 0.12i) \approx -8 + i.$$

Alternatively, we redo this multiplication in euler representation far more efficiently as

$$z \cdot w \approx \sqrt{13}\,e^{\,i56.31°}\,\sqrt{5}\,e^{\,i116.57°} = \sqrt{65}e^{\,i172.88°}.$$

EXPONENTIATION

We extend the complex multiplication logically to complex exponentiation by successively multiplying a complex number $z$ by itself. The modulus of the exponentiation calculates as $|z^n| = |z|^n$. The argument of the exponentiation simplifies as $\arg(z^n) = n\arg z$. The complex exponentiation of a number $z = |z|(\cos\theta + i\sin\theta)$ in trigonometric representation leads to **de Moivre's formula**

$$z^n = |z|^n(\cos n\theta + i\sin n\theta), \quad \text{for all } n \in \mathbb{Z}. \tag{17.6}$$

The complex exponentiation of a number $z = |z|e^{i\theta}$ in euler representation simplifies to

$$z^n = \left(|z|e^{i\theta}\right)^n = |z|^n e^{i(n\theta)}, \quad \text{for all } n \in \mathbb{Z}. \tag{17.7}$$

*Example*: We calculate $z^4$, given $z = 1 + i\sqrt{3}$. The modulus of $z$ equals 2 and its argument equals $\frac{\pi}{3}$. We summarise $z$ trigonometrically as $z = 2\left(\cos\frac{\pi}{3} + i\sin\frac{\pi}{3}\right)$. The complex exponentiation of $z$ to the power 4, exponentiates its modulus to $2^4$ and multiplies its argument to $4\frac{\pi}{3}$. We summarise this exponentiation trigonometrically as

$$z^4 = 16\left(\cos\frac{4\pi}{3} + i\sin\frac{4\pi}{3}\right).$$

Again representing this result algebraically via $\cos\frac{4\pi}{3} = \frac{-1}{2}$ and $\sin\frac{4\pi}{3} = \frac{-\sqrt{3}}{2}$, results finally into $z^4 = -8 - 8i\sqrt{3}$.

### DIVISION

At a first glance, complex division looks troublesome. For instance, dividing 1 by $i$ results in the fraction $\frac{1}{i}$, which does not fit the definition $a + bi$ of complex numbers. This looks similar to for instance dividing the integers $-3$ by 5 which results in the fraction $\frac{-3}{5}$ no longer being an integer. But at a second glance, we can simplify a complex fraction using the complex conjugate of its denominator. To simplify complex fractions, we need to multiply both their numerator and denominator with the conjugate of the denominator. As a try-out, our above example simplifies to

$$\frac{1}{i} = \frac{1}{i} \cdot \left(\frac{-i}{-i}\right) = \frac{-i}{-i^2} = \frac{-i}{1} = -i.$$

*Example*: To simplify the complex fraction $\frac{1}{2+2.5i}$ we need to multiply both its numerator and denominator with the conjugate of the denominator $(2 + 2.5i)^* = 2 - 2.5i$.

$$\frac{1}{2+2.5i} = \left(\frac{1}{2+2.5i}\right) \cdot \left(\frac{2-2.5i}{2-2.5i}\right) = \frac{2-2.5\,i}{2^2 + 2.5^2}$$

$$= \frac{2-2.5i}{10.25} = \frac{2}{10.25} - \frac{2.5}{10.25}i$$

$$\approx 0.20 - 0.24i.$$

Generalising the above simplification method proves it can never fail:

$$\frac{a+bi}{c+di} = \left(\frac{a+bi}{c+di}\right) \cdot \left(\frac{c-di}{c-di}\right) = \frac{(a+bi)(c-di)}{c^2+d^2}$$

$$= \frac{ac+bd}{c^2+d^2} + \left(\frac{-ad+bc}{c^2+d^2}\right)i. \tag{17.8}$$

We advise you to forget this algebraic formula, and instead to remember its procedure practically: 'multiply both numerator and denominator of complex fractions with the conjugate of its denominator'.

Similarly to complex multiplication, the division can be performed more efficiently in the trigonometrical representation. Dividing the complex number $z = |z|(\cos\alpha + i\sin\alpha)$ by $w = |w|(\cos\beta + i\sin\beta)$, we simplify the complex division, applying the trigonometrical Sum Identities (see paragraph 3.7).

$$\frac{z}{w} = \frac{|z|(\cos\alpha + i\sin\alpha)}{|w|(\cos\beta + i\sin\beta)}$$

$$= \frac{|z|}{|w|} \frac{(\cos\alpha + i\sin\alpha)}{(\cos\beta + i\sin\beta)} \frac{(\cos\beta + i\sin\beta)^*}{(\cos\beta + i\sin\beta)^*}$$

$$= \frac{|z|}{|w|} \frac{(\cos\alpha\cos\beta - i\cos\alpha\sin\beta + i\sin\alpha\cos\beta + \sin\alpha\sin\beta)}{\cos^2\beta + \sin^2\beta}$$

$$= \underbrace{\frac{|z|}{|w|}}_{\text{division}} (\cos \underbrace{(\alpha - \beta)}_{\text{subtraction}} + i\sin \underbrace{(\alpha - \beta)}_{\text{subtraction}})$$

The complex division in trigonometric representation summarises as:

▷ the modulus of a complex quotient $\frac{z}{w}$ equals the quotient of the moduli of $z$ and $w$.

$$\left|\frac{z}{w}\right| = \frac{|z|}{|w|}$$

▷ the complex division subtracts the argument of $w$ from the argument of $z$.

$$\arg\left(\frac{z}{w}\right) = \arg z - \arg w$$

The complex division in euler representation summarises as

$$\frac{z}{w} = \frac{|z|e^{i\alpha}}{|w|e^{i\beta}} = \frac{|z|}{|w|}e^{i(\alpha - \beta)}.$$

*Example*:    We divide the complex number $z = 2 + 3i$ by $w = -1 + 2i$. Firstly, we convert both complex numbers $z$ and $w$ into their trigonometrical representation. We determine their moduli as $|z| = \sqrt{13}$ and $|w| = \sqrt{5}$. We retrieve their arguments as $\arg z = \text{atan2}(3, 2)$ and $\arg w = \text{atan2}(2, -1)$.

The modulus of the quotient $\frac{z}{w}$ equals

$$\left|\frac{z}{w}\right| = \frac{|z|}{|w|} = \frac{\sqrt{13}}{\sqrt{5}} = \sqrt{\frac{13}{5}}.$$

The argument of the quotient $\frac{z}{w}$ equals

$$\arg\left(\frac{z}{w}\right) = \arg z - \arg w = \text{atan2}\,(3,\ 2) - \text{atan2}\,(2,\ -1) \approx 299.74°.$$

We approximate the complex quotient of $z$ and $w$ as

$$\frac{z}{w} \approx \sqrt{\frac{13}{5}}\,(\cos 299.74° + i\sin 299.74°).$$

Alternatively, we redo this complex division in euler representation way more efficiently

$$\frac{z}{w} \approx \frac{\sqrt{13}\,e^{\,i56.31°}}{\sqrt{5}\,e^{\,i116.57°}} = \sqrt{\frac{13}{5}}\,e^{\,i(56.31°-116.57°)} = \sqrt{\frac{13}{5}}\,e^{\,i(-60.26°)} = \sqrt{\frac{13}{5}}\,e^{\,i299.74°}.$$

## 17.3 Complex numbers and transformations

### TRANSLATION

We recall the head to tail vector addition, which here allows us to realise translations by the use of complex numbers (see page 127).

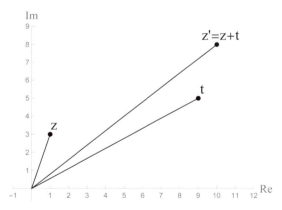

*Figure 17.7*: Translation by the use of complex numbers

Figure 17.7 shows the effect of the 2D translation by a displacement $\vec{t} = \begin{pmatrix} 9 \\ 5 \end{pmatrix}$ as we encountered similar ones before (see figure 14.2). We calculate the image point $z'$ caused by the translation number $t = 9 + 5i$ applied to the original point $z$ by the complex number addition

$$z' = z + t = (1 + 3i) + (9 + 5i) = 1 + 3i + 9 + 5i = 1 + 9 + 3i + 5i = 10 + 8i.$$

STANDARD ROTATION

We recall that multiplication with the factor $i$ geometrically corresponds to a standard rotation over 90°. This allows us to realise standard rotations by the use of complex numbers.

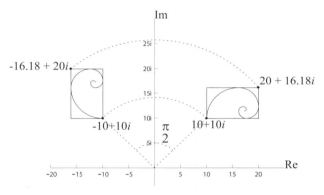

*Figure 17.8*: Rotation by the use of complex numbers

Figure 17.8 recalls the effect of a standard 2D rotation over 90° on the golden rectangle (see page 281). To rotate this golden rectangle by the use of complex numbers requires the multiplication of its vertices in the complex plane by the factor $i$. The argument of the imaginary unity $i$ is $\frac{\pi}{2}$ and its modulus is 1. Due to the trigonometrical representation of the complex multiplication, the modulus of the vertices remains invariant and $\frac{\pi}{2}$ is added to their arguments:

$$|z \cdot i| = |z| \cdot |i| = |z| \cdot 1 = |z| \quad \text{and} \quad \arg(z \cdot i) = \arg(z) + \arg(i) = \arg(z) + \frac{\pi}{2}.$$

We calculate the image points of the golden rectangle's defining vertices as:

$$(10 + 10i)i = -10 + 10i \text{ and } (20 + 16.18i)i = -16.18 + 20i.$$

We realise that the multiplication with a factor featuring modulus 1 and argument $\theta$ geometrically corresponds to a standard rotation over $\theta$. For instance, the multiplication with $\frac{1}{2} + i\frac{\sqrt{3}}{2}$ corresponds to the standard rotation over 60°. The modulus of $\frac{1}{2} + i\frac{\sqrt{3}}{2}$ is 1 and its argument is 60°.

STANDARD SCALING

Multiplying each point of an object with a positive real factor, rescales our object according to the factor's value. Positive scale factors smaller than 1 shrink the original object and positive scale factors larger than 1 enlarge the original object. For instance, the complex multiplication by the factor $\frac{1}{2}$ geometrically corresponds to a standard scaling that halves our golden rectangle uniformly, featuring scale factors $s_x = s_y = \frac{1}{2}$.

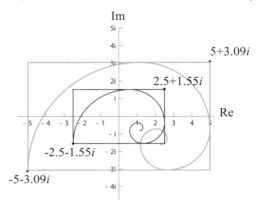

*Figure 17.9*: Scaling by the use of complex numbers

COMPOSITE TRANSFORMATION

Multiplication with an arbitrary complex number $z = |z|(\cos\theta + i\sin\theta)$ corresponds to a transformation consisting of a uniform standard scaling by the modulus $|z| > 0$ and a standard rotation by the argument $\theta$. For instance, the complex multiplication by the factor $\frac{i}{2}$ rescales the object to half its size and rotates it over the right angle around the origin. We reveal this effect by representing $\frac{i}{2}$ trigonometrically:

$$\frac{i}{2} = 0 + \frac{1}{2}i = \frac{1}{2}(0+i) = \frac{1}{2}(\cos 90° + i\sin 90°).$$

As a final example, multiplication with the complex number $z = 1 + i$ corresponds to a transformation combined of an enlargement by the scale factor $|z| = \sqrt{2}$ and a standard rotation by the argument $\theta = 45°$. We apply this composite transformation to the golden rectangle by multiplying its vertices $10 + 10i$ and $20 + 16.18i$ by the factor $1 + i$.

$$(10 + 10i) \cdot (1+i) = 20i \text{ and } (20 + 16.18i) \cdot (1+i) = 3.82 + 36.18i$$

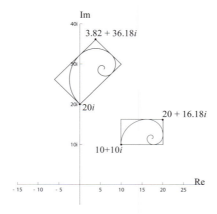

*Figure 17.10*: Composite transformation by the use of complex numbers

## 17.4 Complex continuation of the Fibonacci numbers

Based upon the complex plane, we extend Binet's formula to generate Fibonacci numbers (see formula (11.5)) beyond its conventional range of Fibonacci numbers in $\mathbb{N}$. We achieve this in two steps by interpreting Binet's formula as a function $f : k \mapsto f(k)$ dictated by its recipe

$$f(k) = \frac{(\Phi)^k - (\Phi')^k}{\sqrt{5}}.$$

In the first step, we evaluate this function $f$ as a mapping from the domain $\mathbb{Z}$ to the range $\mathbb{Z}$. In the second step, we extend this function as a mapping from the domain $\mathbb{Q}$ to the range $\mathbb{C}$.

### INTEGER FIBONACCI NUMBERS

We firstly consider the integer function $f : \mathbb{Z} \to \mathbb{Z} : k \mapsto f(k)$, dictated by the above recipe. Evaluating its first negative argument $k = -1$ returns the Fibonacci number $f(-1)$, which we calculate as

$$f(-1) = \frac{1}{\sqrt{5}} \left( \left( \frac{1 + \sqrt{5}}{2} \right)^{-1} - \left( \frac{1 - \sqrt{5}}{2} \right)^{-1} \right)$$

$$= \frac{1}{\sqrt{5}} \left( \frac{2}{1 + \sqrt{5}} - \frac{2}{1 - \sqrt{5}} \right)$$

$$= \frac{1}{\sqrt{5}} \left( \frac{2(1 - \sqrt{5}) - 2(1 + \sqrt{5})}{(1 + \sqrt{5})(1 - \sqrt{5})} \right)$$

$$f(-1) = \frac{1}{\sqrt{5}} \left( \frac{-4\sqrt{5}}{-4} \right) = 1.$$

Proceeding with its return values $f(-2) = \frac{1}{\sqrt{5}} \cdot \left( \left( \frac{1+\sqrt{5}}{2} \right)^{-2} - \left( \frac{1-\sqrt{5}}{2} \right)^{-2} \right) = -1$ and $f(-3) = 2$ and so on, we extrapolate the Fibonacci sequence below zero. We note that this extended function $f$ remains governed by the defining property of a Lucas sequence: $f(n+1) = f(n) + f(n-1)$. This allows us for each decremented argument $n-1$ to calculate $f_{n-1}$ faster via $f_{n-1} = f_{n+1} - f_n$. Therefore, the **countable** continuation of the Fibonacci sequence to $\mathbb{Z}$ equals $\dots, -8, 5, -3, 2, -1, 1, 0, 1, 1, 2, 3, 5, 8, 13, \dots$.

### COMPLEX FIBONACCI NUMBERS

We now extend this function to rational arguments $f : \mathbb{Q} \to \mathbb{C} : k \mapsto f(k)$ by applying the complex exponentiation (see formula (17.6)) on the trigonometric representations of $\Phi$ and $\Phi'$. Therefore, we can consequently calculate complex return values such as

$$f\left( \frac{1}{2} \right) = \frac{1}{\sqrt{5}} \left( \left( \frac{1+\sqrt{5}}{2} \right)^{\frac{1}{2}} - \left( \frac{1-\sqrt{5}}{2} \right)^{\frac{1}{2}} \right)$$

$$= \frac{1}{\sqrt{5}} \left( \left( \left| \frac{1+\sqrt{5}}{2} \right| (\cos 0 + i \sin 0) \right)^{\frac{1}{2}} - \left( \left| \frac{1-\sqrt{5}}{2} \right| (\cos \pi + i \sin \pi) \right)^{\frac{1}{2}} \right)$$

$$f\left( \frac{1}{2} \right) = \frac{1}{\sqrt{5}} \left( \left( \frac{1+\sqrt{5}}{2} \right)^{\frac{1}{2}} \left( \cos \frac{1}{2} 0 + i \sin \frac{1}{2} 0 \right) - \left( \frac{-1+\sqrt{5}}{2} \right)^{\frac{1}{2}} \left( \cos \frac{1}{2} \pi + i \sin \frac{1}{2} \pi \right) \right)$$

$$= \frac{1}{\sqrt{5}} \left( \sqrt{\frac{1+\sqrt{5}}{2}} (1 + 0i) - \sqrt{\frac{-1+\sqrt{5}}{2}} (0 + 1i) \right)$$

$$= \sqrt{\frac{1}{10}(1+\sqrt{5})} - i\sqrt{\frac{1}{10}(-1+\sqrt{5})}.$$

Or we calculate return values such as $f(\frac{5}{2}) = 1.48931 - 0.134291i$, which also resides in the complex plane. This complex range of Fibonacci numbers remains countable, because the rational domain $\mathbb{Q}$ of the function $f$ is countable.

We finally extend Binet's function to real arguments $f : \mathbb{R} \to \mathbb{C} : k \mapsto f(k)$, sustaining the above approach. In this case $f$ returns an **uncountable** complex continuation of the Fibonacci sequence. Plotting for each real argument $k \in \mathbb{R}$ its return value $f(k) \in \mathbb{C}$ visualises the range of the function $f$ graphically in the complex plane.

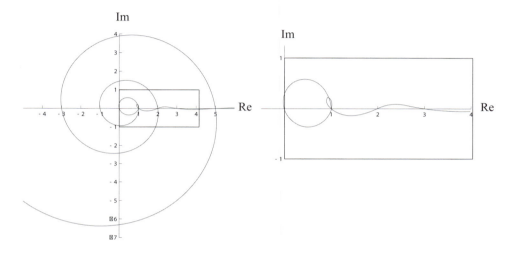

*Figure 17.11*: Complex continuation of the Fibonacci sequence

The complex graph of this Fibonacci continuation illustrates perfectly the way $f$ contains all integer Fibonacci numbers, governed by $f(n+1) = f(n) + f(n-1)$ as a Lucas sequence. We observe graphically that return value 1 occurs three times, which was indeed confirmed by $f(1) = f(2) = f(-1) = 1$. The numbers $2, 5, 13, \ldots$ on the odd positions in the ordinary Fibonacci sequence occur twice.

## 17.5 Quaternions

In the former paragraph, we interpreted complex numbers as points in the 2D complex plane, which allowed us to multiply these points. During the 19<sup>th</sup> century, mathematicians sought to multiply points similarly in a 3D complex space, concluding it required 4D.

It was the Irish mathematician **Wiliam Rowan Hamilton** (1805–1865) who discovered these 4D complex numbers also known as **quaternions**. Initially Hamilton tried to extend complex arithmetic by adopting only one extra imaginary unity $j$. For years, he investigated complex arithmetic using one real part and two imaginary parts, completely in vain. It was not until 1843 that Hamilton tried his idea of complex arithmetic based on one real part and three imaginary unities $i, j$ and $k$. Meanwhile, it is a famous narrative in mathematical history how Hamilton was walking over the Brougham Bridge in Dublin when this genius idea urged him to jag the fundamental rules for $i, j$ and $k$ in its stone. Today the Dubliners have a plaque on the Brougham Bridge, in memory of the quaternions and their inventor.

However quaternions also emerge as well spontaneously within certain *geometric algebra* spaces (for which we kindly refer you to the specialised literature [19], [50]).

To express a quaternion takes, apart from the symbol $i$, two extra symbols $j$ and $k$ to typeset its three imaginary components. We define a **quaternion** as an expression $a + bi + cj + dk$, given $a, b, c$ and $d$ are real coefficients and $i, j$ and $k$ are governed by Hamilton's rules

$$i^2 = j^2 = k^2 = ijk = -1. \qquad (17.9)$$

The imaginary unities $i, j$ and $k$ are arithmetically interrelated by

$$ij = k, \quad jk = i, \quad ki = j, \quad ji = -k, \quad kj = -i, \text{ and } ik = -j. \qquad (17.10)$$

For instance, the numbers $7, 3 + i, 1 + j, 2 - 3i + 5k$ and $9 - i + 2j + 7k$ are quaternions. The quaternion set is a subset of the **hypercomplex number** set. We typeset the quaternion set as $\mathbb{H}$, in honour of Hamilton.

Note that $ij \neq ji, ik \neq ki$ and $jk \neq kj$ which implies the quaternion multiplication is non-commutative. Applying Hamilton's rules and the imaginary unities' interrelations allows for quaternion arithmetic.

For instance, multiplying the number $1 + k$ with $1 + i$ illustrates its non-commutativity:

$$(1+k)(1+i) = 1 + i + k + ki = 1 + i + k + j,$$

$$(1+i)(1+k) = 1 + k + i + ik = 1 + i + k - j.$$

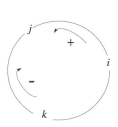
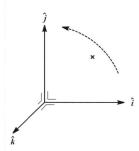

Given their one real and three imaginary components, we can typeset quaternions componentwise as **row vectors**. We express a quaternion $q = q_0 + q_1 i + q_2 j + q_3 k$ as the 4D row vector $[q_0, (q_1, q_2, q_3)] = [q_0, \vec{q}\,]$, separating its scalar component from its imaginary 3D vector $\vec{q} = (q_1, q_2, q_3)$. As an example the other way around, the 4D vector $[1, (1, 1, -1)]$ corresponds to the quaternion $1 + i + j - k$.

Identically to vectors and complex numbers, we define the **quaternion norm** as

$$\|q\| = \sqrt{q_0^2 + q_1^2 + q_2^2 + q_3^2}.$$

We define the **unit quaternions** as quaternions named $u$ with a norm $\|u\|$ equal to 1. We define the **pure quaternions** as quaternions named $q$ with the scalar component $q_0$ equal to 0. Consequently, we define the **pure unit quaternions** as quaternions $[0, (u_1, u_2, u_3)]$ featuring the real component $u_0 = 0$ and a norm $\sqrt{u_0^2 + u_1^2 + u_1^2 + u_1^2} = 1$ and keep typesetting them with the letter $u$ referring to *unit*.

## 17.6 Quaternion arithmetic

Applying Hamilton's rules and interrelationships to ordinary algebra on $i$, $j$ and $k$, yields quaternion arithmetic.

### ADDITION AND SUBTRACTION

We add two quaternions $p = [p_0, (p_1, p_2, p_3)]$ and $q = [q_0, (q_1, q_2, q_3)]$ by adding their corresponding components, similarly to vector addition.

$$p + q = [p_0 + q_0, (p_1 + q_1, p_2 + q_2, p_3 + q_3)]$$
$$= (p_0 + q_0) + (p_1 + q_1)i + (p_2 + q_2)j + (p_3 + q_3)k$$

Consequently, the **opposite quaternion** $-q$ of the quaternion $q = [q_0, (q_1, q_2, q_3)]$ equals $-q = [-q_0, (-q_1, -q_2, -q_3)]$. We define quaternion subtraction as addition with the opposite quaternion.

$$p - q = p + (-q) = [p_0 - q_0, (p_1 - q_1, p_2 - q_2, p_3 - q_3)]$$
$$= (p_0 - q_0) + (p_1 - q_1)i + (p_2 - q_2)j + (p_3 - q_3)k$$

## SCALAR MULTIPLICATION

We base the scalar multiplication of quaternions on the repeated addition of the quaternion to itself. Extending this idea, we scalar multiply a quaternion $q = [q_0, (q_1, q_2, q_3)]$ with a number $\lambda \in \mathbb{R}$ by multiplying each component of $q$ with $\lambda$, identically to scalar multiplication of vectors.

$$\lambda q = [\lambda q_0, (\lambda q_1, \lambda q_2, \lambda q_3)] = \lambda q_0 + (\lambda q_1)i + (\lambda q_2)j + (\lambda q_3)k$$

## NORMALISATION

We define the **normalisation** of a quaternion $q$ as the division of $q$ by its length $\|q\|$. Therefore, $q$ turns into its unit quaternion which typeset with the subscript $n$ referring to *normalised*. For instance, normalising the quaternion $q = [1, (4, 4, -4)]$ by its length $\|q\| = \sqrt{1 + 16 + 16 + 16} = \sqrt{49} = 7$ scales this $q$ into its corresponding unit quaternion $q_n = [\frac{1}{7}, (\frac{4}{7}, \frac{4}{7}, \frac{-4}{7},)]$.

## QUATERNION MULTIPLICATION

We multiply two quaternions $p = [p_0, (p_1, p_2, p_3)]$ and $q = [q_0, (q_1, q_2, q_3)]$ applying Hamilton's rules and interrelationships.

$$p \cdot q = (p_0 + p_1 i + p_2 j + p_3 k) \cdot (q_0 + q_1 i + q_2 j + q_3 k)$$

$$= p_0 q_0 + p_0 q_1 i + p_0 q_2 j + p_0 q_3 k + p_1 q_0 i + p_1 q_1 i^2 + p_1 q_2 ij + p_1 q_3 ik$$
$$+ p_2 q_0 j + p_2 q_1 ji + p_2 q_2 j^2 + p_2 q_3 jk + p_3 q_0 k + p_3 q_1 ki + p_3 q_2 kj + p_3 q_3 k^2$$

$$= p_0 q_0 + p_0 q_1 i + p_0 q_2 j + p_0 q_3 k + p_1 q_0 i - p_1 q_1 + p_1 q_2 k - p_1 q_3 j$$
$$+ p_2 q_0 j - p_2 q_1 k - p_2 q_2 + p_2 q_3 i + p_3 q_0 k + p_3 q_1 j - p_3 q_2 i - p_3 q_3$$

$$= (p_0 q_0 - p_1 q_1 - p_2 q_2 - p_3 q_3) + (p_0 q_1 + p_1 q_0 + p_2 q_3 - p_3 q_2)i$$
$$+ (p_0 q_2 - p_1 q_3 + p_2 q_0 + p_3 q_1)j + (p_0 q_3 + p_1 q_2 - p_2 q_1 + p_3 q_0)k$$

We recall the dot product of the imaginary vectors $\vec{p} = (p_1, p_2, p_3)$ and $\vec{q} = (q_1, q_2, q_3)$ as the number $\vec{p} \cdot \vec{q} = p_1 q_1 + p_2 q_2 + p_3 q_3 \in \mathbb{R}$. We recall the cross product of $\vec{p} = (p_1, p_2, p_3)$ and $\vec{q} = (q_1, q_2, q_3)$ as the vector $\vec{p} \times \vec{q} = (p_2 q_3 - p_3 q_2, p_3 q_1 - p_1 q_3, p_1 q_2 - p_2 q_1)$. Hence, we summarise the quaternion product as

$$p \cdot q = [p_0 q_0 - \vec{p} \cdot \vec{q}, \ (p_0 \vec{q} + q_0 \vec{p} + \vec{p} \times \vec{q})]. \tag{17.11}$$

We note that the non-commutativity of the quaternion product is due to the non-commutativity of the cross product of the imaginary vectors.

*Example:* We multiply the quaternions $p = 1 + 2i + 3j + 4k = [1, (2,3,4)]$ and $q = 5 - 2i + 6j - 7k = [5, (-2,6,-7)]$. We perform their quaternion product $p \cdot q$ stepwise.

▷  $p_0 q_0 = 1 \cdot 5 = 5$

▷  $\vec{p} \cdot \vec{q} = \begin{pmatrix} 2 \\ 3 \\ 4 \end{pmatrix} \cdot \begin{pmatrix} -2 \\ 6 \\ -7 \end{pmatrix} = -14$

We calculate its scalar component as $p_0 q_0 - \vec{p} \cdot \vec{q} = 5 - (-14) = 19$.

▷  $p_0 \vec{q} = 1 \begin{pmatrix} -2 \\ 6 \\ -7 \end{pmatrix} = \begin{pmatrix} -2 \\ 6 \\ -7 \end{pmatrix}$

▷  $q_0 \vec{p} = 5 \begin{pmatrix} 2 \\ 3 \\ 4 \end{pmatrix} = \begin{pmatrix} 10 \\ 15 \\ 20 \end{pmatrix}$

▷  $\vec{p} \times \vec{q} = \begin{pmatrix} 2 \\ 3 \\ 4 \end{pmatrix} \times \begin{pmatrix} -2 \\ 6 \\ -7 \end{pmatrix} = \begin{pmatrix} -45 \\ 6 \\ 18 \end{pmatrix}$

We calculate its imaginary vector as $p_0 \vec{q} + q_0 \vec{p} + \vec{p} \times \vec{q} = \begin{pmatrix} -37 \\ 27 \\ 31 \end{pmatrix}$.

We summarise the product quaternion as

$$p \cdot q = [19, (-37, 27, 31)] = 19 - 37i + 27j + 31k.$$

We illustrate the non-commutativity of the quaternion product by also calculating $q \cdot p$.

▷  $q_0 p_0 = 5 \cdot 1 = 5$

▷  $\vec{q} \cdot \vec{p} = \begin{pmatrix} -2 \\ 6 \\ -7 \end{pmatrix} \cdot \begin{pmatrix} 2 \\ 3 \\ 4 \end{pmatrix} = -14$

We calculate its scalar component as $q_0 p_0 - \vec{q} \cdot \vec{p} = 5 - (-14) = 19$.

▷  $q_0 \vec{p} = 5 \begin{pmatrix} 2 \\ 3 \\ 4 \end{pmatrix} = \begin{pmatrix} 10 \\ 15 \\ 20 \end{pmatrix}$

▷ $p_0\vec{q} = 1 \begin{pmatrix} -2 \\ 6 \\ -7 \end{pmatrix} = \begin{pmatrix} -2 \\ 6 \\ -7 \end{pmatrix}$

▷ $\vec{q} \times \vec{p} = \begin{pmatrix} -2 \\ 6 \\ -7 \end{pmatrix} \times \begin{pmatrix} 2 \\ 3 \\ 4 \end{pmatrix} = \begin{pmatrix} 45 \\ -6 \\ -18 \end{pmatrix}$

We calculate its imaginary vector as $q_0\vec{p} + p_0\vec{q} + \vec{q} \times \vec{p} = \begin{pmatrix} 53 \\ 15 \\ -5 \end{pmatrix}$.

We summarise the product quaternion as

$$q \cdot p = [19, (53, 15, -5)] = 19 + 53i + 15j - 5k.$$

## QUATERNION CONJUGATE

We define the **conjugate quaternion** $q^*$ of a quaternion $q$ analogously to the conjugation of complex numbers. The conjugation of a quaternion keeps its scalar component and takes the opposite of its imaginary vector. Given $q = [q_0, (q_1, q_2, q_3)]$, we define its conjugate quaternion as

$$q^* = [q_0, -\vec{q}] = [q_0, -(q_1, q_2, q_3)] = q_0 - (q_1 i + q_2 j + q_3 k).$$

*Example*:  We multiply a general quaternion $q = [q_0, (q_1, q_2, q_3)]$ with its conjugate quaternion $q^* = [q_0, -(q_1, q_2, q_3)]$.

▷ $q_0 q_0 = q_0^2$

▷ $\vec{q} \cdot (-\vec{q}) = -(q_1^2 + q_2^2 + q_3^2)$

We calculate its scalar component as

$$q_0 q_0 - \vec{q} \cdot (-\vec{q}) = q_0^2 - (-(q_1^2 + q_2^2 + q_3^2)) = q_0^2 + q_1^2 + q_2^2 + q_3^2 = \|q\|^2$$

▷ $q_0(-\vec{q}) = -q_0\vec{q}$

▷ $q_0\vec{q}$

▷ $\vec{q} \times (-\vec{q}) = \begin{pmatrix} q_1 \\ q_2 \\ q_3 \end{pmatrix} \times \begin{pmatrix} -q_1 \\ -q_2 \\ -q_3 \end{pmatrix} = \begin{pmatrix} 0 \\ 0 \\ 0 \end{pmatrix}$

We calculate its imaginary vector as $-q_0\vec{q} + q_0\vec{q} + \vec{q} \times (-\vec{q}) = \begin{pmatrix} 0 \\ 0 \\ 0 \end{pmatrix}$.

We summarise the product quaternion as

$$q \cdot q^* = [\|q\|^2, (0,0,0)] = \|q\|^2 \in \mathbb{R}.$$

### INVERSE QUATERNION

We define the inverse quaternion $q^{-1}$ of the nonzero quaternion $q \neq [0, (0,0,0)]$ as the unique factor for which $q \cdot q^{-1} = q^{-1} \cdot q = [1, (0,0,0)]$, given this product $[1, (0,0,0)] = 1$ also known as the **identity quaternion**. Hence, we can determine the inverse of a quaternion via its conjugate quaternion

$$\underbrace{q^{-1} \cdot q}_{1} \cdot q^* = q^*$$

an equation which we then solve for the inverse as

$$q^{-1} = \frac{q^*}{q \cdot q^*} \Longleftrightarrow$$

$$q^{-1} = \frac{q^*}{\|q\|^2}. \tag{17.12}$$

In case of a unit quaternion $u$ for which $\|u\| = 1$, the above calculation simplifies to $u^{-1} = u^*$, meaning its inverse is equal to its conjugate (which is far easier to determine).

## 17.7 Quaternions and rotations

Contemporary interactive 3D computer graphics thrive on quaternions to handle fast changing location vectors. Multiplying with unit quaternions $u$ features exactly the same properties as applying 3D rotators (see page 298). Therefore, we may use unit quaternions $u$ to rotate location vectors around an axis of rotation in 3D space.

## TRIGONOMETRICAL REPRESENTATION OF QUATERNIONS

Identifying both parts of the algebraic representation of q expressed as
$$q = q_0 + q_1 i + q_2 j + q_3 k = [q_0, (q_1, q_2, q_3)]$$ or in shorthand

$$q = [q_0, \vec{q}]$$

to both parts of its equivalent trigonometrical representation

$$q = \|q\|[\cos\theta, \hat{q}\sin\theta]$$
$$= [\underbrace{\|q\|\cos\theta}_{q_0}, \underbrace{\|q\|\hat{q}\sin\theta}_{\vec{q}}]$$

leads to the conversion formulas

$$\cos\theta = \frac{q_0}{\|q\|} \Rightarrow \sin\theta = \frac{\sqrt{\|q\|^2 - q_0^2}}{\|q\|} = \frac{\sqrt{q_1^2 + q_2^2 + q_3^2}}{\|q\|}, \quad \text{which confirms}$$

$$\sin\theta = \frac{\|\vec{q}\|}{\|q\|}.$$

*Example 1:* Given the quaternion $q = \frac{1}{2} + \frac{1}{2}i + \frac{1}{2}j + \frac{1}{2}k = [\frac{1}{2}, (\frac{1}{2}, \frac{1}{2}, \frac{1}{2})]$, we determine both parts of its equivalent trigonometrical representation via

$$\|q\| = \sqrt{\left(\frac{1}{2}\right)^2 + \left(\frac{1}{2}\right)^2 + \left(\frac{1}{2}\right)^2 + \left(\frac{1}{2}\right)^2} = 1,$$

$$\cos\theta = \frac{1}{2} \Rightarrow \theta = \arccos\left(\frac{1}{2}\right) = 60° \text{ and}$$

$$\hat{q} = \frac{\vec{q}}{\|\vec{q}\|} = \frac{2}{\sqrt{3}}(\frac{1}{2}, \frac{1}{2}, \frac{1}{2}) = (\frac{1}{\sqrt{3}}, \frac{1}{\sqrt{3}}, \frac{1}{\sqrt{3}})$$

which yields

$$q = 1[\cos 60°, \hat{q}\sin 60°] \text{ given } \hat{q} = (\frac{1}{\sqrt{3}}, \frac{1}{\sqrt{3}}, \frac{1}{\sqrt{3}}).$$

## EULER REPRESENTATION OF QUATERNIONS

We abbreviate the second factor of the trigonometric representation applying Euler's formula as

$$e^{\hat{q}\theta} = [\cos\theta, \hat{q}\sin\theta]$$

which leads to the polar form or the euler representation of a quaternion

$$q = \|q\|e^{\hat{q}\theta}.$$

Concluding all previous natural exponentiation flavours (see formulas (4.5), (11.6) and (17.2)) makes $\exp(x) = e^x$ the universal transformator in geometric algebra (see specialised literature [19], [50]), instead of using the conventional transformation matrices.

## QUATERNION EXPONENTIATION

We extend the complex exponentiation in trigonometric representation (see formula (17.6)) without further proof to the quaternion exponentiation

$$q^{\lambda} = \|q\|^{\lambda} [\cos(\lambda\theta), \hat{q}\sin(\lambda\theta)], \tag{17.13}$$

for all exponents $\lambda \in \mathbb{R}$.

Alternatively, the exponentiation of a quaternion $q = \|q\|e^{\hat{q}\theta}$ in euler representation simplifies to

$$q^{\lambda} = \left(\|q\|e^{\hat{q}\theta}\right)^{\lambda} = \|q\|^{\lambda}e^{\hat{q}(\lambda\theta)}.$$

For instance, the inverse $q^{-1} = \|q\|^{-1}e^{\hat{q}(-\theta)} = \dfrac{\|q\|e^{\hat{q}(-\theta)}}{\|q\|^2} = \dfrac{q^*}{\|q\|^2}$  see formula (17.12).

### THE QUEST FOR THE QUATERNION ROTATION

We recall complex multiplication can be interpreted as a transformation featuring a uniform scaling as well as a rotation effect, in standard (with respect to the complex origin $0 + 0i$). Given the complex transformator $r = |r|(\cos\theta + i\sin\theta)$ and any original complex position $z = x + iy$ within $\mathbb{C}$, we derive

$$
\begin{aligned}
z' &= r \cdot z \\
&= |r|(\cos\theta + i\sin\theta) \cdot (x + iy) \\
&= |r| \cdot ((\cos\theta + i\sin\theta)x + i(\cos\theta + i\sin\theta)y) \\
&= |r| \cdot (\cos\theta x + i\sin\theta x + i\cos\theta y + i^2\sin\theta y) \\
&\quad\Downarrow \\
x' + iy' &= |r| \cdot ((\cos\theta x - \sin\theta y) + i(\sin\theta x + \cos\theta y)).
\end{aligned}
$$

This complex multiplication in trigonometric representation can therefore be expressed by the matrix product:

$$
\begin{pmatrix} x' \\ y' \end{pmatrix} = \underbrace{\begin{pmatrix} |r| & 0 \\ 0 & |r| \end{pmatrix}}_{S_O} \cdot \underbrace{\begin{pmatrix} \cos\theta & -\sin\theta \\ \sin\theta & \cos\theta \end{pmatrix}}_{R_{O,\theta}} \cdot \begin{pmatrix} x \\ y \end{pmatrix}.
$$

Inspired by the latter insight, we take off with a similar approach within $\mathbb{H}$. We augment the location vector $\vec{p}$ which heads the point $P$ by one extra dimension to the pure quaternion $p = [0, \vec{p}\,]$. Given the quaternion transformator $r = \|r\|[\cos\theta, \hat{r}\sin\theta\,]$ and any original position $p = [0, \vec{p}\,]$, we apply the *left* quaternion product (17.11):

$$
\begin{aligned}
r \cdot p &= \\
\|r\|[\cos\theta, \; \hat{r}\sin\theta\,] \cdot [0, \; \vec{p}\,] &= \\
\|r\|[(\cos\theta)0 - (\hat{r}\sin\theta) \cdot \vec{p}, \; \cos\theta\vec{p} + (\hat{r}\sin\theta)0 + \hat{r}\sin\theta \times \vec{p}\,] &= \\
\|r\|[\underbrace{-\sin\theta\hat{r} \cdot \vec{p}}_{\text{nonzero}}, \; \cos\theta\vec{p} + \sin\theta(\hat{r} \times \vec{p})\,].
\end{aligned}
$$

We realise this first attempt does not apply for an image position $\vec{p}'$, given its result is no longer a pure quaternion.

We will later try to cancel out the occured nonzero real component $-\sin\theta\hat{r} \cdot \vec{p}$ in our final attempt. Therefore, we first investigate multiplying the other way around by means of the inverse of $r$ separately. Applying the former quaternion exponentiation for exponent value $\lambda = -1$ (see formula (17.13)):

$$
r^{-1} = \|r\|^{-1}[\cos(-\theta), \; \hat{r}\sin(-\theta)\,] = \frac{1}{\|r\|}[\cos\theta, \; -\hat{r}\sin\theta\,].
$$

Given this inverse quaternion transformator $r^{-1}$ and any original position $p = [0, \vec{p}\,]$ and in order not to cancel our previous action on $p$, we hereby apply the *right* quaternion product (17.11):

$$p \cdot r^{-1} =$$

$$[0, \vec{p}\,] \cdot \frac{1}{\|r\|}[\cos\theta, \ -\hat{r}\sin\theta\,] =$$

$$\frac{1}{\|r\|}[0(\cos\theta) - \vec{p} \cdot (-\hat{r}\sin\theta), \ 0(-\hat{r}\sin\theta) + \vec{p}\cos\theta + \vec{p} \times (-\hat{r}\sin\theta)\,] =$$

$$\frac{1}{\|r\|}[\underbrace{+\sin\theta\hat{r} \cdot \vec{p}}_{\text{nonzero}}, \ \cos\theta\vec{p} + \sin\theta(\hat{r} \times \vec{p})\,].$$

Neither does this second attempt apply for an image position $\vec{p}\,'$ interpretation, given its result again is no longer a pure quaternion. However, since its norm is the inverse from our first attempt and its real component is the opposite of the previous product, our hopes in chaining both multiplications into the sandwich product $r \cdot p \cdot r^{-1}$ are justified.

So let us try to cancel out the occurred nonzero real components from both previous attempts. Considering the inverse transformator $r^{-1}$, an original pure position $p = [0, \vec{p}\,]$ and in order not to cancel our previous action by $r$, we chain both previous attempts into:

$$p' = r \cdot p \cdot r^{-1}$$

$$= \|r\|[\cos\theta, \ \hat{r}\sin\theta\,] \cdot [0, \vec{p}\,] \cdot \frac{1}{\|r\|}[\cos\theta, \ -\hat{r}\sin\theta\,]$$

$$= \frac{\|r\|}{\|r\|}[\cos\theta, \ \hat{r}\sin\theta\,] \cdot [0, \vec{p}\,] \cdot [\cos\theta, \ -\hat{r}\sin\theta\,]$$

$$= 1\,[-\sin\theta\hat{r} \cdot \vec{p}, \ \cos\theta\vec{p} + \sin\theta(\hat{r} \times \vec{p})\,] \cdot [\cos\theta, \ -\hat{r}\sin\theta\,]$$

$$= [-\sin\theta\cos\theta\hat{r} \cdot \vec{p} - (-\hat{r}\sin\theta) \cdot (\cos\theta\vec{p} + \sin\theta(\hat{r} \times \vec{p})),$$

$$(-\sin\theta\hat{r} \cdot \vec{p})(-\hat{r}\sin\theta) + \cos\theta(\cos\theta\vec{p} + \sin\theta(\hat{r} \times \vec{p})) + (\cos\theta\vec{p} + \sin\theta(\hat{r} \times \vec{p})) \times (-\hat{r}\sin\theta)\,]$$

$$= [-\sin\theta\cos\theta\hat{r} \cdot \vec{p} + \sin\theta\cos\theta\hat{r} \cdot \vec{p} + (\sin\theta)^2\hat{r} \cdot (\hat{r} \times \vec{p}),$$

$$(\sin\theta)^2(\hat{r} \cdot \vec{p})\hat{r} + (\cos\theta)^2\vec{p} + \sin\theta\cos\theta(\hat{r} \times \vec{p}) + \sin\theta\cos\theta(-(\vec{p} \times \hat{r})) + (\sin\theta)^2(-(\hat{r} \times \vec{p}) \times \hat{r})\,]$$

Properties of the cross product allow for a first round in simplifying terms (see page 141).

$$= [0 + (\sin\theta)^2 0, \ (\sin\theta)^2(\hat{r} \cdot \vec{p})\hat{r} + (\cos\theta)^2\vec{p} + 2\sin\theta\cos\theta(\hat{r} \times \vec{p}) + (\sin\theta)^2(\hat{r} \times (\hat{r} \times \vec{p}))\,]$$

Because the real component of $p'$ simplifies to zero, we will indeed be able to interprete its imaginary vector $\vec{p}\,'$ upon flattening $\hat{r} \times (\hat{r} \times \vec{p})$ by applying the **stacked cross product** identity:

$$\vec{a} \times (\vec{b} \times \vec{c}) = \vec{b}(\vec{c} \cdot \vec{a}) - \vec{c}(\vec{a} \cdot \vec{b}). \tag{17.14}$$

Whilst we kindly refer you to the literature on vector analysis for a proof of this identity, we do emphasise its cyclic pattern (reading the identity through from left to right) as a handy mnemonic for it.

$$
\begin{aligned}
p' &= [\,0,\ (\sin\theta)^2(\hat{r}\cdot\vec{p})\hat{r} + (\cos\theta)^2\vec{p} + \sin 2\theta(\hat{r}\times\vec{p}) + (\sin\theta)^2(\hat{r}(\vec{p}\cdot\hat{r}) - \vec{p}(\hat{r}\cdot\hat{r}))\,] \\
&= [\,0,\ (\sin\theta)^2(\hat{r}\cdot\vec{p})\hat{r} + (\cos\theta)^2\vec{p} + \sin 2\theta(\hat{r}\times\vec{p}) + (\sin\theta)^2(\hat{r}\cdot\vec{p})\hat{r} - (\sin\theta)^2\vec{p}(\hat{r})^2\,] \\
&= [\,0,\ 2(\sin\theta)^2(\hat{r}\cdot\vec{p})\hat{r} + (\cos\theta)^2\vec{p} + \sin 2\theta(\hat{r}\times\vec{p}) - (\sin\theta)^2\vec{p}\ 1\,] \\
&= [\,0,\ (1 - \cos 2\theta)(\hat{r}\cdot\vec{p})\hat{r} + \cos 2\theta\,\vec{p} + \sin 2\theta(\hat{r}\times\vec{p})\,] \\
&= [\,0,\ \underbrace{(\hat{r}\cdot\vec{p})\hat{r}}_{\text{invariant part}} + \underbrace{\cos 2\theta(\vec{p} - (\hat{r}\cdot\vec{p})\hat{r})}_{\text{reference part}} + \underbrace{\sin 2\theta(\hat{r}\times\vec{p})}_{\text{perpendicular part}}\,]
\end{aligned}
$$

Interpreting this pure quaternion's $p'$ imaginary vector's component $\vec{p}'$ for its possible rotation of vector $\vec{p}$ around $\hat{r}$ over an angle $2\theta$, we distinguish three different vector parts (see figure 17.12). Upon discussing each of these parts, let us combine them.

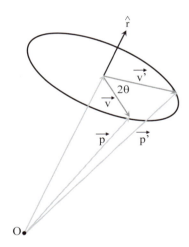

*Figure 17.12*: Rodrigues 3D-rotation theorem for rotation around $\hat{r}$ (here by angle $2\theta$)

Firstly, any spacious rotation of vector $\vec{p}$ around an axis of rotation through unit vector $\hat{r}$ takes the orthogonal projection $\vec{p}\cdot\hat{r}$ of vector $\vec{p}$ onto this unit vector $\hat{r}$, yielding the **invariant vector** $(\hat{r}\cdot\vec{p})\hat{r}$ for all rotations around $\hat{r}$.

Secondly, we establish a **reference vector** $\vec{v}$ perpendicular to the axis of rotation via solving the (head to tail) vector addition $(\hat{r}\cdot\vec{p})\hat{r} + \vec{v} = \vec{p}$ for this vector $\vec{v} = \vec{p} - (\hat{r}\cdot\vec{p})\hat{r}$.

Thirdly, we construct a **perpendicular vector** $\vec{w} = \hat{r} \times \vec{v}$ being both orthogonal to the previous reference vector $\vec{v}$ and orthogonal to the axis of rotation $\hat{r}$. We simplify this vector $\vec{w} = \hat{r} \times (\vec{p} - (\hat{r} \cdot \vec{p})\hat{r}) = \hat{r} \times \vec{p}$ based on the properties of the cross product.

Finally, rotating vector $\vec{p}$ to its image $\vec{p}'$ drags its reference $\vec{v}$ in the plane of rotation to its image $\vec{v}'$, which we decompose in the perpendicular frame $(\vec{v}, \vec{w})$ straightforwardly as $\vec{v}' = \cos 2\theta \vec{v} + \sin 2\theta \vec{w}$, in case of the rotation angle $2\theta$. Vector addition of the invariant part to the rotated part yields the rotated image vector $\vec{p}'$, which we summarise for *double* angles as

$$\vec{p}' = (\hat{r} \cdot \vec{p})\hat{r} + \cos 2\theta (\vec{p} - (\hat{r} \cdot \vec{p})\hat{r}) + \sin 2\theta (\hat{r} \times \vec{p}) \tag{17.15}$$

This idea of decomposing and recomposing about the axis of rotation is the achievement of and named after the French mathematician **Benjamin Olinde Rodrigues** (1795–1851) as the well-known Rodrigues Rotation Theorem.

In conclusion, to achieve pure quaternion rotations around a spatious axis $\hat{r}$ by a *single* angle $\theta$ we need to halve the rotation angle in our quaternion operator $r$, leading to its more practical version. Therefore, we define the **rotation quaternion** $r$ for rotating around a unit direction vector $\hat{r}$ over an angle $\theta$ as

$$r = [r_0, \vec{r}] = \|r\| \left[ \cos \frac{\theta}{2}, \hat{r} \sin \frac{\theta}{2} \right].$$

## UNIT ROTATION QUATERNION

We define the **unit rotation quaternion** $r$ for rotating around a unit direction vector $\hat{r}$ over an angle $\theta$ as

$$r = [r_0, \vec{r}] = \left[ \cos \frac{\theta}{2}, \hat{r} \sin \frac{\theta}{2} \right].$$

We prove that rotation quaternions $r$ are unit quaternions, as $\|r\|$ equals 1.

$$\|r\| = \sqrt{\left( \cos \frac{\theta}{2} \right)^2 + \|\hat{r}\|^2 \left( \sin \frac{\theta}{2} \right)^2} = \sqrt{\left( \cos \frac{\theta}{2} \right)^2 + 1 \cdot \left( \sin \frac{\theta}{2} \right)^2}$$

$$= \sqrt{1} = 1$$

We augment the location vector $\vec{p}$ which heads the point $P$ by one extra dimension to the pure quaternion $p = [0, \vec{p}]$. We express the quaternion rotation of a point $P$ around a unit direction vector $\hat{n}$ over an angle $\theta$, yielding the image point $P'$, as

$$p' = r \cdot p \cdot r^{-1}$$
$$[0, \vec{p}'] = r \cdot [0, \vec{p}] \cdot r^{-1}.$$

This rotation expression is not valid for arbitrary quaternions $u$! It only suits the formerly defined rotation quaternion $r = \left[\cos\frac{\theta}{2}, \hat{r}\sin\frac{\theta}{2}\right]$ given its axis of rotation $\hat{r}$.

Given $\|\hat{r}\| = 1$, our defined rotation quaternion $r$ is also a unit quaternion, which simplifies the above by $r^{-1} = r^*$. In conclusion, we realise the rotation of the point $P$ by the quaternion $r$ to the image point $P'$ as

$$[0, \vec{p}'] = r \cdot [0, \vec{p}] \cdot r^*$$

$$= \left[\cos\frac{\theta}{2}, \hat{r}\sin\frac{\theta}{2}\right] \cdot [0, \vec{p}] \cdot \left[\cos\frac{\theta}{2}, -\hat{r}\sin\frac{\theta}{2}\right]. \qquad (17.16)$$

*Example 1*: We rotate the point $P(6,2,4)$ around an axis of rotation along the direction

vector $\begin{pmatrix} 1 \\ 1 \\ 1 \end{pmatrix}$ over an angle of $120°$, applying quaternions. For this purpose, we need to

normalise $\begin{pmatrix} 1 \\ 1 \\ 1 \end{pmatrix}$ to $\hat{r} = \frac{1}{\sqrt{3}}\begin{pmatrix} 1 \\ 1 \\ 1 \end{pmatrix}$. Consequently, we find the rotation quaternion $r$ as

$$r = \cos 60° + i\,\frac{\sin 60°}{\sqrt{3}} + j\,\frac{\sin 60°}{\sqrt{3}} + k\,\frac{\sin 60°}{\sqrt{3}}$$

$$= \frac{1}{2} + \frac{1}{2}\,i + \frac{1}{2}\,j + \frac{1}{2}\,k$$

$$= \left[\frac{1}{2}, \left(\frac{1}{2}, \frac{1}{2}, \frac{1}{2}\right)\right].$$

We verify that $r$ is a unit quaternion by calculating its norm $\|r\| = 1$. The conjugate quaternion of $r$ equals $r^* = \left[\frac{1}{2}, -\left(\frac{1}{2}, \frac{1}{2}, \frac{1}{2}\right)\right]$. We augment the point $P(6,2,4)$ to its pure quaternion $p = [0, (6,2,4)]$. We perform the specified rotation by the quaternion expression

$$p' = r \cdot p \cdot r^*$$

$$= \left[\frac{1}{2}, \left(\frac{1}{2}, \frac{1}{2}, \frac{1}{2}\right)\right] \cdot [0, (6,2,4)] \cdot \left[\frac{1}{2}, \left(-\frac{1}{2}, -\frac{1}{2}, -\frac{1}{2}\right)\right]$$

$$= [-6, (4,2,0)] \cdot \left[\frac{1}{2}, \left(-\frac{1}{2}, -\frac{1}{2}, -\frac{1}{2}\right)\right]$$

$$= [0, (4,6,2)].$$

In conclusion, the rotation of the point $P(6,2,4)$ around the direction vector $\begin{pmatrix} 1 \\ 1 \\ 1 \end{pmatrix}$ over

an angle of $120°$ results in the image point $P'(4,6,2)$.

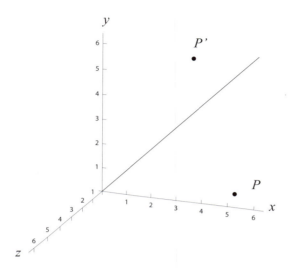

*Figure 17.13*: Rotation by quaternions

*Example 2*: We rotate the general point $P(x,y,z)$ around the $z$-axis over a general angle $\theta$, applying quaternions. For this purpose, we do not need to normalise $\hat{e}_z = \begin{pmatrix} 0 \\ 0 \\ 1 \end{pmatrix}$ since it is already a unit vector. Consequently, we find the rotation quaternion $r$ as

$$r = \left[\cos\frac{\theta}{2}, \hat{e}_z \sin\frac{\theta}{2}\right]$$

$$= \left[\cos\frac{\theta}{2}, (0,0,1)\sin\frac{\theta}{2}\right]$$

$$= \left[\cos\frac{\theta}{2}, \left(0,0,\sin\frac{\theta}{2}\right)\right].$$

We verify that $r$ is a unit quaternion by calculating its norm $\|r\| = 1$. The conjugate quaternion of $r$ equals $r^* = \left[\cos\frac{\theta}{2}, -\left(0,0,\sin\frac{\theta}{2}\right)\right]$. We augment the point $P(x,y,z)$ to its pure quaternion $p = [0,(x,y,z)]$. We perform the specified rotation by the quaternion expression (see exercise 158)

$$p' = r \cdot p \cdot r^*$$

$$= \left[\cos\frac{\theta}{2}, \left(0,0,\sin\frac{\theta}{2}\right)\right] \cdot [0,(x,y,z)] \cdot \left[\cos\frac{\theta}{2}, \left(0,0,-\sin\frac{\theta}{2}\right)\right]$$

$$= \left[-z\sin\frac{\theta}{2}, \left(x\cos\frac{\theta}{2} - y\sin\frac{\theta}{2}, x\sin\frac{\theta}{2} + y\cos\frac{\theta}{2}, z\cos\frac{\theta}{2}\right)\right] \cdot \left[\cos\frac{\theta}{2}, \left(0,0,-\sin\frac{\theta}{2}\right)\right]$$

$$= [0,(x\cos\theta - y\sin\theta, x\sin\theta + y\cos\theta, z)].$$

The resulting pure quaternion $p'$ contains the image point

$$P'(x\cos\theta - y\sin\theta, x\sin\theta + y\cos\theta, z).$$

We found the same result after applying the matrix rotator for a roll in 3D of point $P$ around the $z$-axis (see formula (13.10)).

$$\begin{pmatrix} \cos\theta & -\sin\theta & 0 & 0 \\ \sin\theta & \cos\theta & 0 & 0 \\ 0 & 0 & 1 & 0 \\ 0 & 0 & 0 & 1 \end{pmatrix} \begin{pmatrix} x \\ y \\ z \\ 1 \end{pmatrix} = \begin{pmatrix} x\cos\theta - y\sin\theta \\ x\sin\theta + y\cos\theta \\ z \\ 1 \end{pmatrix}$$

Quaternion rotation is popular in computer applications because it needs only one move for the rotation around a direction vector $\vec{n}$ over an angle $\theta$. Whereas the same effect takes three moves (around the major axes) in matrix rotation, which can cause complications amongst which is gimbal lock.

In this table, we list some standard 3D rotations around the major axes and their corresponding rotation quaternion $r \in \mathbb{H}$. Again note that all rotation quaternions are unit quaternions.

| rotation around the x-axis | rotation quaternion | rotation around the y-axis | rotation quaternion | rotation around the z-axis | rotation quaternion |
|---|---|---|---|---|---|
| over 180° | $[0,(1,0,0)]$ | over 180° | $[0,(0,1,0)]$ | over 180° | $[0,(0,0,1)]$ |
| over 90° | $\left[\frac{\sqrt{2}}{2}, \left(\frac{\sqrt{2}}{2},0,0\right)\right]$ | over 90° | $\left[\frac{\sqrt{2}}{2}, \left(0,\frac{\sqrt{2}}{2},0\right)\right]$ | over 90° | $\left[\frac{\sqrt{2}}{2}, \left(0,0,\frac{\sqrt{2}}{2}\right)\right]$ |
| over −90° | $\left[\frac{\sqrt{2}}{2}, \left(-\frac{\sqrt{2}}{2},0,0\right)\right]$ | over −90° | $\left[\frac{\sqrt{2}}{2}, \left(0,-\frac{\sqrt{2}}{2},0\right)\right]$ | over −90° | $\left[\frac{\sqrt{2}}{2}, \left(0,0,-\frac{\sqrt{2}}{2}\right)\right]$ |
| over 60° | $\left[\frac{\sqrt{3}}{2}, \left(\frac{1}{2},0,0\right)\right]$ | over 60° | $\left[\frac{\sqrt{3}}{2}, \left(0,\frac{1}{2},0\right)\right]$ | over 60° | $\left[\frac{\sqrt{3}}{2}, \left(0,0,\frac{1}{2}\right)\right]$ |

## 17.8 Exercises

**Exercise 153**  Represent the following four complex numbers algebraically and trigono-metrically.

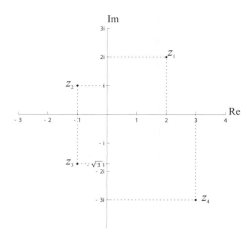

**Exercise 154**  Calculate the following expressions, given $z_1 = -4i$, $z_2 = 3 - 2i$ and $z_3 = -1 + i$.

1)  $z_1 - 2z_2 + 3z_3$

2)  $2z_1 z_2^*$

3)  $z_2 z_3$

4)  $\frac{z_1^* z_2}{z_3}$

5)  $z_1(2z_2^* - z_1) + z_3^*$

6)  $\frac{z_1 - z_2^*}{3z_3^*}$

**Exercise 155**  Consider the rectangle spanned by the vertices $A(1,1), B(\sqrt{3},1), C(\sqrt{3},-1)$ and $D(1,-1)$.

1)  Give the complex number for a standard rotation around the origin over an angle of $120°$. Apply this complex number to calculate the rectangle's image vertices after rotation.

2)  Give the complex number for a standard rotation around the origin over an angle of $80°$ and a standard scaling with scale factor 3.

**Exercise 156**  Apply the function $f : \mathbb{Z} \to \mathbb{Z} : k \mapsto f(k) = \frac{(\Phi)^k - (\Phi')^k}{\sqrt{5}}$ to calculate the elements $f(-1), f(-2)$ and $f(-3)$ of the continued Fibonacci sequence. Both $\Phi = \frac{1+\sqrt{5}}{2}$ and $\Phi' = \frac{1-\sqrt{5}}{2}$ are known, respectively as the golden number and its variant.

**Exercise 157**  Calculate the following expressions, for $q = 2 - 4i + 2k$, $w = 1 - 2i + j - 2k$, $s = -1 + i + j + k$ and $t = i + 2j - k$.

1)  $q + w^*$

4)  $q - 2s + t^*$

2)  $\|s\|$

5)  $q \cdot w$

3)  $s \cdot t^*$

6)  $t^{-1}$

**Exercise 158**  Calculate the quaternion product

$$\left[\cos\frac{\theta}{2}, \left(0, 0, \sin\frac{\theta}{2}\right)\right] \cdot [0, (x, y, z)] \cdot \left[\cos\frac{\theta}{2}, \left(0, 0, -\sin\frac{\theta}{2}\right)\right].$$

**Exercise 159**

1)  Construct the rotation matrix for a roll over $90°$ and apply this roll to the point $P(1, 0, 0)$.

2)  Repeat this rotation applying a rotation quaternion.

**Exercise 160**

1)  Construct the rotation matrix for a pitch over $240°$ and apply this pitch to the point $P(2, 3, 1)$.

2)  Repeat this rotation applying a rotation quaternion.

**Exercise 161**  Calculate the image coordinates of the point $P(1, 1, 0)$ after the rotation over $180°$ around the direction vector $\begin{pmatrix} 1 \\ 2 \\ 1 \end{pmatrix}$, applying quaternion arithmetic.

**Exercise 162**  The given matrix expression $Z$ behaves identically to the complex number $z = a + bi$. It even adopts a 'complex' conjugate expression $Z^*$. We define

$$Z = a \begin{pmatrix} 1 & 0 \\ 0 & 1 \end{pmatrix} + b \begin{pmatrix} 0 & -1 \\ 1 & 0 \end{pmatrix},$$

given $a, b \in \mathbb{R}$ and its conjugate expression as

$$Z^* = a \begin{pmatrix} 1 & 0 \\ 0 & 1 \end{pmatrix} - b \begin{pmatrix} 0 & -1 \\ 1 & 0 \end{pmatrix}.$$

Calculate $Z + Z^*$, $Z - Z^*$ and $Z \cdot Z^*$.

# Annex A · Real numbers in computers

Real numbers are stored identically into the computer. From the irrational numbers such as $\pi$ to giant integers such as $10^{10}$, radicals and negative fractions, they all fit into the machine in the same way. Let us also add $-1$ billion to our example list.

## A.1 Scientific notation

The storage of real numbers into computers is based on their **scientific notation** which separates the sign and the precision from the order of magnitude of each exact number $x$, arranged into the product

$$x = (-1)^s \times N_{10} \times 10^{E_{10}}.$$

The first factor $(-1)^s$ shows the **sign** of $x$, the second factor $N_{10}$ is the decimal **normalised significand** lying between 1 and 10 and finally the exponent $E_{10}$ indicates the **decimal order of magnitude** of $x$.

| exact value $x$ | decimally displayed | decimally scientifically displayed |
|---|---|---|
| $\frac{1}{10}$ | 0.1 | $+1. \times 10^{-1}$ |
| $\pi$ | $3.141592653\ldots$ | $+3.141592653\ldots \times 10^0$ |
| 0.00001234 | 0.00001234 | $+1.234 \times 10^{-5}$ |
| $-1$ billion | $-1000000000.$ | $-1.000000000 \times 10^9$ |

This normalised scientific notation allows us to simulate the storage of our real examples into a decimal machine which allocates a standardised digit sequence for each of them.

## A.2 The decimal computer

Let us straightforwardly consider a decimal computer which stores one digit denoting the sign, one digit indicating the order of magnitude and stores four **significant digits** of the original value $x$.

| scientific notation | uniform machine precision | stored **machine number** $x'$ |
|---|---|---|
| $+1. \qquad\qquad \times 10^{-1}$ | $(-1)^0 \times 1. \qquad \times 10^{-1}$ | $(-1)^0 \times 1.000 \times 10^{-1}$ |
| $+3.141592653\ldots \times 10^0$ | $(-1)^0 \times 3.142 \times 10^0$ | $(-1)^0 \times 3.142 \times 10^0$ |
| $+1.234 \qquad\qquad \times 10^{-5}$ | $(-1)^0 \times 1.234 \times 10^{-5}$ | $(-1)^0 \times 1.234 \times 10^{-5}$ |
| $-1.000000000 \quad \times 10^9$ | $(-1)^1 \times 1.000 \times 10^9$ | $(-1)^1 \times 1.000 \times 10^9$ |

Our simplified decimal computer stores exact values $x \in \mathbb{R}$ systematically in a fixed digit sequence $x'$ containing the sign (1 digit), the exponent (1 digit) and the normalised significand (4 digits). This computer is limited to storing only four significant digits and consequently standardises its stored numbers $x'$ with fixed **machine precision**. We call the finite subset of real numbers $x'$ which are inevitably **rounded** to fit into the computer, **machine numbers**. The accompanying figure shows all positive machine numbers, from the smallest to the largest one in $\mathbb{R}^+$ in case of 8-bit (which means 2-decimal digit) numbers.

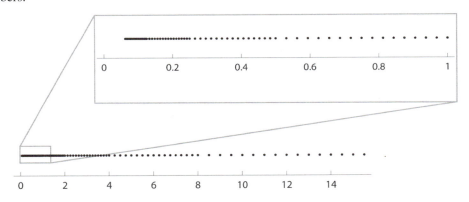

Figure A.1: The subset of (fictitious) 8-bit machine numbers $x'$ in $\mathbb{R}^+$

## A.3 Special values

Calculations which result in numbers smaller than the smallest machine number, suffer **real underflow**. Arithmetical outputs which are larger than the largest machine number feature **real overflow**. For instance, storing the real number *zero* is an issue, since it would require the exponent $E_{10} = -\infty$. To be able to store the number zero, and similarly the *infinities* requiring exponent $E_{10} = +\infty$ and the indeterminate such as $\frac{0}{0}$ in our decimal machine, we predefine these exceptions respectively as **NULL**, **INFINITY** and **NAN** abbreviating 'Not A Number'.

# Annex B · Notations and Conventions

## B.1 Alphabets

| meaning | symbol |
|---|---|
| constants and coefficients | $a, b, c, \ldots$ |
| unknown quantities and variables | $x, y, z, \ldots$ |
| points | $P, Q, R, \ldots$ |
| lines | $r, s, t, \ldots$ |
| planes | $v_R, v_P, \ldots$ |
| vectors | $\vec{v}, \vec{w}, \ldots$ |
| unit vectors | $\hat{v}, \hat{w}, \ldots$ |
| matrices | $A, B, C, \ldots$ |
| angles | $\hat{A}, \hat{B}, \hat{C}, \ldots$ |
| (angles alternatively in Greek) | $\alpha, \beta, \theta, \ldots$ |
| Bezier segment | $\vec{b}_{012\ldots n}$ |
| B-spline | $\vec{s}_{012\ldots n}$ |
| translations | $T_{\vec{c}}$ |
| standard rotations | $R_O$ |
| standard scalings | $S_O$ |
| composite action transformations | $A$ |
| pivot transformation | $P_B$ |
| conventional composite transformations | $TRS$ |
| (nonconventional) orbit transformation | $O_B$ |
| embedding transformation | $E_i$ |
| camera transformation | $F_{\vec{c}}$ |
| view transformation | $V_{\vec{c}}$ |
| quaternions | $q, p, \ldots$ |
| normalized quaternions | $q_n, p_n, \ldots$ |
| unit quaternions | $u$ |
| rotation quaternions | $u, r, \ldots$ |

GREEK ALPHABET

Traditionally, we use Greek characters to denote angles (especially in trigonometry). We also choose Greek characters for typesetting mathematical and physical constants.

| name | Greek character | name | Greek character |
|------|----------------|------|----------------|
| alpha | $\alpha$ | nu | $\nu$ |
| beta | $\beta$ | xi | $\xi$ |
| gamma | $\gamma$ | omicron | $o$ |
| delta | $\delta, \Delta$ | pi | $\pi$ |
| epsilon | $\varepsilon$ | rho | $\rho$ |
| zeta | $\zeta$ | sigma | $\sigma$ |
| eta | $\eta$ | tau | $\tau$ |
| theta | $\theta$ | upsilon | $\upsilon$ |
| iota | $\iota$ | phi | $\phi, \Phi$ |
| kappa | $\kappa$ | chi | $\chi$ |
| lambda | $\lambda$ | psi | $\psi$ |
| mu | $\mu$ | omega | $\omega$ |

# B.2 Mathematical symbols

SETS

| number sets including zero | symbol |
|----------------------------|--------|
| natural numbers (unsigned integers) | $\mathbb{N}$ |
| integer numbers (integers) | $\mathbb{Z}$ |
| rational numbers or fractions | $\mathbb{Q}$ |
| real numbers (floating points) | $\mathbb{R}$ |
| complex numbers | $\mathbb{C}$ |
| hypercomplex numbers or quaternions | $\mathbb{H}$ |

We embed these number sets as

$$\mathbb{N} \subset \mathbb{Z} \subset \mathbb{Q} \subset \mathbb{R} \subset \mathbb{C} \subset \mathbb{H}.$$

## MATHEMATICAL SYMBOLS

| name | symbol |
|---|---|
| empty set | $\{\}$ |
| set minus | $\backslash$ |
| element of | $\in$ |
| cardinality (number of elements) | $\#$ |
| factorial | $!$ |
| equal to | $=$ |
| equivalent with | $\Leftrightarrow$ |
| implies | $\Rightarrow$ |
| distance | $d$ |
| difference | $\Delta$ |
| degrees | $\circ$ |
| infinity (unbound large value) | $\infty$ |
| summation | $\Sigma$ |
| dot product | $\cdot$ |
| cross product | $\times$ |
| transpose | $T$ |
| conjugate | $*$ |
| imaginary unities | $i, j, k$ |
| cartesian coordinates | $(\ )_{cc}$ |
| polar coordinates | $(\ )_{pc}$ |
| tiny error | $\varepsilon$ |

MATHEMATICAL KEYWORDS

| name | symbol |
|---|---|
| logarithm in base $b$ | $\log_b$ |
| exponential in base $e$ | exp |
| radian | rad |
| sine | sin |
| cosine | cos |
| tangent | tan |
| cotangent | cot |
| arcsine | arcsin |
| arccosine | arccos |
| arctangent | arctan |
| extended arctangent | atan2 |
| determinant | det |
| absolute value | abs |

NUMBERS

| name | symbol, (rounded) value |
|---|---|
| pi | $\pi \approx 3.1416$ |
| radian | $1 \text{ rad} \approx 57.30°$ |
| silver number | $\delta = 1 + \sqrt{2} \approx 2.4142$ |
| golden number | $\Phi = \frac{1+\sqrt{5}}{2} \approx 1.6180$ |
| paired golden number | $\Phi' = \frac{1-\sqrt{5}}{2} \approx -0.6180$ |
| imaginary unities (quaternions) | $i^2 = j^2 = k^2 = -1$ and $ij = k$ |
| natural base | $e \approx 2.7183$ |
| acceleration due to gravity (average) | $g \approx 9.8067$ |

# Annex C · The International System of Units (SI)

## C.1 SI Prefixes

We may use the international default prefixes to specify decimal orders of magnitude.

| name | symbol | factor |
|------|--------|--------|
| yotta | Y | $10^{24}$ |
| zetta | Z | $10^{21}$ |
| exa | E | $10^{18}$ |
| peta | P | $10^{15}$ |
| tera | T | $10^{12}$ |
| giga | G | $10^{9}$ |
| mega | M | $10^{6}$ |
| kilo | k | $10^{3}$ |
| hecto | h | $10^{2}$ |
| deca | da | $10^{1}$ |
| | | |
| deci | d | $10^{-1}$ |
| centi | c | $10^{-2}$ |
| milli | m | $10^{-3}$ |
| micro | $\mu$ | $10^{-6}$ |
| nano | n | $10^{-9}$ |
| pico | p | $10^{-12}$ |
| femto | f | $10^{-15}$ |
| atto | a | $10^{-18}$ |
| zepto | z | $10^{-21}$ |
| yocto | y | $10^{-24}$ |

*Examples:*

$$12 \text{ k}m = 12 \times 10^{3} \; m = 12\,000 \; m$$
$$34 \text{ m}m = 34 \times 10^{-3} \; m = 0.034 \; m$$

In our modern world, we quantify measures standardised by the **SI** (the International System of Units). Nature's base measures length $l$, mass $m$ and time $t$ are measured in metres $m$, kilograms $kg$ and seconds $s$ respectively. We may typeset units by putting *square brackets* around their corresponding measures.

## C.2 SI Base measures

We mainly use just these three base measures throughout this book; for the few remaining base measures we refer you to the physics literature.

| measure | symbol | SI-unit | |
|---------|--------|---------|---|
| length | $l$ | $[l] = m$ | metre |
| mass | $m$ | $[m] = kg$ | kilogram |
| time | $t$ | $[t] = s$ | second |

## C.3 SI Supplementary measure

Unlike the real physics units, expressing plane angles in radian is typeset by the supplementary measure or mathematical tag 'rad'.

| measure | symbol | SI-unit | |
|---------|--------|---------|---|
| plane angle | $\alpha$ | $[\alpha] = \text{rad}$ | radian |

## C.4  SI Derived measures

Derived measures are composed of the above measures. We mainly use the following derived measures throughout this book; for the remaining ones with special names we refer you to the physics literature.

| measure | symbol | SI-unit | |
|---|---|---|---|
| width | $b$ | $[b] = m$ | metre |
| height | $h$ | $[h] = m$ | metre |
| radius | $r$ | $[r] = m$ | metre |
| diameter | $d$ | $[d] = m$ | metre |
| distance | $d$ | $[d] = m$ | metre |
| norm of a location vector | $\|\vec{s}\|$ | $[\|\vec{s}\|] = m$ | metre |
| area | $area$ | $[area] = m^2$ | square metre |
| volume | $volume$ | $[volume] = m^3$ | cubic metre |
| speed | $v$ | $[v] = \frac{m}{s}$ | metre per second |
| magnitude of acceleration | $a$ | $[a] = \frac{m}{s^2}$ | metre per second squared |
| acceleration due to gravity | $g$ | $[g] = \frac{m}{s^2}$ | metre per second squared |
| frequency | $f$ | $[f] = s^{-1}$ | Hertz |
| angular location | $\theta$ | $[\theta] = \text{rad}$ | radian |
| angular speed | $\omega$ | $[\omega] = \frac{\text{rad}}{s}$ | radians per second |
| angular acceleration | $\alpha$ | $[\alpha] = \frac{\text{rad}}{s^2}$ | radians per second squared |

# Bibliography

BOOKS

1) L. Ammeraal (1998). *Computer Graphics for Java Programmers*, John Wiley & Sons Ltd.

2) P.A. Egerton, W.S. Hall (1999). *Computer Graphics, Mathematical First Steps*, Pearson Education.

3) A.J. Hanson (2006). *Visualizing Quaternions. Series in interactive 3D technology*, Morgan Kaufmann.

4) M. Kamminga-van Hulsen, P.E.J.M. Gondrie, G.A.T.M. van Alst (1994). *Toegepaste wiskunde met computeralgebra*, Academic Service.

5) W. Kleijne, T. Konings (2000). *De Gulden Snede*, Epsilon Uitgaven.

6) P. Lothar (1993). *Wiskunde voor het hoger technisch onderwijs deel 1*, Academic Service.

7) P. Lothar (1993). *Wiskunde voor het hoger technisch onderwijs deel 2*, Academic Service.

8) W. Stahler, D. Clingman, K. Kaveh (2004). *Beginning Math and Physics for Game Programmers*, Pearson Education.

9) B. Langerock (2011). *Wiskunde voor ontwerpers*, LannooCampus.

10) T. Crilly (2008). *Vijftig inzichten wiskunde*, Veen Magazines.

11) J. Gielis (2001). *De uitvinding van de cirkel*, Geniaal.

12) D. Marsh (2005). *Applied Geometry for Computer Graphics and CAD*, Springer.

13) G. Farin (2002). *Curves and Surfaces for CAGD*, Morgan Kauffman.

14) K. Samyn (2016). *Applied Physics*, Digital Arts & Entertainment (Howest BE).

15) J. M. Van Verth, L. M. Bishop (2008). *Essential Mathematics for Games and Interactive Applications*, Morgan Kauffman.

16) D. C. Giancoli (2013). *Physics: Principles with Applications*, Pearson Education.

17) R. Wolfson (2007). *Essential University Physics*, Pearson Education.

18) R.C. Hill, K. Plantenberg (2013). *Conceptual Dynamics*, SDC Publications.

19)   L. Dorst, D. Fontijne, S. Mann (2007). *Geometric Algebra for Computer Science*, Morgan Kaufmann.

20)   J.-P. Ottoy, (2014). *Differentiaal- en integraalrekening*, Academia Press.

21)   J. Schwichtenberg (2018). *Physics from Symmetry*, Springer.

WEBSITES

22)   Wolfram: Corporate website (http://www.wolfram.com/) (access June 2013).

23)   Wolfram MathWorld: The Web's Most Extensive Mathematics Resource (http://mathworld.wolfram.com/) (access June 2013).

24)   Wolfram Demonstrations Project: Powered by CDF Technology (http://demonstrations.wolfram.com) (access June 2021).

25)   Wolfram Application Server: successor to webMathematica (https://www.wolfram.com/application-server) (access June 2021).

26)   Wolfram|Alpha: computational knowledge engine (http://www.wolframalpha.com/) (access June 2013).

27)   D. Klingens: Gulden snede en Fibonacci (http://www.pandd.demon.nl/sectioaurea.htm) (access April 2009).

28)   University of Surrey, Faculty of Engineering and Physical Sciences: (http://www.mcs.surrey.ac.uk) (access April 2009).

29)   J. Wassenaar: Two dimensional curves (http://www.2dcurves.com) (access April 2009).

30)   NVvW – Nederlandse Vereniging van Wiskundeleraren: (http://www.nvvw.nl) (access April 2009).

31)   Dimensions: Chapters 5 and 6 (http://www.dimensions-math.org/Dim_CH5_NL.htm) (access February 2009).

32)   Startpagina | regionaal steunpunt Eindhoven: (http://www.win.tue.nl/wiskunded/files/public/Complexe_getallen/complex-2007-06.1s.pdf) (access February 2009).

33)   Graphics: Programma (http://staff.science.uva.nl/~jose/graphics/matrix.pdf) (access April 2009).

34)   Point in triangle test: (http://www.blackpawn.com/texts/pointinpoly/default.html) (access April 2009).

35) Senocular tutorial: Understanding the Transformation Matrix in Flash 8 (`http://www.senocular.com/flash/tutorials/transformmatrix/`) (access April 2009).

36) Genicap: Corporate website (`http://www.genicap.com/`) (access July 2011).

37) Online Math Learning: complete from Kindergarten to University (`http://www.onlinemathlearning.com/`) (access June 2013).

38) Mathwords: Terms and Formulas from Beginning Algebra to Calculus (`http://www.mathwords.com/`) (access January 2021).

39) Wikipedia: free online Encyclopedia (`http://en.wikipedia.org/`) (access January 2021).

40) CLEAR at Rice: Introduction to Computer Graphics (`http://www.clear.rice.edu/comp360/lectures/`) (access June 2013).

41) Illuminations: Online Teaching Resources (`http://illuminations.nctm.org/`) (access June 2013).

42) Kairos: Academic Repository (`http://kairos.laetusinpraesens.org/`) (access June 2013).

43) The Physics Classroom: Courseware (`http://www.physicsclassroom.com/`) (access July 2015).

44) The Physical Markup Language: Academic standard (`http://web.mit.edu/mecheng/pml/`) (access July 2015).

45) Unity Script Reference: Documentation (`http://docs.unity3d.com/ScriptReference/`) (access July 2016).

46) Conceptual Dynamics [18]: Companion site (`http://www.conceptualdynamics.com/`) (access August 2015).

47) MIT Open Courseware: Academic repository (`http://ocw.mit.edu/`) (access August 2015).

48) GeoGebra: Maths tool (`http://www.geogebra.org/`) (access February 2021).

49) CK12: Interactive maths and science (`http://www.ck12.org/`) (access April 2016).

50) Bivector: the one-stop resource on geometric algebra (`https://www.bivector.net/`) (access November 2020).

# Index